Microcontroller Prototypes with
Arduino and a 3D Printer

Microcontroller Prototypes with Arduino and a 3D Printer

Learn, Program, Manufacture

Dimosthenis E. Bolanakis

Department of Air Force Science
Hellenic Air Force Academy
Athens
Greece

This edition first published 2021

Registered Offices
John Wiley & Sons, Inc., 111 River Street, Hoboken, NJ 07030, USA
John Wiley & Sons Ltd, The Atrium, Southern Gate, Chichester, West Sussex, PO19 8SQ, UK

Editorial Office
The Atrium, Southern Gate, Chichester, West Sussex, PO19 8SQ, UK

For details of our global editorial offices, customer services, and more information about Wiley products visit us at www.wiley.com.

Wiley also publishes its books in a variety of electronic formats and by print-on-demand. Some content that appears in standard print versions of this book may not be available in other formats.

Library of Congress Cataloging-in-Publication Data

Names: Bolanakis, Dimosthenis E., 1978– author.
Title: Microcontroller prototypes with Arduino and a 3D printer : learn, program, manufacture / Dimosthenis E. Bolanakis.
Description: Hoboken, NJ : Wiley, 2021. | Includes bibliographical references and index.
Identifiers: LCCN 2021003270 (print) | LCCN 2021003271 (ebook) | ISBN 9781119782612 (cloth) | ISBN 9781119782674 (adobe pdf) | ISBN 9781119782681 (epub)
Subjects: LCSH: Microcontrollers. | Arduino (Programmable controller) | Three-dimensional printing.
Classification: LCC TJ223.P76 B665 2021 (print) | LCC TJ223.P76 (ebook) | DDC 006.2/2–dc23
LC record available at https://lccn.loc.gov/2021003270
LC ebook record available at https://lccn.loc.gov/2021003271

Cover Design: Wiley
Cover Image: © Rasi Bhadramani/iStock/Getty Images

Set in 9.5/12.5pt STIXTwoText by SPi Global, Pondicherry, India
Printed and bound by CPI Group (UK) Ltd, Croydon, CR0 4YY

C9781119782612_190321

I have never seen any of the 7 wonders of the world. I believe there are more though!

∞ To my wife and my three kids ∞

Contents

About the Author

Dimosthenis (Dimos) E. Bolanakis was born in Crete, Greece (1978) and graduated in Electronic Engineering (2001) from ATEI Thessalonikis, Greece. He received the MSc degree (2004) in Modern Electronic Technologies and the PhD degree (2016) in Education Sciences (focusing on Remote Experimentation), both from University of Ioannina, Greece. He has (co)authored more than 30 papers (mainly on Research in Engineering Education) and 3 books. He has held positions in both industry and education and his research interests focus on *μC-based and FPGA-based Hardware Design* and *Research in Education*. He currently lives in Athens (Greece) together with his wonderful wife Katerina and their three delightful kids, Manolis, Eugenia, and Myronas.

List of Figures

List of Tables

Preface

This book provides a guide to learning *microcontrollers (μCs)*, appropriate for educators, researchers, and makers. Microcontrollers constitute a popular type of embedded computers. This technology has been experiencing, in the last decade, the widespread dissemination of *do it yourself* (DIY) culture and gradually shifting away from *electronic engineering* (EE) discipline (where it was originally meant to be used). The today's wave of the *ready-to-use* and *stackable* μC board systems and the *shareable* – over the internet – libraries, render feasible the rapid development of microcontroller-based applications. Furthermore, the modern μC programming methods have managed to abstract the low-level tasks, and consequently, today's technology can also be utilizable by the *computer science* (CS) discipline. However, to learn, in in-depth, how to develop microcontroller-based products, one has to pay particular attention to practices related to the hardware domain as well. The current effort exploits the modern development tools and programming methods, while also providing a scholastic examination of the critical, hardware-related practices on microcontrollers. Through a series of carefully designed example codes and explanatory figures, the book provides to the reader a unique experience on establishing a clear link between the software and hardware. The latter constitutes perhaps the most challenging area of study, which could be considered as the basis for making the passage from "*have knowledge of*" to "*mastering*" the design of microcontroller-based projects and applications. The book also features a theoretical background appropriate for instructors and educational researchers, which provides to the reader a unique perspective on the educational aspects related to the process of microcomputer programming and application development.

Chapter 1 is structured with scientific rigorousness and is primarily addressed for educators and educational researchers. The chapter first introduces a novel classification of today's embedded computer systems, in terms of the tasks linked either to the CS or EE discipline. This interdisciplinary technology between the two disciplines aims at helping readers clarify the possibilities, limitations, and learning difficulties of each category. Then the chapter provides a unique perspective on the educational aspects of microcomputer programming and application development. The original analysis applies to the *technological pedagogical content knowledge* (TPACK) model, and attempts to clarify why the programming language should be considered as the technology integration, toward helping students create their knowledge about the subject matter. It also justifies why the employed technology may arrange the tutoring more appropriate for either the CS, or the EE, discipline. Subsequent to that analysis, the chapter explores the additional endeavor required to understand the capabilities of microcomputer technology and addresses the coined *micro-computational thinking* (μCT) term in order to describe the thought processes involved in the solutions, carried out by a microcomputer-based system. The μCT term is not differentiated from the existing and well-known *computational thinking* (CT) concept, but is rather addressed to reveal the thought processes related to the application development with embedded computers. Because of the *maker* movement in education (and the fact that microcontrollers constitute a low-cost and easily accessible technology that can be straightforwardly incorporated within any *makerspace*), it would be wise to consider an upcoming turn to the educational research efforts related to microcomputer programming, at expense of (or complementary to) the conventional computer programming courses. This attempt would raise questions such as: "What is the difference between programming a regular computer and a microcomputer? How could we arrange the content knowledge of a technical subject matter without too much focus on the specified technology?" This chapter applies to such issues and provides information that can be used as a reference guide for further educational research in microcontrollers, and embedded computers in general. Moreover, in order to understand the today's impact of microcontroller technology on the maker industry, the reader follows the advancement of microcontroller programming and application development and identifies the (i) *long-cycle* and (ii) *short-cycle* development eras (in terms of the requisite time needed to learn, program, and develop a microcontroller-based application), as well as the recent trends in sensor devices and how they pave the way for creativity and new solutions in embedded computing devices.

Chapter 2 is structured with less scientific rigorousness and is addressed for engineers and makers. The reader explores the Arduino software and hardware tools, which have become a viral technology for microcontrollers as they provide a quick jump-start and flexibility in the implementation of a microcontroller-based project. The chapter summarizes the fundamental aspects of sequential programming, in consideration of a relative compatibility between the Arduino and C language programming. The authoring strategies of the chapter apply to the motto *"less is more"* and address the *minimalist* principles in the design of the software. Additionally to the task of reducing the software development to its necessary elements, the chapter exploits the familiar and simplified board called the Arduino Uno, so as to reduce the use of hardware when there is a need to explore the results occurred by the execution of an example code. Makers and engineers may directly start from this particular chapter, in order to make quick jump-start into the practical part of learning microcontroller.

Chapter 3 applies to practices related to the hardware interface with the outside world. The chapter starts from the familiarization and utilization of the *ready-to-use* Arduino libraries, and gradually moves to more advanced issues, such as interrupting the regular execution of the code, building libraries from scratch, and so forth. Hence, the chapter incorporates information that intends to satisfy the curiosity of makers, but also engineers and researchers who work with microcontroller devices. The purpose of this chapter is to provide a thorough examination of the hardware interface topics, which are considered of vital importance when building projects around microcontrollers. Through a scholastic examination of the critical, hardware-related practices on microcontrollers (arranged around carefully designed example codes and explanatory figures) the book provides to the reader a unique experience in establishing a clear link between the software and hardware.

This area of study is essential in building the solid background required to make the passage from "*have knowledge of*" to "*mastering*" the design of microcontroller-based projects and applications.

Chapter 4 applies to sensors (used in microcontroller projects) and to the data acquisition process. Because modern sensor devices constantly pave the way for creativity and innovation in embedded solutions, this chapter aims to inspire the interest and curiosity of the reader, through examples that apply to the detection of orientation, motion, gesture, distance, and color sensing. The examples are implemented with some of the most popular and contemporary boards in the worldwide market, that is, Teensy 3.2, TinyZero, and Micro:bit. The examples are designed in a way so that they direct readers in achieving *simplicity in design*. The process of interfacing with mobile phone through Bluetooth technology is also explored. Once again, the explanatory figures of the chapter are conducted with particular devotion in order to help readers achieve a deep understanding of the explored topics.

Chapter 5 applies to the *tinkering* practices of μC-based electronic products using Arduino-related hardware and software tools, as well as *prototyping* techniques using 3D printing technology. Having dealt with the theoretical and practical topics covered by the previous chapters (mainly by Chapters 2–4), the reader should be ready to proceed to practices related to real-world projects, as those covered by this particular chapter. *Creativity* and *simplicity in design* are two of main features that are addressed by the carefully thought examples of this chapter.

The freeware tools used by the current project are the following:

- **Arduino integrated development environment (**IDE**)** Software tool to develop (and upload) μC code.
- **Termite** RS 232 terminal console (data write/read to/from serial port).
- **Free Serial Port Monitor** Software tool to spy data exchanged through the serial port.
- **Notepad++** Editor for the source code development.
- **Minimalist GNU for Windows (**MinGW**)** Open source programming tool.
- **Gnuplot** Command-line driven graphing utility (can be invoked by C code).
- **OpenGL and GLUT** Open graphics library and toolkit for 2D and 3D vector graphics.
- **FreeCAD** General-purpose parametric 3D computer-aided design (CAD) modeler.
- **Ultimaker Cura** 3D printing software for preparing prints.

Additionally, the book offers a variety of supplementary resources – including source codes and examples – hosted on an accompanying website to be maintained by the author: www.mikroct.com.

Acknowledgments

I would like to thank God for blessing me with the passion of creation. I am truly thankful for my family, and I wish to thank my sweet wife Katerina who has always been there, supporting me in pursuing my dreams. I would like to sincerely thank Wiley for the opportunity to publish this project with one of the world's leading publishers. I thank from the bottom of my heart my kind commissioning editor, Sandra Grayson, for her trust and all the support with this project, as well as the rest of the team that worked to complete this book. Lastly, I am eternally grateful to my generous Professor Georgios A. Evangelakis for his immense and contribution during my maiden authoring venture.

Abbreviations

Greek-English Alphabet:

μC	microcontroller
μCT	micro-computational thinking
μP	microprocessor
μS	microsecond(s)
kΩ	kiloohm

English Alphabet:

ACC	accumulator
ACCLRM	accelerometer
ACK	acknowledge
ACK(M)	acknowledge (master)
ACK(S)	acknowledge (slave)
ADC	analog-to-digital converter
API	application programming interface
APP	application
ASCII	American standard code for information interchange
ASIC	application-specific integrated circuit
B	byte
BAT	battery
BCD	binary-coded decimal
BIN	binary
BIT	binary digit
BLE	Bluetooth low energy
BOM	bill of material
CAD	computer-aided design
CAN	controller area network
CE	computer engineering
CSG	constructive solid geometry

CK	content knowledge
cm	centimeter
CNC	computer numerical control
Coeffs	coefficients
CPHA	clock phase
CPLD	complex programmable logic device
CPOL	clock polarity
CPU	central processing unit
CS	computer science
CT	computational thinking
DAC	digital-to-analog converter
DAQ	data acquisition
DC	duty cycle or direct current
DEC	decimal
DIY	do it yourself
DOF	degrees of freedom
EE	electronic engineering
EEPROM	electrically erasable programmable read-only memory
EN	enable
F	frequency
FDM	fused deposition modeling
FPGA	field-programmable gate array
FW	firmware
GCC	GNU compiler collection
GND	ground
GPIO	general-purpose input/output
H	height
HDL	hardware description language
HEX	hexadecimal
HW	hardware
Hz	hertz
I2C	inter-integrated circuit
IC	integrated circuit
IDE	integrated development environment
IO	input/output
ISP	in-system programming
ISR	interrupt service routine
kB	kilobyte
L	length
LED	light-emitter diode
LiPo	lithium polymer
LSB	least significant bit
LSD	least significant digit
MEMS	micro-electro-mechanical systems
MinGW	minimalist GNU for Windows

MISO	master input slave output
MIT	Massachusetts Institute of Technology
MOSI	master output slave input
m	meter(s)
mm	millimeter(s)
ms	millisecond(s)
MSB	most significant bit
MSD	most significant digit
NACK	no-acknowledge
NACK(M)	no-acknowledge (master)
NACK(S)	no-acknowledge (slave)
ns	nanosecond(s)
ODR	output data rate
OEM	original equipment manufacturer
OpenGL	open graphics library
OS	operating system
Pa	Pascal
PBL	project-based learning
PC	program counter or personal computer
PCB	printed-circuit board
PCK	pedagogical content knowledge
PEI	polyetherimide
PETG	polyethylene terephthalate glycol
PLA	polylactic acid
PLD	programmable logic device
PK	pedagogical knowledge
PRNG	pseudo-random number generator
PTFE	polytetrafluoroethylene
PWM	pulse width modulation
quo	quotient
R	radius
RAM	random-access memory
RC	remote control
reg	register
rem	remainder
RF	radio frequency
RGB	red, green, blue
ROM	read-only memory
RS-232	recommended standard 232
RTC	real-time clock
R/W	read/write
s	second(s)
SCL	serial clock
SCLK	serial clock
SDA	serial data

SDK	software development kit
SiP	system in package
SMT	surface-mount technology
SoC	system on chip
SPI	serial peripheral interface
SPLD	simple programmable logic device
SS	slave select
SRAM	static random-access memory
SW	software
T	time period (or period)
TCK	technological content knowledge
ToF	time-of-flight
TK	technology knowledge
TPACK	technological pedagogical content knowledge
TPK	technological pedagogical knowledge
UART	universal asynchronous receiver/transmitter
USB	universal serial bus
UVPROM	ultraviolet programmable read-only memory
v	volts
VPL	visual programming language
W	width

Syllabus

The book provides a series of example codes to help learners explore, in practice, the critical topics related to microcomputer programming as well as to establish a clear link between the firmware and hardware. The carefully designed examples intend to inspire readers' *critical thinking* and *creativity*, and help them to gradually make the journey from "*have knowledge of*" to "*mastering*" the design of microcontroller-based projects. Hereafter is the recommended syllabus. To provide a sense of compatibility between the Arduino and Embedded C programming, only a limited set of Arduino-specific functions is utilized, apart from the requisite – in Arduino sketches – *setup()* and *loop()* functions.

The first column of Table P.1, presented below, incorporates each lesson of the course. The second column incorporates the examples codes examined by each lesson. The third column employs the hardware board where each example runs. The fourth column provides information about the accompanying software needed to explore each particular example. The final column provides information about what readers can expect to learn by each example code. In addition, Table P.2 summarizes the Arduino-specific (i.e. non-C language) functions used by the book. Then Table P.3 presents the Arduino-original, third-party, as well as the custom-designed libraries. Finally, Table P.4 introduces the custom-designed software running on the host personal computer (PC).

Table P.1 Recommended syllabus (description of example codes of each lesson).

Lesson	Example	Hardware	Software	Description
1	Ex2_1	Arduino Uno	Arduino serial monitor and temite	**Data output from the μC to a host PC.** The reader explores: (i) printable and nonprintable ASCII characters and *strings*; (ii) storage of stings in either *Data Memory* or *Program Memory* of Arduino Uno.
	Ex2_2	Arduino Uno	Arduino serial monitor	***Datatypes* and *typecasting* tasks.** The reader explores: (i) signed/unsigned data, datatypes of *bool*, *char*, *short*, *int*, *long*, and *float*; (ii) *typecasting* conversions between datatypes; (iii) number of bytes reserved in Arduino Uno memory by each particular datatype (*sizeof()* operator).
2	Ex2_3	Arduino Uno	Arduino serial monitor	**Program flow of control.** The reader explores: (i) *unconditional branching* with *while()*; (ii) repetition of a code part with *for()*; (iii) the *variable scope* coding style in *for()* loops. Examples are applied to *printable* and *nonprintable* ASCII and to *typecasting*, while the *assignment (=)*, *comparison (<)*, and *compound (++)* operators are used by the example.
	Ex2_4	Arduino Uno	Arduino serial monitor	**Boolean operators in program flow of control.** The reader explores the *Logical OR(II)* and *AND(&&)* operators along with the *if()* and *continue* statements, so as to implement *conditional branch*.
	Ex2_5	Arduino Uno	Temite	**Data input to the μC from a host PC.** The reader explores how to access the elements of an array via either *subscripts* when using loops of *predefined* iterations, or *pointers* when using loops of *undefined* iterations. In the latter case the reader explores the meaning of a null-terminated string, as well as the influence of a *unary pre-increment/decrement* or *post-increment/decrement* operation to a pointer. The *Indirection (*)*, *Address-of (&)*, and *Logical NOT (!)* operators are utilized by the example.

Table P.1 (Continued)

Lesson	Example	Hardware	Software	Description
3	Ex2_6	Arduino Uno	Temite	**Convert *numeric string* to *binary* value.** The reader explores further the issues learned before (i.e. inputting data to the µC, accessing arrays with pointers, exploiting logical operations in the program flow of control) and, in addition, he/she applies *arithmetic* operations toward converting a *numeric string* to *binary* value. The *Arithmetic* operators of *addition (+)*, *subtraction (−)*, and *multiplication (*)* are utilized by the example.
	Ex2_6b	Arduino Uno	Temite	**Convert *binary* value to *numeric string*.** The reader explores further the *arithmetic* operations in the µC's memory via the opposite operation than before. That is, by converting a *binary* value to *numeric string*. The conversion is performed in two steps: (i) by applying *BCD encoding* to the binary value and (ii) by converting the *BCD encoding* outcome to ASCII character. The latter task is performed via a *switch()...case* statement. The *Arithmetic* operators of *division quotient (/)* and *division remainder (%)* are utilized by the example.
	Ex2_7	Arduino Uno	Temite	**2's complement representation of signed numbers.** The reader exploits the entry of a numeric string from the host PC, which represents a single-byte decimal value (i.e. from 0 to 255). Through particular bitwise and arithmetic operations performed by the µC, the reader explores the signed number representation in 2's complement arrangement. The method of extracting bit information from a variable using *bitmasks*, along with bitwise tasks, is also explored.
	Ex2_7b	Arduino Uno	Temite	**Multiply/Divide binary numbers to a power of 2.** The reader explores further the *bitwise* operations, and in particular the *bit-shift* left (<<) and right (>>) operators, toward multiplying a binary number with a power of 2 (e.g. 2^3, 2^7, and so on).

(*Continued*)

Table P.1 (Continued)

Lesson	Example	Hardware	Software	Description
	Ex2_8	Arduino Uno	Temite	**Code decomposition with *functions*.** The reader explores further how to decompose the tasks performed by the previous example codes, with *functions*.
	Ex2_8b	Arduino Uno	Temite	**Code decomposition with *macros*.** The reader explores further the code decomposition techniques with the use of *macros* as well as the difference between *macros* and *functions*.
4	Ex3_1	Arduino Uno	—	**Blinking LED.** The reader explores how to provide, from the μC device, a *digital pin interface* with the outside world. Through the simplest example of a blinking LED, the reader realizes the rapid instruction execution by the μC's processor, as well as the need to slow down particular processes, through *delay()* routines, so as to support the user interface.
	Ex3_1b	Arduino Uno	—	**Blinking LED.** The reader explores further how to generate an easily upgradable code with the use of *global variables*.
	Ex3_1c	Arduino Uno	—	**Blinking LED.** The reader explores further how to generate an easily upgradable code with the use of *object-like macros* as well as the space in memory reserved by *macros* compared to the *global variables*.
	Ex3_1d	Arduino Uno	—	**Blinking LED.** The reader explores further how to generate a more readable code with the use of *functions*.
	Ex3_1e	Arduino Uno	—	**Blinking LED.** The reader explores further how to generate a more readable code with the use of *function-like macros*.
	Ex3_2	Arduino Uno	—	**Digital PORT interface.** The reader explores how the digital IO pins of the Arduino Uno are grouped together and how to concurrently access all pins of a PORT without using Arduino-specific functions.
	Ex3_2b	Arduino Uno	—	**Digital PORT interface.** The reader explores further how to concurrently access some, but not all pins of a PORT without using the Arduino-specific functions. To this end, the reader exploits *bitwise* operators along with the proper *bitmasks*.

Table P.1 (Continued)

Lesson	Example	Hardware	Software	Description
	Ex3_2c	Arduino Uno	Temite	**Digital PORT interface.** The reader explores further the direct manipulation of a PORT via *bitwise* operators and *bitmasks*, which are now incorporated within custom *function-like macros*. The reader explores the delay generated by the coding differences in the examples Ex3_2, Ex3_2b, Ex3_2c, when removing the long delay of 1 s.
	Ex3_3	Arduino Uno	—	**Digital PORT interface.** The reader explores further the latency generated when controlling two digital pins of the same PORT *sequentially*, compared to the instant response of the pins when they are controlled *concurrently*, via the PORT registers.
	Ex3_4	Arduino Uno, push-button	—	**Digital data input to the μC with a push-button.** The reader explores the data input process from a digital pin, using a common *push-button* as input unit. The connection of the push-button to one of the μC pins via a pull-up resistor, is also analyzed. (Data output using a LED is also carried out.)
	Ex3_5	Arduino Uno, push-button	—	**Debounce delay.** The reader explores the *switch-bounce phenomenon* and how it can affect the regular functionality of a system, and address a *debounce delay* to resolve that issue.
	Ex3_5b	Arduino Uno, push-button	—	**Digital data input via direct port manipulation.** The reader explores the data input process from a digital pin, via direct manipulation of the PORT's registers using proper *bitwise operations* and *bitmasks*.
5	Ex3_6	Arduino Uno, push-button	Arduino serial monitor and plotter	**Analog data input.** The reader explores how to read analog data from the μC when the latter encompasses an ADC.
	Ex3_7	Arduino Uno, short-circuit cable	Arduino serial monitor and plotter	**Analog data output via PWM.** The reader explores how to indirectly generate an analog signal, via the mean value occurred by a *pulse width modulation*.

(Continued)

Table P.1 (Continued)

Lesson	Example	Hardware	Software	Description
	Ex3_8	Arduino Uno, push-button	Arduino serial monitor	**Interrupt.** The reader explores how to enable *external interrupt* to a special-function digital pin and how the processor invokes an *interrupt service routine (ISR),* as well as resumes the control flow after servicing the ISR.
6	Ex3_9	Arduino Uno	Arduino serial monitor	**Built-in UART serial interface.** The reader explores, in detail, the functionality of the UART interface through a simple example, which repeatedly transmits the ASCII character 'A' to the host PC. An analysis of the quite known *8N1* type of communication, through the observation of the serial data acquired by an oscillator, is performed.
	Ex3_10	Arduino Uno, Uno click shield, FTDI click	Temite	**External UART module.** The reader explores how to implement a UART serial interface using an external module. (That particular task is useful when incorporating a μC of no built-in UART.)
	Ex3_11	Arduino Uno, short-circuit cable	Temite	**Software-implemented UART.** The reader explores how to implement a software-based UART serial interface via regular IO pins (particularly useful when no *built-in* or *external* UART is available).
7	Ex3_12	Arduino Uno, Uno click shield, BME280 sensor (SPI config.)	Temite	**Built-in SPI serial interface (read).** The reader is introduced, (through a simplified example) to the operating principles of the SPI serial interface. The example performs an SPI read to the *chip id* of BME280 environmental sensor and sends data to the host PC.
	Ex3_12b	Arduino Uno, Uno click shield, BME280 sensor (SPI config.)	Temite	**Built-in SPI serial interface (write).** The reader explores further the SPI operating principles, by applying an SPI write to one of the registers of BME280 sensor, used for the configuration of the device.
	Ex3_12c	Arduino Uno, Uno click shield, BME280 sensor (SPI config.)	Temite	**Built-in SPI serial interface (multiple-byte read).** The reader explores further how to speed up the SPI reading transaction, through the possibility of applying multiple-byte read to the adjacent registers of an SPI peripheral device.

Table P.1 (Continued)

Lesson	Example	Hardware	Software	Description
	Ex3_13	Arduino Uno, Uno click shield, BME280 sensor (SPI config.)	Arduino serial monitor	**Built-in SPI (environmental measurements).** The reader explores the first real-world project, which obtains environmental measurements from BME280 sensor. Through this particular example, the reader explores how to build an SPI driver for the sensor device as well as perform advanced arithmetic and bitwise operation, which are needed to obtained the compensated *Temperature*, *Pressure*, and *Humidity* values. (The requisite operations are described, in detail, in the BME280 datasheet.)
	Ex3_14	Arduino Uno, Uno click shield, BME280 sensor (SPI config.)	Temite	**Software-implemented/custom-designed SPI.** As before, the reader (through a simplified example) performs an SPI read to the *chip id* of BME280 environmental sensor. However, this time we address a custom-defined (i.e. software-implemented) SPI library. The reader explores how to build custom-designed Arduino (i.e. C++) libraries. This library is quite useful for μCs of no built-in SPI, as well for the case where pins other than the dedicated SPI pins are required (for a particular reason) to drive the SPI module.
	Ex3_15	Arduino Uno, Uno click shield, BME280 sensor (SPI config.)	Arduino serial monitor	**Software-implemented/custom-designed SPI.** As before, the reader obtains environmental measurements from BME280 sensor, but using the custom-designed SPI library of the previous example.
8	Ex3_16	Arduino Uno, Uno click shield, BME280 sensor (I2C config.)	Termite	**Built-in I2C serial interface (read).** The reader is introduced, (through a simplified example) to the operating principles of the I2C serial interface. The example performs an I2C read to the *chip id* of BME280 environmental sensor and sends data to the host PC.
	Ex3_16b	Arduino Uno, Uno click shield, BME280 sensor (I2C config.)	Temite	**Built-in I2C serial interface (write).** The reader explores further the I2C operating principles, by applying an I2C write to one of the registers of BME280 sensor, used for the configuration of the device.

(*Continued*)

Table P.1 (Continued)

Lesson	Example	Hardware	Software	Description
	Ex3_16c	Arduino Uno, Uno click shield, BME280 sensor (I2C config.)	Temite	**Built-in I2C serial interface (multiple-byte read).** The reader explores further how to speed up the I2C reading transaction, through the possibility of applying multiple-byte read to the adjacent registers of an I2C peripheral device.
	Ex3_17	Arduino Uno, Uno click shield, BME280 sensor (I2C config.)	Arduino serial monitor	**Built-in I2C (environmental measurements).** As in the SPI example, the reader explores how to build an I2C driver for the acquisition of environmental measurements by BME280 sensor.
	Ex3_18	Arduino Uno, Uno click shield, BME280 sensor (I2C config.)	Temite	**Software-implemented/custom-designed I2C.** As in the SPI example, the reader (through a simplified example) performs an I2C read to the *chip id* of BME280 environmental sensor. However, this time we address a custom-defined (i.e. software-implemented) I2C library. The reader explores how to build the custom-designed Arduino (i.e. C++) library for the I2C control of a peripheral module. This library is quite useful for μCs of no built-in I2C, as well for the case where pins other than the dedicated I2C pins are required (for a particular reason) to drive the I2C module.
	Ex3_19	Arduino Uno, Uno click shield, BME280 sensor (I2C config.)	Arduino serial monitor	**Software-implemented/custom-designed I2C.** As before, the reader obtains environmental measurements from BME280 sensor, but using the custom-designed I2C library of the previous example.
9	Ex4_1	Arduino Uno, Uno click shield, BME280 sensor (I2C config.)	Arduino serial monitor, free serial port monitor, *Serial_Ex4_01.c*	**Real-time monitoring of atmospheric pressure.** The reader exploits BME280 sensor (which has been used in the previous chapter) toward implementing a *data acquisition system (DAQ)*. The reader explores, in depth, the critical timings that determine the hardware's sampling interval (i.e. the dead time among adjacent air pressure samples). Then he/she learns how to build a custom-designed C language code for acquiring the pressure samples that arrive (from the μC device) to the serial port (i.e. USB-to-UART) of a host PC. The samples are stored, through the C code, to a *text (.txt)* file for additional *offline* analysis. To build the *executable (.exe)* file from the C source code, the reader exploits the freeware *Minimalist GNU for Windows (MinGW)* programming tool, while the source code development is performed by the freeware *Notepad++* editor.

Table P.1 (Continued)

Lesson	Example	Hardware	Software	Description
	Ex4_1b	Arduino Uno, Uno click shield, BME280 sensor (I2C config.)	Arduino serial monitor, *Serial_ Ex4_01.c*	**Data acquisition of atmospheric pressure.** The reader implements the same example code but this time, the μC acquires 100 samples temporarily stored to its data memory, and then, the μC transmits the 100 samples to the host PC (*non-real-time monitoring*).
	Ex4_1c	Arduino Uno, Uno click shield, BME280 sensor (I2C config.)	Arduino serial monitor, *Serial_ Ex4_01-gnuplot.c*	**Data acquisition of atmospheric pressure.** The reader explores further how to provide a graphical monitoring feature to the C code running on the host PC. For that particular purpose, the reader exploits the freeware *Gnuplot* graphing utility, which can be invoked by C code.
	Ex4_2	Arduino Uno, Uno click shield, 2× BME280 sensor (I2C config.)	Arduino serial monitor and plotter, *Serial_ Ex4_02-gnuplot.c*	**Data acquisition of atmospheric pressure.** The reader upgrades Ex4_1 (firmware and software) so that air pressure measurements are concurrently obtained from two identical sensors. This particular setup can be used for calculating the absolute height in between the two sensor devices (i.e. a regular implementation in indoor navigation systems, and other creative applications).
10	Ex4_3	Teensy 3.2	Arduino serial monitor	**Getting started with Teensy 3.2.** This simple example, which blinks a LED and sends a *string* to the USB-to-Serial port of a host PC, attempts to help the reader in getting started with the 32-bit Teensy 3.2 board. To download the code into the Teensy board, the user needs to install (the available for free) *Teensyduino* add-on software.
	Ex4_4	Teensy 3.2, Pesky's BNO055 + BMP280	Arduino serial monitor, *Teensy-duino, Serial_ Ex4_01-gnuplot-TEENSY.c*	**DAQ of atmospheric pressure with Teensy 3.2.** The reader learns how to obtain air pressure samples from BMP280 sensor found on the *Pesky's BNO055 + BMP280* module. The reader exploits the compatibility between BMP280 and BME280 sensors and addresses the same driver, used earlier for the BME280 device. Thereby, the reader stays focused on the differentiations found in the application code between *Teensy* and *Arduino Uno* boards.
	Ex4_5	Teensy 3.2, Pesky's BNO055 + BMP280	Arduino serial plotter	**Orientation detection.** The reader continuous the involvement with contemporary sensor devices and in this example, he/she exploits BNO055 *system in package (SiP)* toward the *orientation detection* expressed with *Euler* angles (i.e. *Pitch*, *Roll*, *Yaw*).

(*Continued*)

Table P.1 (Continued)

Lesson	Example	Hardware	Software	Description
	Ex4_5b	Teensy 3.2, Pesky's BNO055 + BMP280	*BNO055-gnuplot.c and BNO055-openGL.c*	**Orientation detection.** The previous firmware has been updated to work with the custom-designed DAQ software (i.e. *BNO055-gnuplot.c*) running on the host PC, as well as with the *OpenGL*-based software explained at the end of this subchapter (i.e. *BNO055-openGL.c*). The latter software requires installing *GLUT* toolkit and generates a real-time rotation of 3D box on host PC, where rotation is in agreement with the *Euler* angels obtained from the hardware board.
	Ex4_6	Teensy 3.2, Pesky's BNO055 + BMP280	Arduino serial plotter	**Gesture recognition and motion detection.** The reader exploits BNO055 SiP toward the *Gravity vector* as well as *Linear acceleration* detection. Then the reader explores further how to utilize *Gravity vector* toward applying *gesture recognition*, and how to utilize *Linear acceleration* toward *motion detection*.
	Ex4_6b	Teensy 3.2, Pesky's BNO055 + BMP280	*BNO055-gnuplot.c*	**Gesture recognition and motion detection.** The previous firmware has been updated to work with the custom-designed DAQ software (i.e. *BNO055-gnuplot.c*) running on the host PC.
11	Ex4_7	TinyZero	Arduino serial monitor	**Getting started with TinyZero.** This simple example, which blinks a LED and sends a *string* to the USB-to-Serial port of a host PC, attempts to help the reader in getting started with the 32-bit TinyZero board. To download the code into the TinyZero board, the user should follow the directions provided by the book.
	Ex4_8	TinyZero, wireling adapter TinyShield, 2× TOF sensor wireling	Arduino serial monitor and plotter, *VL53L0X.c*	**Distance detection and 1D gesture recognition.** The reader is introduced to the distance (ToF) sensor VL53L0X and develops a firmware, which applies either to distance detection (when using a single sensor) or to 1 dimension (1D) gesture recognition (when using two identical sensors). The accompanying DAQ software *VL53L0X.c* applies to *distance detection*.

Table P.1 (Continued)

Lesson	Example	Hardware	Software	Description
12	Ex4_9	Micro:bit	Arduino serial monitor	**Getting started with Micro:bit.** This simple example, which read two push-buttons, blinks a LED and sends a *string* to the USB-to-Serial port of a host PC, attempts to help the reader in getting started with the 32-bit Micro:bit board. To download the code into the Micro:Bit board, the user should follow the instructions provided by the book.
	Ex4_10	Micro:bit, RGB sensor board	Arduino serial monitor and plotter, *TCS34725-openGL.c*	**RGB color sensing.** The reader is introduced to color sensing techniques and develops firmware as well as software for the host PC, which recognizes *red (R)*, *green (G)*, and *blue (B)* colors (via the Adafruit's board employing TCS3472 sensor). It is here noted that the software code for the PC constitutes *OpenGL* code and, hence, *GLUT* toolkit is required for this particular example.
	Ex4_11	Micro:bit, RGB sensor board, mobile phone with Bluetooth 4.2	Arduino serial monitor, *Bluefruit Connect app*	**RGB color sensing via Bluetooth.** The reader upgrades the previous example code so that data are forwarded to a mobile phone, as well, via *Bluetooth low energy (BLE)*. To obtain the data to the mobile device (of Bluetooth version 4.2 or higher) the user should install the Adafruit *Bluefruit Connect app*.
	Ex4_11b	Micro:bit, RGB sensor board, *mobile phone with Bluetooth 4.2*	Arduino serial monitor	**RGB color sensing via Bluetooth.** The reader upgrades the previous example code so that data are forwarded to the mobile phone, as well, as soon as the user transmits any character to the Micro:bit board.
13	Ex5_1	Teensy 3.2, Pesky's nRF52 add-on for Teensy, *mobile phone with Bluetooth 4.2*	Arduino serial monitor, *Bluefruit Connect app*	**Tinkering an RC car.** The user explores a trial code so as to test the Bluetooth connectivity between the Teensy 3.2 board and a smartphone.
	Ex5_1b	Teensy 3.2, Pesky's nRF52 add-on for Teensy, *mobile phone with Bluetooth 4.2*	Arduino serial monitor, *Bluefruit Connect app*	**Tinkering an RC car.** The user explores how to decode the commands available at the *Control Pad* user interface of *Bluefruit Connect app*.
	Ex5_2	See Table 5.1	Arduino serial monitor, *Bluefruit Connect app*	**Tinkering an RC car.** The user develops the final firmware code for the control of an RC car through a smartphone device.

(*Continued*)

Table P.1 (Continued)

Lesson	Example	Hardware	Software	Description
14	Ex5_3	Micro:bit, Zip Halo	—	**Prototyping an interactive game for sensory play.** The reader starts the bottom-up design method of the prototype system via the control of a LED ring (found on the Kitronik's *Zip Halo* board), while also implementing a *random number generator*. (This code part is meant for triggering the children's *sight* sense.)
	Ex5_4	Micro:bit, Zip Halo, BMP280, DC blowing fan	Arduino serial plotter	**Prototyping an interactive game for sensory play.** The reader continues the bottom-up design method of the prototype system and explores how to generate a virtual button with an air pressure sensor and a DC fan. (This code part is meant for triggering the children's *proprioception* sense.)
	Ex5_5	Micro:bit, speaker module	—	**Prototyping an interactive game for sensory play.** The reader continues the bottom-up design method of the prototype system and this time, he/she explores how to reproduce sound from a μC and a speaker. (This code part is meant for triggering the children's *hearing* sense.)
	Ex5_6	See Table 5.2	Arduino serial monitor	**Prototyping an interactive game for sensory play.** The reader concludes the bottom-up design method with the implementation of the overall firmware of the proposed game, which engages the children's *sight*, *hearing*, and *proprioception* senses.
15	Explore 3D printing technology and prepare the apparatus of the prototype interactive game.			

Table P.2 Arduino-specific (i.e. non-C) functions used by the book examples.

Arduino-specific functions/macros	Examples using each function/macro
setup()	All examples
loop()	All examples
Serial.begin()	Ex2_1, Ex2_2, Ex2_3, Ex2_4, Ex2_5, Ex2_6, Ex2_6b, Ex2_7, Ex2_7b, Ex2_8, Ex2_8b, Ex3_6, Ex3_7, Ex3_8, Ex3_9, Ex3_10, Ex3_11, Ex3_12, Ex3_12b, Ex3_12c, Ex3_13, Ex3_15, Ex3_16, Ex3_17, Ex3_18, Ex3_19, Ex4_1, Ex4_1b, Ex4_1c, Ex4_2, Ex4_3, Ex4_4, Ex4_5, Ex4_5b, Ex4_6, Ex4_6b, Ex4_8, Ex4_10, Ex4_11, Ex4_11b, Ex5_1, Ex5_1b, Ex5_4, Ex5_6
Serial.print()	Ex2_1, Ex2_2, Ex2_3, Ex2_4, Ex2_5, Ex2_6, Ex2_6b, Ex2_7, Ex2_7b, Ex2_8, Ex2_8b, Ex3_10, Ex3_11, Ex3_12, Ex3_12b, Ex3_12c, Ex3_13, Ex3_14, Ex3_15, Ex3_16, Ex3_17, Ex3_18, Ex3_19, Ex4_3, Ex4_8
Serial.println()	Ex2_1, Ex2_2, Ex2_6, Ex3_6, Ex3_8, Ex3_12c, Ex3_15, Ex3_17, Ex3_19, Ex4_1, Ex4_1b, Ex4_1c, Ex4_2, Ex4_4, Ex4_5, Ex4_5b, Ex4_6, Ex4_6b, Ex4_10, Ex4_11, Ex4_11b, Ex5_1b, Ex5_4, Ex5_6
F()	Ex2_1, Ex2_2, Ex3_6
delay()	Ex2_1, Ex2_2, Ex2_3, Ex2_4, Ex2_5, Ex2_6, Ex2_6b, Ex3_1, Ex3_1b, Ex3_1c, Ex3_1d, Ex3_1e, Ex3_2, Ex3_2b, Ex3_2c, Ex3_5, Ex3_5b, Ex3_6, Ex3_7, Ex3_9, Ex3_10, Ex3_11, Ex3_12, Ex3_12b, Ex3_12c, Ex3_13, Ex3_14, Ex3_15, Ex3_16, Ex3_17, Ex3_18, Ex3_19, Ex4_1, Ex4_1b, Ex4_1c, Ex4_2, Ex4_3, Ex4_4, Ex4_5, Ex4_5b, Ex4_6, Ex4_6b, Ex4_7, Ex4_8, Ex4_8, Ex4_10, Ex4_11, Ex4_11b, Ex5_1, Ex5_1b, Ex5_2, Ex5_3, Ex5_4, Ex5_5, Ex5_6
Serial.available()	Ex2_5, Ex2_6, Ex2_6b, Ex2_7, Ex2_7b, Ex2_8, Ex2_8b, Ex4_1, Ex4_1b, Ex4_1c, Ex4_2, Ex4_4, Ex4_5b, Ex4_6b
Serial.read()	Ex2_5, Ex2_6, Ex2_6b, Ex2_7, Ex2_7b, Ex2_8, Ex2_8b, Ex4_1, Ex4_1b, Ex4_1c, Ex4_2, Ex4_4, Ex4_5b, Ex4_6b
Serial.write()	Ex2_6, Ex2_6b, Ex2_7, Ex2_7b, Ex2_8, Ex2_8b, Ex3_9, Ex5_1
pinMode()	Ex3_1, Ex3_1b, Ex3_1c, Ex3_1d, Ex3_1e, Ex3_4, Ex3_5, Ex3_5b, Ex3_8, Ex3_12, Ex3_12b, Ex3_12c, Ex3_13, Ex3_14, Ex3_15, Ex3_18, Ex3_19, Ex4_1, Ex4_1b, Ex4_1c, Ex4_2, Ex4_3, Ex4_7, Ex4_8, Ex5_2, Ex5_5, Ex5_6
digitalWrite()	Ex3_1, Ex3_1b, Ex3_1c, Ex3_1d, Ex3_1e, Ex3_4, Ex3_12, Ex3_12b, Ex3_12c, Ex3_13, Ex3_14, Ex3_15, Ex3_18, Ex3_19, Ex4_1, Ex4_1b, Ex4_1c, Ex4_2, Ex4_3, Ex4_7, Ex4_8, Ex5_2, Ex5_5, Ex5_6
digitalRead()	Ex3_4, Ex3_5, Ex3_11, Ex3_18, Ex3_19, Ex4_8
analogRead()	Ex3_6, Ex3_7, Ex5_3, Ex5_5, Ex5_6
analogWrite()	Ex3_7, Ex3_8, Ex5_2
attachInterrupt()	Ex3_8
SPI.begin()	Ex3_12, Ex3_12b, Ex3_12c, Ex3_13
SPI.beginTransaction()	Ex3_12, Ex3_12b, Ex3_12c, Ex3_13
SPI.transfer()	Ex3_12, Ex3_12b, Ex3_12c, Ex3_13
delayMicroseconds()	Ex3_14, Ex3_15, Ex3_18, Ex3_19, Ex5_5, Ex5_6

(*Continued*)

Table P.2 (Continued)

Arduino-specific functions/macros	Examples using each function/macro
Wire.begin()	Ex3_16, Ex3_17, Ex4_1, Ex4_1b, Ex4_1b, Ex4_2, Ex4_4, Ex4_5, Ex4_5b, Ex4_6, Ex4_6b, Ex4_8
Wire. beginTransmission()	Ex3_16, Ex3_17, Ex4_1, Ex4_1b, Ex4_1c, Ex4_2, Ex4_4, Ex4_5, Ex4_5b, Ex4_6, Ex4_6b
Wire.write()	Ex3_16, Ex3_17, Ex4_1, Ex4_1b, Ex4_1c, Ex4_2, Ex4_4, Ex4_5, Ex4_5b, Ex4_6, Ex4_6b
Wire.endTransmission()	Ex3_16, Ex3_17, Ex4_1, Ex4_1b, Ex4_1c, Ex4_2, Ex4_4, Ex4_5, Ex4_5b, Ex4_6, Ex4_6b
Wire.requestFrom()	Ex3_16, Ex3_17, Ex4_1, Ex4_1b, Ex4_1c, Ex4_2, Ex4_4, Ex4_5, Ex4_5b, Ex4_6, Ex4_6b
Wire.available()	Ex3_16, Ex3_17, Ex4_1, Ex4_1b, Ex4_1c, Ex4_2, Ex4_4, Ex4_5, Ex4_5b, Ex4_6, Ex4_6b
Wire.read()	Ex3_16, Ex3_17, Ex4_1, Ex4_1b, Ex4_1c, Ex4_2, Ex4_4, Ex4_5, Ex4_5b, Ex4_6, Ex4_6b
Wire.setSDA()	Ex4_4, Ex4_5, Ex4_5b, Ex4_6, Ex4_6b
Wire.setSCL()	Ex4_4, Ex4_5, Ex4_5b, Ex4_6, Ex4_6b
SerialUSB.begin()	Ex4_7, Ex4_8
SerialUSB.print()	Ex4_7
SerialUSB.println()	Ex4_8
Serial1.begin()	Ex5_1, Ex5_1b, Ex5_2
Serial1.available()	Ex5_1, Ex5_1b, Ex5_2
Serial1.read()	Ex5_1, Ex5_1b, Ex5_2
randomSeed()	Ex5_3, Ex5_5, Ex5_6
random()	Ex5_3, Ex5_5, Ex5_6

Table P.3 Arduino-original, third-party, and custom-designed libraries.

Arduino original, third-party, and user-defined libraries	Description (and examples invoking the library)
SoftwareSerial.h	Arduino library to implement a software-based UART (Ex3_11)
SPI.h	Arduino library of the built-in SPI (Ex3_12, Ex3_12b, Ex3_12c, Ex3_13)
__CH3_BME280spi.h	User-defined library applying to the built-in SPI driver of BME280 (Ex3_13)
swSPI.h and swSPI.cpp	Custom-designed Arduino (C++) library applying to SPI interface (Ex3_14)
Wire.h	Arduino library of the built-in I2C (Ex3_16, Ex3_16b, Ex3_16c, Ex3_17, Ex4_4, Ex4_5, Ex4_5b, Ex4_6, Ex4_6b, Ex4_8, Ex4_10, Ex4_11, Ex4_11b)
__CH3_BME280i2c.h	User-defined library applying to the built-in I2C driver of BME280 (Ex3_17)
swWire.h and swWire.cpp	Custom-designed Arduino (C++) library applying to I2C interface (Ex3_18)
__CH4_BME280i2c_revA.h	Revised (user-defined) driver for BME280 (Ex4_1, Ex4_1b, Ex4_1c, Ex4_2, Ex4_4)
__CH4_BNO055i2c_revA.h	User-defined library applying to the built-in I2C driver of BNO055 (Ex4_5, Ex4_5b, Ex4_6, Ex4_6b)
Wireling.h	TinyZero library for Arduino code which controls *Wireling adapter*(Ex4_8)
VL53L0X.h	TinyZero library for Arduino code which controls VL53L0X distance sensor
Adafruit_TCS34725.h	Adafruit library for Arduino code which controls TCS3472 color (RGB) sensor
Adafruit_Microbit.h	Adafruit library for the programming of Micro:bit with Arduino code (Ex4_11, Ex4_11b)
Adafruit_TCS34725.h	Adafruit library for the control of TCS34725 RGB color sensor (Ex4_11, Ex4_11b)
Adafruit_NeoPixel.h	Adafruit library for the control of a LED ring (Ex5_3, Ex5_4, Ex5_5, Ex5_6)
SlowSoftWire.h	Third-party library for Arduino code which emulates *Wire.h* library (Ex5_4, Ex5_6)
DFRobot_BMP280.h	DFRobot library for the control of BMP280 barometric sensor (Ex5_4, Ex5_6)
__CH5_MusicNotes.h	User-defined library associating music notes to the semi-period of each note (Ex5_5, Ex5_6)

Table P.4 Custom-designed software running on the host PC.

Custom-designed software running on the host PC	Description (and examples invoking the library)
Serial_Ex4_01.c and Serial.h	DAQ software acquiring air pressure samples via USB-to-UART port (Ex4_1, Ex4_1b)
Serial_Ex4_01-gnuplot.c and Serial.h	DAQ software acquiring air pressure samples via USB-to-UART port (Ex4_1c)
Serial_Ex4_02-gnuplot.c and Serial.h	DAQ software acquiring air pressure samples via USB-to-UART port (Ex4_2)
Serial_Ex4_01-gnuplot-TEENSY.c and Serial.h	DAQ software acquiring air pressure samples (via USB-to-UART port) from Teensy 3.2 board (Ex4_4)
BNO055-gnuplot.c and Serial.h	DAQ software acquiring either *Euler angles* or *Gravity-vector/Linear-acceleration* (Ex4_5b, Ex4_6b)
BNO055-openGL.c and Serial.h	DAQ software acquiring *Euler angles* and portraying 3D object graphics (Ex4_5b)
VL53L0X.c and Serial.h	DAQ software acquiring distance from a single ToF sensor (EX4_8)
TCS34725-openGL.c and Serial.h	DAQ software detecting *red (R)*, *green (G)*, and *blue (B)* colors

1

The Art of Embedded Computers

The rapid evolution of embedded computers, along with the abundant educational possibilities they offer, has attracted the interest of instructors at all levels of education. This chapter recommends five distinct categories of embedded computers, in terms of the tasks linked either to *computer science (CS)* or *electronic engineering (EE)* discipline. An overview of this interdisciplinary technology between the two disciplines aims at helping instructors clarify the possibilities and limitations, as well as learning difficulties of each category. Then the chapter provides readers with a unique perspective on the educational aspects of microcomputer programming and application development. The analysis applies to the *technological pedagogical content knowledge* (*TPACK*) model and attempts to clarify why the programming language should be considered as the technology integration toward helping students create their knowledge about the subject matter. It also justifies why the employed technology may arrange the tutoring more appropriate for CS or EE students.

Subsequent to this analysis, the author explores the additional endeavor required to understand the capabilities of microcomputer technology and addresses the coined *micro-computational thinking* (*μCT*) term to describe the thought processes involved in the solutions, carried out by a microcomputer-based system. The author does not intend to coin a new term entirely differentiated from the existing *computational thinking* (*CT*) concept, but rather to reveal the thought processes related to the application development with embedded computers.

To understand the today's impact of microcontroller technology on the *maker* industry, the chapter also follows the advancement of microcontroller programming and application development and identifies the *long-cycle and short-cycle* development era.

Overview of Embedded Computers and Their Interdisciplinarity

Computer programming teaching and learning has been thoroughly researched throughout the years and at all levels of education [1–3]. Lately, instructors experience the widespread dissemination of embedded computer systems and the challenge to enhance students' perspective in the field of embedded computer programming and application

Microcontroller Prototypes with Arduino and a 3D Printer: Learn, Program, Manufacture, First Edition.
Dimosthenis E. Bolanakis.

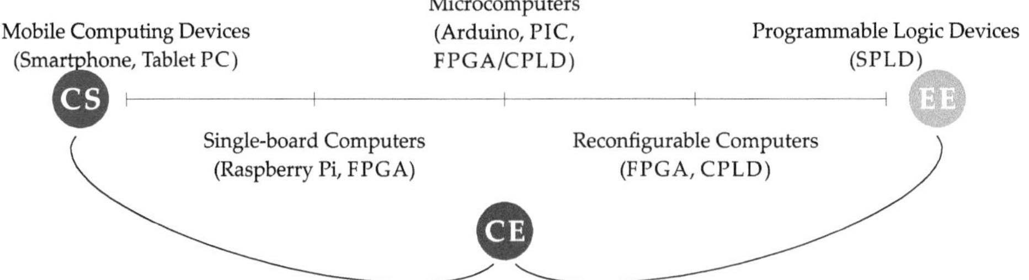

Figure 1.1 Interdisciplinarity of embedded computers.

development [4–6]. The questions raised are "What is the difference between the programming of a regular computer and an embedded computer system? What are the challenges for an educator who wishes to get involved with the second approach?"

It would be wise to start with a definition of the term *embedded computer*. Yet, it is sometimes essential to compromise with an informal definition of a complex and multifaceted theme that cannot be straightforwardly expressed within a single phrase. In consideration of the reader who is introduced to a field of study that does not necessarily fall within his/her area of expertise, the term is primarily addressed as follows. *Embedded computers* encompass any electronic device (contained in a hardware system) that can be programmed (with some kind of code) to carry out some computing. By definition, the process of developing programming code for an electronic device draws one's attention to an interdisciplinary task between the disciplines of CS and EE.

Figure 1.1 distinguishes five categories of embedded computers, in terms of the tasks that are more closely linked with the discipline of either CS or EE. To understand the position that each type of embedded computer holds on the proposed scheme, it is important to make reference to the basic features of conventional computer programming (which is originally rooted to CS).

Computer vs. Embedded Computer Programming and Application Development

The application software that is designed by a computer scientist or engineer to run on a personal computer interacts with the computer hardware through a system software known as the *operating system* (*OS*). The programming language used by the developer who builds custom-designed software incorporates utilities that are considered part of the OS. This set is referred to as the *application programming interface* (*API*) and encrypts the underlying hardware operations from the developer. For instance, the API in C programming language is declared in the header files, such as the "stdio.h," which embeds input and output functions. Hence, the developer learns how to exploit functions in order to *input/output* (*IO*) data to/from the computer system. Starting with the design of the simplest application, students are introduced to functions and syntax rules toward inputting/outputting data from/to the outside world. To do so, the students make use of two regular IO units of the computer, that is, the (input) keyboard and (output) monitor.

According to the aforementioned information, a computer program does not only direct the computer hardware toward computing tasks, it is also in charge of handling some IO units. Despite the standard keyboard and monitor units, the developer may perform more advanced IO operations through today's dominant computer interfaces, such as *universal serial bus* (*USB*), Ethernet, and so forth. In computer programming, advanced IO operations demand merely the calling of (perhaps) more complicated ready-to-use functions. The absence of these libraries would demand tremendous endeavor by the software developer just for the control of such IO units. One would have to go through several low-level tasks in order to access the computer hardware.

According to their complexity, an embedded system may or may not use an OS. If it does, then the emphasis is placed on software design tasks and the application development constitutes (more likely) a distinctive procedure for the computer scientist. Otherwise, the application development necessitates serious involvement with the hardware, and therefore it can be considered a more familiar territory for an electronic engineer. Due to the requisite accessibility at the machine level, the firmware development process has a need for special treatment of the incorporated programming language (compared to the software development for a personal computer, even if we use the exact same programming language). Moreover, an embedded computer application regularly incorporates nonstandard IO units that render the learning process of a novice designer more difficult.

All these aspects of complexity and interdisciplinarity of embedded computers are a case study of an interdisciplinary curriculum, which integrates partial training from CS and EE. This discipline is known as *computer engineering* (*CE*) and bears the major responsibility on the embedded computer programming and application development. However, due to the rapid evolution of this technology along with the abundant educational possibilities they offer, embedded computers often transcend the boundaries of CS, EE, and CE classes and migrate to several diverse disciplines [7, 8]. Some researchers have attempted to introduce embedded computers in K12 education, as well [9, 10]. In the following, the author proposes and overviews five distinct categories of embedded computers in order to help instructors clarify the possibilities and limitations and the learning difficulties of each category.

Group 1: Programmable Logic Devices

Adjacent to the discipline of EE discipline we could place the *programmable logic devices (PLDs)*. While this technology is nowadays considered obsolete, it is useful to be further explored in order to identify the (embedded computer) practices related to the discipline of electronics.

Twenty years ago, the development of an embedded computer system was not trivial. The process was regularly launched by the design of a *printed-circuit board* (*PCB*) for holding a set of carefully selected and interconnected components, outlining the functionality of the overall electronic system. If the system carried out some kind of digital computing, the PCB designer could select a PLD to build a reconfigurable digital circuit. For instance, the incorporation of a single PLD chip of some combinational logic that was determined by the designer, worked as a replacement for a few different logic gates (i.e. one chip replacing a few electronic chips). This option prevented the designer from wasting additional PCB resources, and subsequently extra working time and effort, as the PCB design alone constitutes a particularly time-consuming task.

The programming languages of such devices are called *hardware description languages (HDLs)* as they are used for describing the behavior of an electronic (hardware) circuit. Figure 1.1 uses the term *simple programmable logic devices (SPLDs)* for this technology, after the naming decided by Atmel Corporation (products that nowadays belong to Microchip Inc.). SPLDs are of electrically erasable flash memory technology, while a popular HDL that is used for their configuration is known as CUPL.

Group 2: Reconfigurable Computers

Near to the EE discipline, but with a greater distance from the allied technology of SPLDs, are two comparable technologies known as *complex programmable logic devices (CPLDs)* and *field-programmable gate arrays (FPGAs)*. To avoid too many technical details, we consider CPLDs as an advancement of the SPLDs, with gates, flip-flops, and fast IO pins, where the designer can upload more complicated digital circuits. FPGAs, on the other hand, offer even more IO pins as well as resources for the configuration of particularly sophisticated and high-speed designs. Therefore, FPGAs constitute an ideal solution for testing complicated circuits before proceeding to the production of an error-free *application-specific integrated circuit* (*ASIC*). Because of the possibility of implementing high-speed and complex computations along with the fast IO response, FPGAs has played an important role in fast data acquisition systems, like, for example, the ones that are used in the high energy physics experiments at CERN [11].

The dominant manufacturers of FPGAs and CPLDs are Altera and Xilinx, while the most widely used HDLs for their configuration are Verilog and VHDL. In contrast to the conventional programming method for computers (that is, the sequential programming where statements are executed in their order of occurrence in the code), HDLs are addressed for expressing concurrency. Therefore, the designer must have (during the programming process) a clear sense of the digital logic circuits that describe the structure and behavior of the potential electronic system (to be uploaded into the FPGA/CPLD). This is the reason why these two technologies are placed nearby the EE discipline, in Figure 1.1. We may also acknowledge the sequential programming in FPGAs and CPLDs and in that particular case, FPGAs and CPLDs can be shifted nearer in the direction of CS discipline.

The term *reconfigurable computer* is sometimes addressed to illustrate FPGA technology. Because CPLDs have been advanced to the point that can be exploited by other means, too (rather than just implementing the behavior of a digital circuit), we will use this term to describe both FPGAs and CPLDs (in terms of their position in Figure 1.1). All these issues will be discussed next in the chapter.

Group 3: Microcomputers

The term *microcomputer* has been entirely changed from its original meaning. To thoroughly illustrate this concept we need to identify the fundamental parts that compose a common computer system. These are: (i) the processor (which is responsible for the execution of machine instructions generated by a computer software), (ii) the memory (i.e. program and data memory, where the former holds the machine instructions and the latter, the data that may occur during the code execution), and (iii) the IO devices (that is, the means by which the IO units are connected to the computer so as to insert/pull-out data into/from the application system).

Based on the aforementioned definition, *microcomputer* term is regularly referred to as a complete computer embedded to a single chip. A popular microcomputer in our age is the microcontroller. The *microcontroller* (*μC*) is a chip device that incorporates program and data memory as well as IO devices that end up to the device's pins. Therefore, to build a microcomputer-based application the designer needs to connect some IO units to the microcontroller's pins (e.g. switches, keypads, sensors, liquid-crystal displays) and develop the code that inputs/outputs data to/from the microcontroller, as well as perform some computing. Today, the *do it yourself* (*DIY*) and *maker* cultures have established a wave of ready-to-use board systems for microcontrollers, which, along with the availability of free and shareable libraries over the internet, they have rendered feasible the rapid development of microcontroller-based applications. The designer may purchase a microcontroller-based "motherboard" as well as a separate (and compatible) daughterboard that employs, for instance, a Bluetooth module. Attaching the boards together and then utilizing a shareable library, the designer can straightforwardly implement an application that exchanges data between the microcontroller and a mobile phone.

The programming of a μC has dramatically changed the last decade. In the previous era, the μC was programmed with an assembly-level approach. This option required a deep understanding of the microcontroller's internal structure and a fluent handling of the assembly programming language. This programming method, together with the need for designing and developing a PCB from scratch, would have placed the microcontroller technology closer to the EE discipline. However, the passage to today's era where a higher level of programming is being dominated, and the PCB design is being eliminated (because of the available DIY board systems for microcontrollers), the μC technology could be positioned at the middle of CS and EE disciplines (Figure 1.1).

The most popular and widely used method to program a microcontroller device, nowadays, is the embedded C programming (which is based on the well-known C programming language plus a set of extensions). Because of the absence of an OS, the firmware development for a microcontroller device entails several register access operations in order to perform particular IO tasks with the outside world. This is perhaps the most important differentiation from the regular C programming method of a personal computer. Other extensions of familiar programming languages, such as the Pascal and Python, have been addressed for microcontroller programming, while the DIY culture of our age has engaged hobbyists' interests in the use of Arduino programming. The Arduino programming approach is based on ready-to-use libraries and modules, where the programming process is focused merely on the top-level code. However, the development of a microcontroller-based system cannot be defined by fixed hardware and, hence, learning microcomputer programming and application development entails involvement, to some extent, with some hardware resources [12].

Microcontrollers share this interdisciplinary (and equally distributed in both disciplines) position of Figure 1.1 with the CPLD and FPGA technologies (of the previous group of embedded computers). This is because CPLDs and FPGAs offer the possibility of building a microcontroller core within the reconfigurable computer, such as the PicoBlaze 8-bit microcontroller for the devices of the manufacturer Xilinx. This possibility relieves the demanding process of expressing concurrency of a digital computing system (as the incorporation of a sequential logic is, in many cases, required).

Group 4: Single-Board Computers

Next and near the CS discipline we may identify the single-board computers. Based on the aforementioned analysis on the computer's fundamental parts, as well as the title in this category, we refer to ready-to-use boards that incorporate a processor (or microprocessor), memory, and IO devices. This group of embedded computers is considered to run an OS and, hence, the programming approach looks much like the process followed for a regular computer.

A popular type of embedded computers in this category is the Raspberry Pi. This particular single-board computer incorporates a slot that accepts an SD card for holding the OS, while the designer may connect a monitor and a keyboard to the available HDMI and USB IO connectors, and program the Raspberry Pi as a common computer. The code development can be performed with a regular programming language, such as Python and C. Moreover, the board employs a connector with *general-purpose input/output* (*GPIO*) pins, which can be exploited as per need (and similarly to the pins of a microcontroller device). This additional feature of Raspberry Pi (and of other comparable boards), along with the fact that runs an OS, defines the position they hold on the diagram of Figure 1.1 (influenced more by CS and less by EE discipline).

We observe that single-board computers share this position in the diagram of Figure 1.1 with the FPGA technology (also present in the previous two groups of embedded computers). This sharing arises from the fact that the contemporary technology of FPGAs can be either delivered with a hardwired processor core, or being configured with a software-defined processor (such the Xilinx MicroBlaze 32-bit architecture), and thereafter to be used in association with an OS. It is worth noting that FPGAs constitute a particularly complicated technology, but also flexible in consideration of the way that is being utilized, as revealed by three positions they hold in the diagram of Figure 1.1.

Group 5: Mobile Computing Devices

The final group of embedded computers (specified as mobile computing devices) refers to all portable computers that are delivered with a mobile OS (such as the Android and iOS). Using this technology, the designer has an opportunity to develop software components (also known as *apps*) and exploit the device's IO units in a custom-designed application. Modern handheld and portable computers (such as the smartphones and tablets) embed several sophisticated components; that is, the touch screen, sensors (accelerometer, gyroscopes, barometers, etc.), communication interface modules (e.g. Bluetooth and WiFi), and so forth.

The programming and application development for this kind of embedded computers is limited to the utilization of a standard set of hardware devices, where low-level practices are encrypted by the OS. This is the reason why mobile computing devices are located this close to the CS discipline. However, the cutting-edge technology of this category provides to the educator several experimentation possibilities. For instance, an application development that utilizes the barometer sensor can direct the students toward an inquiry-based learning approach in barometric altimetry [13].

The code development method for a mobile computing device depends on the employed OS. For instance, the software development for an Android OS is usually performed in Java using the Android *software development kit* (*SDK*); the SDK refers to a set of software tools for developing apps for particular software and hardware platforms. There are also other tools for *apps* development, appropriate for novice programmers. A popular one is the *App Inventor* for Android, which is supported by the *Massachusetts Institute of Technology* (*MIT*) and provides a drag-and-drop method of programming.

TPACK Analysis Toward Teaching and Learning Microcomputers

Microcontrollers have become very popular lately. Because of the widespread dissemination of DIY culture in microcontroller technology, it has nowadays been developed an entire industry that is meant for novice designers. This technological evolution has also spread like a virus to the *maker* culture of our age, as the innovation with microcontroller-based electronics can be addressed for several interdisciplinary practices (e.g. innovative experiments in science, wearable electronics with art, and much more). Nowadays, microcontrollers constitute a low-cost and easily accessible technology that can be incorporated by every single *makerspace* and *Fab Lab* [14, 15]. Because of the *maker* movement in education (and today's rapid evolution of embedded computers), it would be wise to consider a possible increase to the educational research efforts related to microcomputer programming, at expense of or complementary to the conventional computer programming learning. Following an educational research on microcomputer practices appropriate for undergraduate students within computing curricula[16–19], the author provides a personal perspective on the teaching and learning aspects of microcomputer programming and application development. A *TPACK* analysis of the interdisciplinary technology of microcontrollers is performed hereafter.

TPACK Analysis of the Interdisciplinary Microcontroller Technology

TPACK constitutes a framework that enhances the initial perception on *pedagogical content knowledge* (*PCK*) [20, 21]; i.e. the intersection between the *content knowledge* (*CK*) and *pedagogical knowledge* (*PK*). The former refers to the instructor's knowledge domain about the subject matter to be learned or taught, and the latter to the knowledge domain of teaching and learning methods.

Building on this concept, TPACK incorporates another component to describe the technology integration in the framework, as defined by Pierson in 2001 [22]. Initially referred to as TPCK, today's structure of the model has been illustrated by Mishra and Koehler, in 2006, as a Venn diagram [23]. The latest component in the model, that is, the *technology knowledge* (*TK*), determines the role of technology in improving education. TK reveals three more intersections: (i) the *technological pedagogical knowledge* (*TPK*) found in between TK and PK, (ii) the *technological content knowledge* (*TCK*) found in between TK and CK, and (iii) the TPACK, which is the intersection of the three components (CK, PK, TK).

The analysis on TCK and TPK interceptions [23, 24] mentions (among others) that teachers should be aware of how the technology application affects the subject matter (i.e. TCK intersection) as well as teaching (i.e. TPK intersection). Based on the aforementioned suggestions, this subsection addresses the TPACK model to share a personal perspective on the three components CK, PK, TK in (i.e. *what* we know, *how* we teach, and *why* the selected technology has a serious effect on) microcontroller-based education. Then, it justifies why the employed technology may cause a shift in microcomputer education between disciplines of EE and CS. When necessary, the parallelization between computer and microcomputer programming and application development is addressed so as to illustrate these concepts (in terms of the sequential method of computer programming, which is the standard programming method for microcontrollers).

Content Knowledge (The What)

In their survey of teaching introductory programming, Pears et al. [2] identify that the traditional views of learning programming (including the associated textbooks) are orientated toward the syntax and structures of the employed programming language. Yet, they also identify the strong movement in CS education and their staunch supporters who consider programming as the application of skills toward solving a problem.

Nowadays, there are several general-purpose programming languages that can be used to direct computers (and microcomputers) toward solving a problem. Some of them support a kind of higher-level approach of programming (such as the Python, where programs are developed in fewer code lines to support code's readability). Others provide a kind of lower-level programming approach (such as the C language, the constructs of which map more effectively with instructions at the machine level). While the approach of programming can be varied from the viewpoint of the code development process (when different languages are addressed), the result in the way generalized mechanics are implemented at the machine level relies on the same concepts.

I argue, too, that learning of computer and microcomputer programming should be considered as the development of skills toward solving a problem. Therefore, the programming language should not be considered as being incorporated in the CK component. Rather, it should be considered as the technology integration toward helping students create their knowledge about the subject matter. The latter should more appropriately be defined as the concepts and theories of the sequential method of computer programming toward solving a problem. Some of today's most popular programming languages may be obsolete in a few years. Yet, the computer programming topics, such as the methods of alternating the normal execution of the code (aka *control flow*), the arrangement and accessing of data in memory, etc., will also be present in tomorrow's popular programming languages.

The aforementioned concepts and theories of CK component, in regard to a potential introductory programming course, are related merely to *software practices* (as hardware practices are encrypted by the computer system). A corresponding introductory course on microcomputer programming would require some hardware practices, as well. The difficulty in arranging a technology-related course, textbook, etc., is in agreement with the fact that the employed technology will most possibly be obsolete in a few years. The question raised is "How could we arrange CK of a technological subject matter without too much focus on the specified technology?"

Technology Knowledge (The Why)

When it comes to microcontroller education, the procedure needed to defocus from a particular technology is not trivial. To build some experiments you need to arrange the examples around a particular microcontroller device. And there is no compliance, even with devices of the same manufacturer. Chip vendors supply different models with more or fewer differentiations in the architecture in order to cover a large number of buyers (according to the needs specified by a custom-designed application). Such differentiations are the memory locations of the internal registers, the available IO subsystems embedded in the microcontroller (such as timers and analog-to-digital converters), and so forth. This is why typical textbooks in microcontrollers regularly incorporate a description of the device's architecture [25–27].

Nevertheless, there are strategies that can be addressed in order to defocus from a particular technology in microcontroller education. One of them could be the incorporation of the external subsystem, instead of the complex internal subsystems available in the employed microcontroller device. In this way, the information related to the architecture of a particular device is limited to the knowledge of issues, such as how to use the microcontroller pins as regular inputs/outputs. Of course, the incorporation of external subsystem increases the PCB requirements; however, the DIY culture of our age facilitates this option.

Another strategy could be the arrangement of experiments around long-lasting technologies. The regularly limited pin resources in a microcontroller device create the need for serial communication interfaces in typical microcomputer applications. Particular examples of long-lasting types of serial protocols are the *universal asynchronous receiver/transmitter (UART)*, *serial peripheral interface (SPI)*, and *inter-integrated circuit (I2C)*, which have been for many years used in microcontroller applications, (and there more to come as there is a tremendous development of chips based on these three protocols).

These serial communication protocols are included as standard IO subsystems available in several microcontroller devices. However, the configuration and control of such subsystems are performed through the accessing of three sorts of registers; that is the *control*, *status*, and *data* registers. The contents and memory addresses of these registers, as well as the pins location of the specified subsystems, differ from device to device. Hence, the learner has no choice but study the architecture of the employed microcontroller device in solving the problem. A possible strategy toward implementing an architecture-independent serial interface is to manually implement the process through the regular IO parallel interface of the microcontroller pins [28, 29].

Figure 1.2 recommends a revision of Kordaki's [30] problem-solving model for beginners learning programming (using C programming language). The revised scheme illustrates how the hardware practices in microcomputer programming are differentiated from the software practices of a regular computer (when addressing a high-level programming approach). The dashed-line arrows depict at which point (of the problem-solving process)

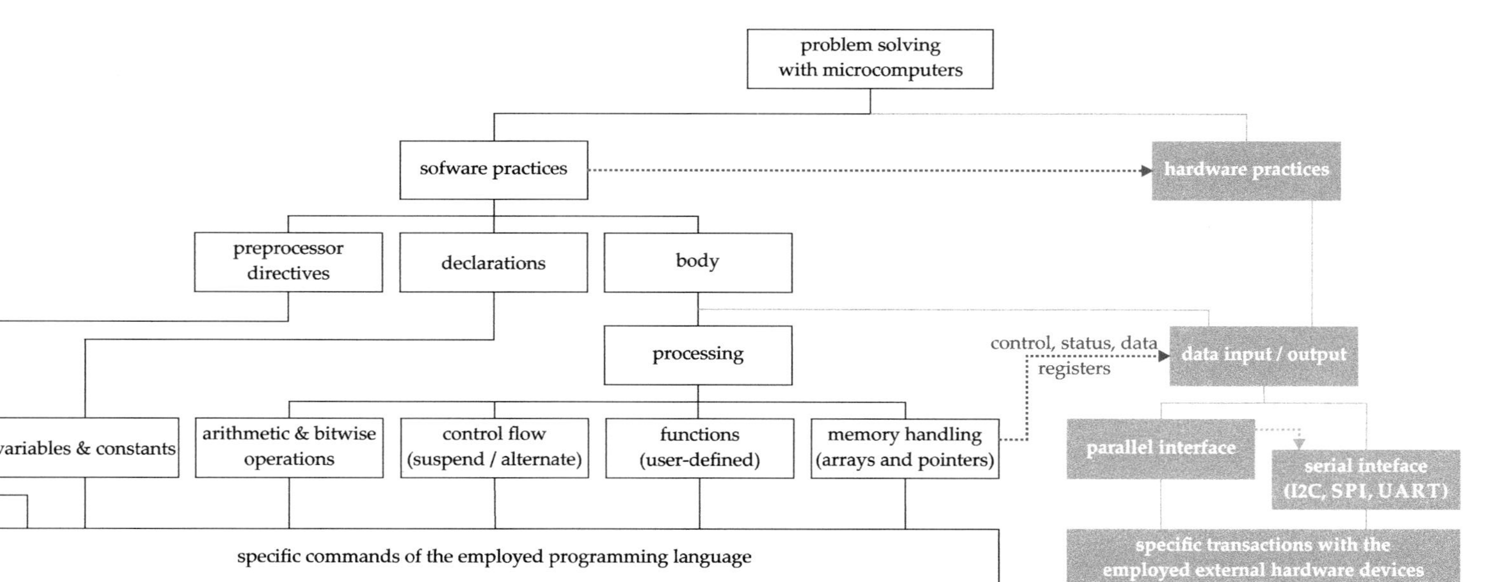

Figure 1.2 Problem-solving with microcomputers (a revision of Kordaki's model).

the differentiation is achieved. That is, the *data* IO processes to/from the microcontroller device, which are handled by the corresponding memory locations.[1] The dashed-line arrow pointing from the *parallel interface* to the *serial interface* block illustrates the alternative choice of implementing architecture-independent serial interfaces in order to defocus from a particular technology [28].

To reply to the question *"why the selected technology has a serious effect on microcontroller education?"* we need to consider what happens if a low-level programming language is addressed for the tutoring system. This choice would place the *memory handling* block immediately after the *processing* block (Figure 1.2). What this means is that all software practices would be related to the microcomputer architecture. For example, to implement an iterative loop the designer should: (i) utilize a register from data memory as counter; (ii) decrease the content of the counter at each repetition of the loop as well as evaluate its content, and (iii) explore the content of the processor's register bits in order to decide the alternation of the program flow. Such processes shift away from the software practices familiar to CS students and are directed toward the hardware practices appropriate for EE.

The assembly language learning requires a particular endeavor, whereas the time for practice within an introductory class is already limited. Subsequently, the possibility of the learner to develop skills in solving problems would be considerably decreased [31]. This would affect the subject matter of CK in the TPACK model; i.e. because of the influence generated by the TCK intersection. In that case, the tutoring would most probably be focused on learning the syntax and structures of the assembly programming language, rather than on problem-solving practices. In addition, the process of defocusing from a particular technology (i.e. from a particular microcontroller device and its identical instruction set architecture) would almost be impossible [16].

Pedagogical Knowledge (The How)

Figure 1.3a presents the generalized scheme of the TPACK model, while Figure 1.3b presents a personal perspective of the model being applied to the recent trends in microcontroller education. Arranging a course (or textbook) around a regular (high-level) programming approach, the subject matter of the CK component can be equally shared in software and hardware practices. This allows addressing efficient pedagogies, such as the *project-based learning* (*PBL*) approach which constitutes one of the most popular research methods in microcontroller education for many years now [32, 33]. This student-centered pedagogy approach cannot be easily addressed with a low-level programming approach. As mentioned earlier, the difficulties in learning (as well as developing code in) assembly language absorb considerable time resources from the students' practice. The process of facilitating the programming language learning by the incorporation of a high-level approach allows us to address such pedagogies (i.e. TPK intersection), which consequently support the arrangement of the subject matter (i.e. PCK intersection) toward engaging students in investigation tasks, collaborative development of the project, etc.

---

1 No process is depicted subsequent to the corresponding blocks of the original hierarchical network (referred to as *data entry* and *output* blocks in Kordaki's scheme); a choice that clearly portrays the encrypted IO operations in learning computer programming. Please refer to Kordaki's original scheme [30] for this particular information.

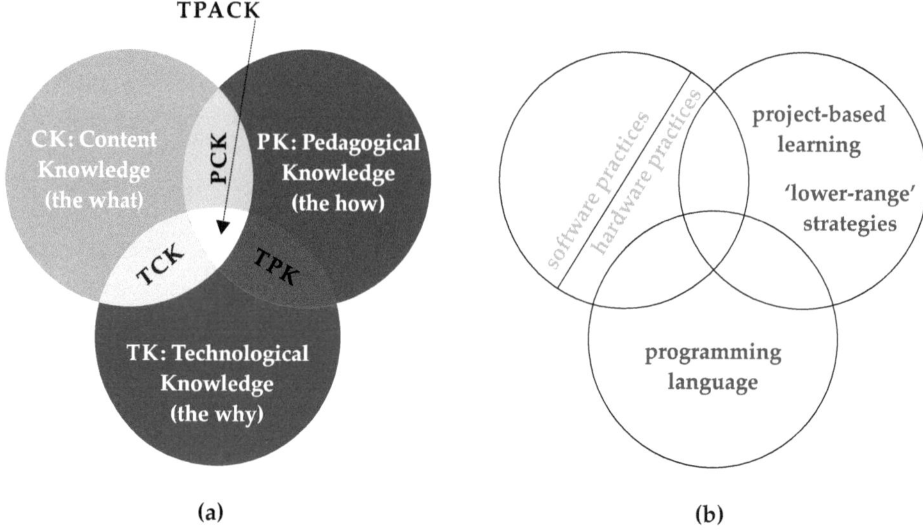

Figure 1.3 (a) TPACK model and (b) its application in microcontroller education.

Other "lower-range," though of particular importance, strategies are also affected by the incorporated technology (i.e. TPK intersection). In detail, the selection of low-level programming approach may create the need for teaching strategies that facilitate the assembly language learning [18]. On the other hand, the arrangement of tutoring toward a higher-level programming approach could rely merely on the utilization of more transparent teaching strategies, such as *flowcharts* and *pseudo-codes* (that also aim at facilitating the language learning). It is worth noting that *flowcharts* are in agreement with Levin's suggestions [34] on the transformational picture function, appropriate for difficult-to-remember information. On the other hand, the *pseudo-code* works as an intermediate step between the programming steps depicted by a flowchart, and the strict syntax required by the development of the final code. While the utilization of *flowcharts* and *pseudo-codes* are taken for granted in computer programming learning, they are of particular importance since the learning endeavor of a programming language, especially for a beginner, is not negligible.

From Computational Thinking (CT) to Micro-CT (μCT)

This subchapter explores the additional endeavor required to understand the capabilities of microcomputer technology and addresses the coined μCT term to describe the thought processes involved in the solutions carried out by a microcomputer-based system. Embedded computers have become an integral part of everyday life as they can be found in almost every single electronic device. If the most effective approach for developing CT is learning computer science, it would be wise to consider today's widespread dissemination of embedded computes and promote μCT concept within the computing curricula. The purpose of this effort is not to coin a new term entirely differentiated from the existing

CT concept, but rather to reveal the thought processes related to the application development with embedded computers.

In addition to the justification and detailed description of the μCT term, the information presented hereafter introduces the terms (i) *pseudo-architecture* and (ii) *pseudo-timing diagram*, and (iii) *pseudo-hardware*, complementary to the well-known and "transparent" teaching method of *pseudo-code* (used to support students' learning processes). Like the *pseudo-code* is addressed to describe the operating principles of a computer code (which is independent of language-specific rules and does not rely on to a standard format), the *pseudo-architecture*, *pseudo-timing diagram*, and *pseudo-hardware* are addressed for illustrating the hardware operating principles of microcomputers, while also making a clear link between the firmware execution and the hardware response.

CT Requirement and Embedded Computers

In 2006, Jeannette M. Wing introduced the term CT to share a vision that everyone can benefit from thinking like a computer scientist [35]. The term has received considerable attention since then, while also raising an ongoing discussion about the definition and essence of CT. Wing revisited the CT term and described it (in 2011 [36]) as "the thought processes involved in formulating problems and their solutions so that the solutions are represented in a form that can be effectively carried out by an information-processing agent." While such a high-level definition kept the debate warm, some educational researchers found the term acceptable and focused on how to promote CT in education [37, 38]. In 2016, the *K–12 Computer Science Framework Steering Committee* [39] recognized CT as the heart of the CS practices. In detail, seven core practices represent what computationally literate students should do to fully engage with the CS core concepts of (i.e. what they should know), where four of the overall seven practices are aspects of CT (that is, *Recognizing and Defining Computational Problems*, *Developing and Using Abstractions*, *Creating Computational Artifacts*, *Testing and Refining Computational Artifacts*). It would be wise to stand aside from this debate on CT definition and focus the reader's attention on what CT requires [39], as defined by the *K–12 Computer Science Framework* (p. 68):

"Computational thinking requires understanding the capabilities of computers, formulating problems to be addressed by a computer, and designing algorithms that a computer can execute. The most effective context and approach for developing computational thinking is learning computer science; they are intrinsically connected. Computational thinking is essentially a problem-solving process that involves designing solutions that capitalize on the power of computers; this process begins before a single line of code is written. Computers provide benefits in terms of memory, speed, and accuracy of execution. Computers also require people to express their thinking in a formal structure, such as a programming language."

Given the aforementioned description and taking into account the widespread dissemination of embedded computers in everyday life, the question raised is: What critical thought processes are involved in the solutions carried out by embedded computers? To answer this question, we first need to explore the different types of embedded computer systems as well as the possibility of being exploited by a broad range of users. As Aho remarks [40], "the term computation means different things to different people depending on the kinds of computational systems they are studying and the kinds of problems they are investigating." According to the proposed categorization of embedded computers presented earlier in this chapter, we may identify two major categories in accordance to the applied method of programming. Accordingly, we may address an HDL (e.g. VHDL) to express concurrency of a digital computing system, or an imperative style of programming that applies to a common programming language (C, Python, etc.). Leaving aside the former type of embedded computers (such as the reconfigurable FGPA technology) as they require advanced skills not suitable for inexperienced learners, the second group (more appropriate for novice designers) can be further separated into two subcategories. Accordingly, in the electronics industry we may identify embedded computers that run an OS (such board systems, like the popular Raspberry Pi, regularly employ microprocessor technology), as well as embedded computers that do not use an OS (such board systems, like the familiar Arduino, incorporate microcontroller technology).

Microcomputers and Abstraction Process

Microprocessor (*μP*) and *μC* devices are regularly referred to as microcomputers, even though it is more precise to associate the second technology with the generalized term of microcomputers. This is because μCs embed the processor unit, the (program and data) memory as well as the IO devices (that interface with the outside world) within the same *integrated circuit* (*IC*) and, hence, they compose single-chip complete computers. The information given hereafter applies to both systems and therefore, μCT term may refer to either a μP-based or μC-based system. In regard to μP technology which is commonly found in boards running an OS, the information applies to the nonstandard hardware interfacing practices. Such practices are related to the endeavor required to plug hardware devices to the GPIO port pins of a board (e.g. a Raspberry Pi single-board computer) and to develop a custom-designed application. This kind of practices is associated with the thought processes the coined μCT term is meant to highlight. To make μCT term even clearer, particular attention is paid to the OS-less μC-based system. Microcontrollers (i.e. a technology that was primarily meant for electrical/electronic engineers) have become very popular the last decade and have been gradually shifting in the direction of CS discipline. The answer key for this reallocation is hidden behind the most common used word in the definition of CT (as depicted by the familiar word-cloud graph of reference [41]); that is, the abstraction.

In [42], Wing clearly explains the significance of abstraction in computing and the importance in deciding what details to highlight or ignore. When working with regular computers the abstraction is generally a clearer process. However, when working with μCs the thought processes become more complicated and the decision on the abstraction is quite critical. The cause is mainly due to the fact that:

a) μC-based applications are arranged around a set of nonstandards IO units (LEDs, switches, keypads, sensors, actuators, etc.) and interfaces (USB, Ethernet, wireless, and so forth);
b) The OS-less system of a μC requires specific configuration, and this configuration might differ a lot from device to device.

The *DIY* culture in μC technology has accomplished a major achievement in the delivery of abundant low-cost stackable hardware platforms, which render the implementation custom-designed systems as easy as one-two-three. In regard to the software domain, the DIY culture is orientated toward the development of drivers of the employed external subsystems that abstract low-level tasks, such as, the complex configuration of an Ethernet module. This is a desirable process of abstraction, which is also in agreement with the directions given in P4.2 description of *K–12 Computer Science Framework* (p. 78) [39]. That is, the students should be able to use well-defined abstractions that hide complexity, and understand that they do not need to know the underlying implementation details of the abstractions they use. Similarly, the well-defined interfaces between the layers of abstractions in a common computer enable developers to interact with components, without needing to know all the details of the component's implementation [42].

The aforementioned description illustrates the contribution of DIY culture in shifting μC technology closer to CS discipline, by placing the emphasis on the abstraction process of the incorporated peripheral devices. However, the developer of a μC-based application works with two distinct elements; that is, (i) the μC device and (ii) the peripheral devices (such as, sensors, interface modules, etc.). *The most critical process in the programming and application development with microcomputers is the signal transaction between the processor and a peripheral device.* Because of the absence of an OS, the modern development environments for writing μC code encompass built-in features and libraries that tend to encrypt such transactions. However, those low-level processes should not be delivered totally encrypted to the learners, but rather the students should be able to develop their own abstractions, as defined by *P4.3 guidelines* of *K–12 Computer Science Framework* (p. 79) [39]. *The most common way to identify and fix errors in a custom-designed system is to explore the signal transaction between the μC and a peripheral device.* For instance, the option a designer regularly holds in order debug and restore the communication between two wirelessly interfaced μC-based setups, is to evaluate the exchanged information between the μC and the incorporated module, which establishes the wireless interface. This might seem like a "scary" process for a novice designer, however, there are only a few dominant hardware interfaces used in microcomputers. This is the critical segment in programming and application development with μCs, as it is in line with what students should be able to compare in results and intended outcomes (P6.1), and subsequently examine and correct their thinking (P6.2) [39]. As Brennan and Resnick remark [43], "Things rarely (if ever) work just as imagined; it is critical for designers to develop strategies for dealing with – and anticipating – problems." The next section aims at demystifying the fundamental and essentially limited resources that are needed to exploit microcomputer technology within computer science education at an introductory level.

The μCT Concept: An Onion Learning Framework

Relative to the program code developed for a personal computer, the fundamental difference in developing microcomputer code consists in the IO interfacing with the outside world. While students within an introductory programming course make use of two standard IO units (that is, the keyboard and monitor) to interface with the outside environment, the students of a microcomputer-based programming class have an abundance of nonstandard units to choose from. Thereby, the introductory examples are commonly arranged around IO interfacing practices. Moreover, while the software code is arranged around ready-to-use functions that control the computer units (e.g. the monitor), the corresponding code for a microcomputer controls the application-specific units through the device's IO pins. The latter are directed by read/write operations of specific memory locations, which is another feature encrypted in the conventional programming.

Simple IO operations shape the microcomputer's parallel interface to the outside world. However, a μC-based application usually incorporates several peripheral modules, a requirement that has a negative consequence to the (limited) pin resources of the μC chip. For that reason, the serial interfaces are of vital importance for the implementation of a microcomputer-based system. Due to this need, typical μC devices embed subsystems that support the implementation of a serial interface (commonly I2C, SPI, and UART) with the outside world. It is here noted that other popular subsystems can also be found inside of a μC chip, such as *analog-to-digital converter* (*ADC*), *pulse width modulation* (*PWM*), USB controllers, and so forth. However, the possibilities and limitations, as well as their configuration and control of such subsystems, differ from device to device.

While the μC programming and application development entails the knowledge of the device's architecture, the μCT should look beyond specific architectures. Rather, the learner should have a clear sense of the generalized architectural framework and operating principles of μC, in the direction of solving a problem with microcomputers. The future designer needs to be able to cross boundaries across diverse chip devices and architectures. Fortunately, the DIY culture (applied to μC technology) sustains this educational possibility with the abundant daughterboard systems, available in today's market. Figure 1.4 recommends on an onion framework toward learning μC programming and application development. If the language practices, common to μC and personal computer programming (e.g. control-flow statements, arrays, subscripts), are placed within the innermost layer then the outer layers are consecutively formed as follows:

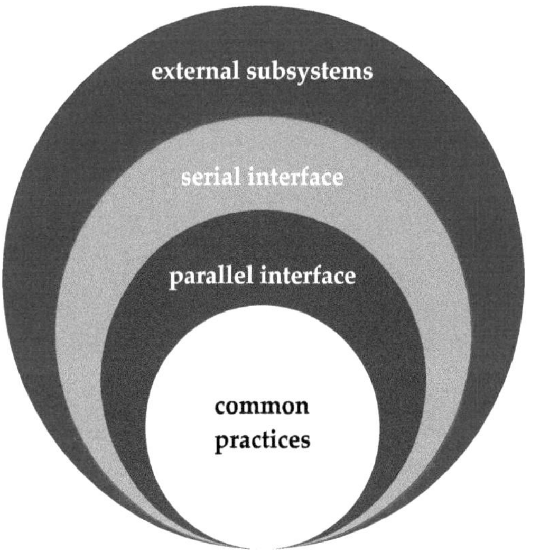

Figure 1.4 Onion learning framework for μC programming and application development.

i) ***Parallel interface:*** refers to the microcontroller's IO pin transactions as directed by the read/write operations of specific memory locations. This is foremost and perhaps the most important difference between programming a μC and a conventional personal computer, as it determines the hardware response according to the code execution. This category can enclose more advanced issues, such that the interrupt mechanisms;

ii) ***Serial interface:*** refers to the hardware interface between the microcontroller and the external peripheral devices. There are three different methods of implementing a serial interface. The most popular is the employment of a μC device with an embedded serial interface module (yet, this is an architecture-specific method). The second is the utilization of ready-to-use functions available within the employed compiler, which direct the regular μC pins to a serial transaction. The final (which is my personal choice because it supports low-level practices [28]) is in agreement with the building of custom-designed functions for this purpose. This alternative is strongly associated with another important concept in μC programming and application development; that is, the timing concept.

iii) ***External subsystems:*** refers to the selection of the appropriate peripheral system as well as the necessary transactions that will contribute to the building of a custom application. The external subsystems could be as simple as an ADC that is used to capture an analog signal and pass this information to the μC device, or a complex interface, such as the Ethernet which meant for outside-the-box communications [44]. It should be noted that the time constraints needed to learn the functional properties of a complex module, can considerably overload the amount of information one introductory class can cover [45].

"Transparent" Teaching Methods

The term transparency has been used to characterize particular pedagogical technologies in consideration of the possibility of being directly related to their function (e.g. a pencil) [46], which is opposed to protean feature of digital technologies (e.g. a computer) in regard to their usability in many different ways [47]. This term is hereafter borrowed to describe the well-known *pseudo-code* teaching method, addressed to illustrate the operating principles of a computer code. There is no standard format for the generation of a *pseudo-code* and no particular dependency on rules defined by a programming language. Yet, it is a dominant teaching approach for computer and microcomputer programming, though of low-range (compared to other approaches in μC education of extended-range such as, the familiar PBL approach [32]). To expand further the utilization of transparent teaching methods (of low-range) as they apply to the aforementioned analysis of μCT concept, we herein introduce the terms *pseudo-architecture*, *pseudo-timing diagram*, and *pseudo-hardware.*

The term *pseudo-architecture* is addressed to illustrate a simplified (and generalized) scheme of the inner workings of a μC. As in the *pseudo-code* approach, the representation depends on strategies determined by the instructor. My personal strategies (as they apply to 8-bit μCs [28]) are in agreement with a limited representation of the basic registers of the *central processing unit* (*CPU*), as well as a part of data memory (i.e. IO and regular registers) and a part of program memory (depicting vectors, the location where the application code is stored, etc.) in a Von Neumann architectural representation for simplifying conceptualization. The *pseudo-architecture* is addressed to illustrate how the internal registers of the μC response to the execution of a *pseudo-code*. A *pseudo-timing* diagram is additionally addressed to initiate the timing concept, which is particularly important in μC programming and application development. It does not refer to the actual clocked processes that take place during the code execution, but rather to the fundamental processes decided by the instructor. A *pseudo-hardware* (referring to a simplified structure of the employed μC and possibly some peripheral units) can be addressed to provide a clear link between the firmware execution and the hardware response.

Figure 1.5 presents a *pseudo-code* example that blinks a *light-emitter diode* (*LED*), which is uploaded to the *pseudo-architecture* of a *pseudo-hardware* device. The *pseudo-timing diagram* in the upper left area of the figure represents the first six successive clock cycles that take place in the execution. The arrow depicts the current execution, which forces the LED to turn ON. This is a result of writing a logical one to the bit 5 of data (IO) register traced in *addr1* memory location. The *program counter* (*PC*) inside of the CPU points to the memory location of the subsequent pseudo-instruction to be executed (i.e. PC = 0x00A). The *accumulator* (*ACC*) register is addressed to illustrate the difference in the length of the

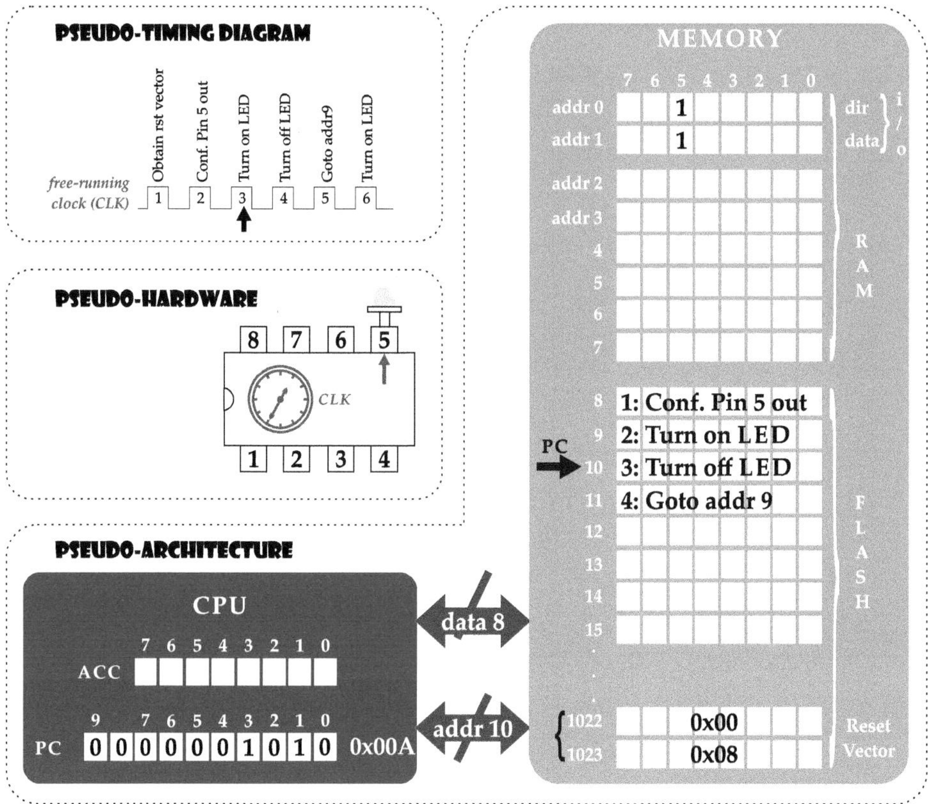

Figure 1.5 An example of pseudo-architecture, pseudo-timing diagram, and pseudo-hardware.

two CPU registers as the memory organization in the current μC is of 8-bit, while the available registers in overall memory cover a range greater than $2^8 = 255$. In detail, the available registers in overall memory are extended to $2^{10} = 1024$ memory locations; a value that is in agreement with the current length of the program counter.

It is worth noting that the underlying hardware mechanisms (as the ones depicted by Figure 1.5) are encrypted during an introductory computer programming course. However, those mechanisms are of particular importance for a course applying to μC programming as they settle the conditions for more advanced topics (e.g. interrupt mechanisms, bootloader in microcontroller's memory, a device reset caused by stack overflow, and much more). Many of these aspects arise from the fact that a microcomputer is of limited abilities, compared to a regular computer system. For instance, the developer will not deal with a situation that the generated code may not fit within the computer memory (which is not an impossible fact for a μC code). Complementary to the transparent (and low-range) *pseudo-code* teaching approach, the *pseudo-architecture* along with the *pseudo-teaching diagram,* and *pseudo-hardware* representations, could be the transparent teaching approaches that sustain the μCT in an introductory microcomputer programming and application development course.

The Impact of Microcontroller Technology on the Maker Industry

From the technological point of view, the μC device constitutes a single-chip computer, incorporating the following parts:

i) ***Central processor unit (CPU):*** that is, the device's processor core, which is responsible for executing the application code (where the *instruction execution* is performed in *sequential order*);
ii) ***Program memory:*** refers mainly to the type of memory that is used to hold the application code (aka *firmware*); that is, a *read-only memory* (*ROM*) the content of which can be updated by the user (i.e. the code developer);
iii) ***Data memory:*** refers to the *random-access memory* (*RAM*) which is used to hold the temporary data generated during the code execution, as well as to the IO RAM registers that interface with the outside world. The microcontroller's IO registers can be used to control the device's IO port pins. The latter can act as simple digital IO pins (for inputting/outputting a digital signal from/to the outside world), or as special function pins that are directed to an embedded subsystem inside the microcontroller device (e.g. an ADC, PWM); It is worth noting that a μC device may incorporate some or more subsystems, which are commonly found in embedded applications.

According to the aforementioned information, the *microcontroller* constitutes a *microcomputer,* which is optimized for control applications; as defined by the identical term of this technology. Microcontrollers constitute a popular type of *embedded computers*, nowadays, while this technology has been experiencing the widespread dissemination of *DIY* culture, the last decade. Today's wave of the *ready-to-use* and *stackable* boards, and the *shareable* – over the internet – libraries, renders feasible the rapid development of microcontroller-based applications. Furthermore, the modern μC programming methods have managed to abstract the low-level tasks with the underlying hardware and, hence, have rendered this technology sensible for the inexperienced *makers*. Nowadays, microcontrollers constitute a low-cost and easily accessible technology that can be straightforwardly incorporated within any *makerspace*.

To understand the impact of microcontroller technology on the maker industry, it would be wise to follow the advancement of microcontroller programming and application development during the years of maturation. The advancement of this technology can be separated into two major eras, in consideration of the requisite time needed to learn, program, and develop a microcontroller-based application [48]. That is, the (i) *long-cycle* development era and (ii) *short-cycle* development era.

Hardware Advancement in μC Technology

A significant reduction of the requisite time needed to program and develop a microcontroller-based application can be reasonably supposed to have arisen from the advancement of the *nonvolatile* memory technology. The *nonvolatile* (contrary to the *volatile* type of) memory is the one that is used to hold the application code, as it retains the stored information when the power supply is removed. The previous generation of μCs where either of an early type of *electrically erasable programmable read-only memory* (*EEPROM*), or of *ultraviolet programmable read-only memory* (*UVPROM*).

Both types of memories required a separate board (regularly referred to as *programming board*) in order to upload new code in the μC. This demand arose from the fact that it was required a different circuit for reading and executing code from μC's memory, and a different circuit for writing new code to the μC's memory. The interface for uploading the updated code to the μC device was regularly based on the *recommended standard 232* (*RS-232*), while the programming board required an additional power supply unit to power the internal components of the *programming board* (Figure 1.6a). In addition, the UVPROM type of memory required an extra UVPROM eraser, where the quartz window on the top of the device allowed the μC's memory to be exposed to ultraviolet light in order to erase previous data before uploading the new code (Figure 1.6b). The common exposure time was approximately 20 minutes and, hence, the code developer could assess the functionality of an error-free code two to three times per hour. In most cases, the user should develop a custom-designed PCB that would hold the application circuit (like the example presented in Figure 1.7).

The aforementioned information reveals the considerable hardware involvement, time, and cost of the requisite occupation just to upload a firmware to the μC's memory (in the *long-cycle* development era). When the later-type of EEPROM memory made its appearance in μC market, it allowed the data to be erased and rewritten within the application circuit. This fact rendered unnecessary the extra *programming board* that was used for the firmware updates and promoted the, so-called, *in-system programming* (*ISP*) capability. In addition, the more recent *Flash* type of memory does not require the μC to be completely erased before rewritten (i.e. it can be read/written in blocks) and, hence, it promotes faster firmware updates that accelerate the debugging process.

The critical time during the μC programming and application development process reduced even more with the replacement of the RS-232 interface. The today's USB industry

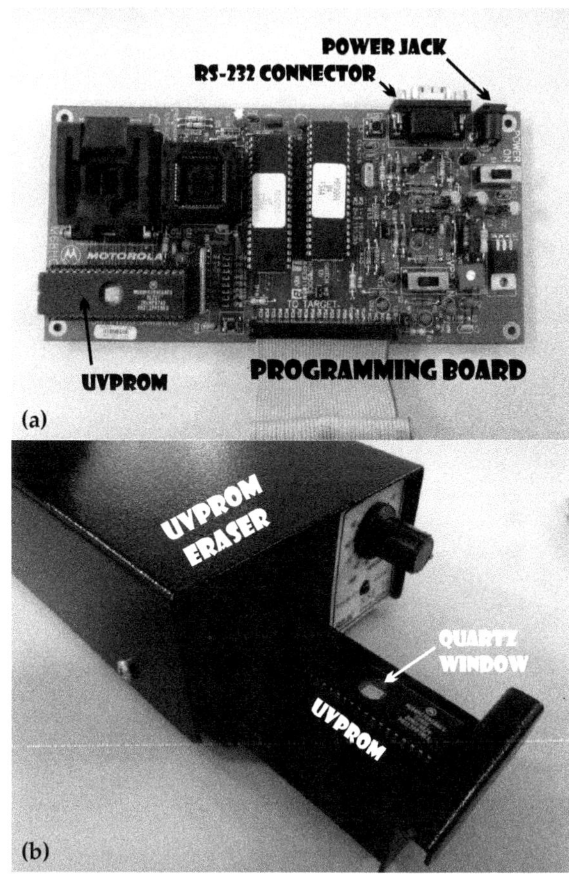

Figure 1.6 (a) μC programming board and (b) UVPROM eraser (from the *long-cycle* era).

Figure 1.7 Custom-designed PCB of the μC's application circuit.

standard not only supports a high-speed communication between the μC and a personal computer, but also provides the ability to supply power to an external device. Hence, the modern microcontroller boards are commonly delivered with a USB connector, which is used to supply electric power, program the microcontroller, as well as to provide a communication interface of the μC board with the personal computer. All the above hardware-related improvements shape the *short-cycle* development era of the modern microcontroller-based applications.

Figure 1.8a depicts a typical microcontroller board of our age; that is, the classic Arduino Uno. Like similar boards on today's market, the Arduino Uno (on the left of Figure 1.8a) is preprogrammed with a *bootloader* code. The bootloader is referred to a small portion of code in μC's memory which provides the ability of updating the user-defined firmware code. With a single press of a button, in the corresponding software running on the user's personal computer, the bootloader receives the firmware information externally from the

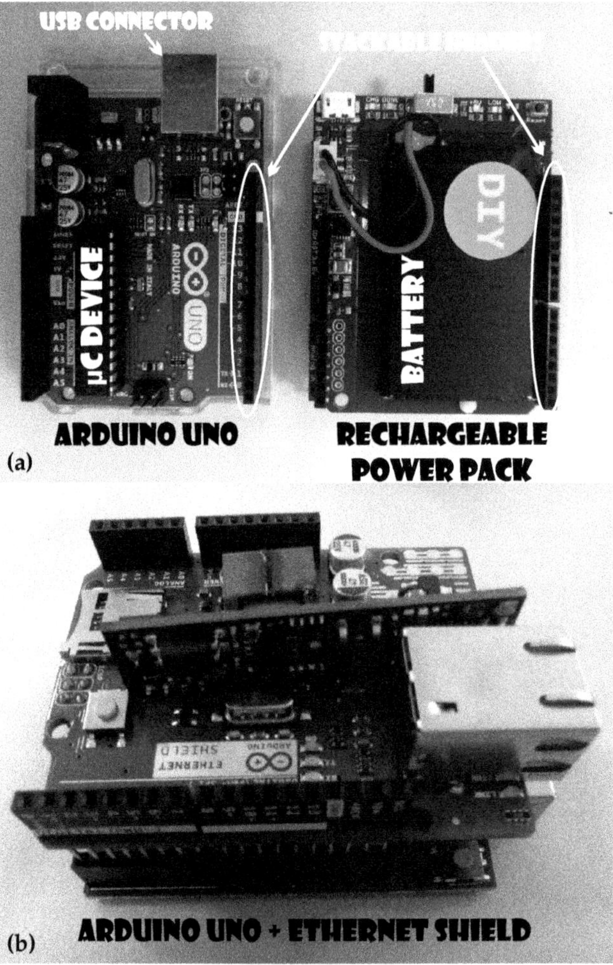

Figure 1.8 (a) Arduino uno board and powerpack shield; (b) Arduino Uno and Ethernet shield.

USB board, and uploads the new application code to the μC's memory. The stackable headers employed in such board systems, allow us to attach one or more daughterboards (aka *shields*) to the main motherboard and arrange the hardware part of a sophisticated microcontroller-based system (more or less) within a minute or so (i.e. in many cases there is no need for a custom-designed PCB in order to implement a μC-based application). For instance, on the right of Figure 1.8a a rechargeable powerpack can be attached to the Arduino Uno board and make available an autonomous μC-based system. Another example is depicted in Figure 1.8b where an Ethernet shield is connected on the top of the Arduino Uno motherboard, through which the μC can connect to the internet.

Software Advancement in μC Technology

The requisite time needed to develop the source code for a microcontroller device considerably reduced with the replacement of the *low level*, with a *higher level* of programming. The dominant μC programming method in the *long-cycle* development era was performed in assembly language. The assembly language constituted a suitable programming method at that time, as the code developer would normally spend considerable effort in simulations before uploading the updated firmware to the device (in order to evaluate its functionality). As mentioned earlier, the successive firmware updates were possible, merely two to three times per hour. Hence, the source code debugging process through simulations, was an unavoidable necessity, at that time.

The replacement of the assembly level with a higher level of programming, along with the hardware advancement related to the successive firmware updates in μC's memory, considerably accelerated the development process an error-free code. Due to the need of getting total control of the underlying hardware, the most popular μC programming method among engineers nowadays is the *embedded C/C++* programming. Other popular programming method of our age is the *MicroPython* interpreted[2] language, which is a *Python* variant optimized for μCs. Lately, there is a tendency of moving toward *visual programming languages (VPLs)*, which run on a web browser and are intended for beginners. An identical example of VPL for μCs, nowadays, is the *MakeCode,* which is used for the code development of the popular *BBC Micro:bit* board system.

The Impact of Arduino on the μC Community

When the *low-level* gave its place to a *higher-level* programming approach, the *embedded C* thrived on μC programming, and maintained a dominant position for many years. The *8-bit* architecture was the standard microcontroller technology of the time, and two vendors that held a substantial share of the world market were the (i) *PIC* microcontrollers of *Microchip Technology* and (ii) *AVR* devices of *Atmel Corporation.* If we try to make an informal classification of the two brands, it could be assumed that the PIC devices were the favorite development tool for *professional engineers*, while the AVRs were (more likely) the best

2 In *interpreted* programming languages, executions are implemented directly without previously compiling the source code into machine language (as it is performed in the *compiled* languages, such as in C/C++).

choice for the *novices* and *enthusiasts*. On one hand, the several embedded subsystems inside the PIC μCs were providing flexibility to the *development engineer (*upon the design of a new product). On the other hand, the availability of the freeware tools for the source-code development, were increasing the motivation of *novices* and *enthusiasts* to make a choice on the AVR μCs. Among other target processor families, the familiar and free *GNU compiler collection* (*GCC*) system supports the AVR architecture, as well.

Apart from these two noble microcontroller manufactures,[3] there were (and still are) many third-party partners providing software and hardware tools for the development of μC-based applications. Each vendor merchandise different development tools without compatibility among different corporations. It could be easily assumed that this is a fair politics to "encourage" users to keep supporting the hardware and software development tools that were originally selected. However, the diverse software and hardware tools give rise to a thriving Tower of Babel, in regard to the selection of the proper development apparatus for a custom-designed system.

Under these circumstances, the Arduino company made a clever move. They created a free-of-charge and particularly simplified *integrated development environment* (*IDE*) that was used to program the company's (rather limited) μC motherboards. The latter were delivered with a preloaded bootloader, which allowed firmware updates (through a USB cable) with merely the press of a button. In Figure 1.9 the corresponding "Upload" button is automatically highlighted in white, when the user scrolls the mouse pointer over it.

The Arduino's μC motherboards, such as the Arduino Uno, use a USB to Serial converter (the drivers of which are installed upon the Arduino IDE installation) and, hence, the only preparation the user should perform before pressing the "Upload" button is to select the

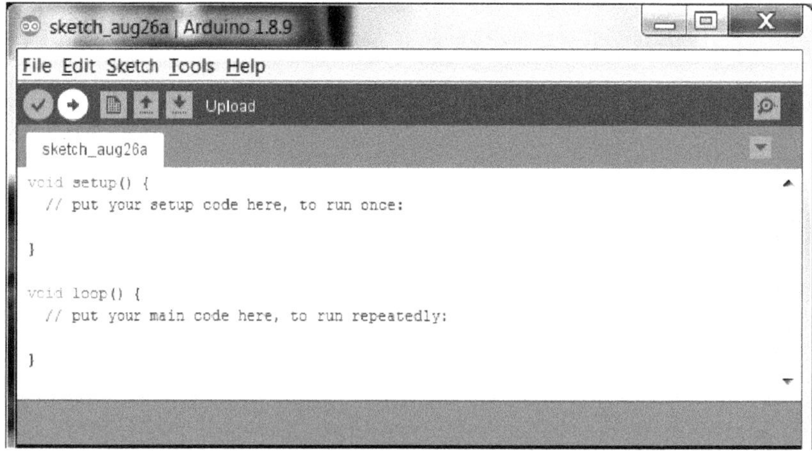

Figure 1.9 Uploading the new firmware code to μC's memory through the Arduino IDE. *Source:* Arduino Software.

3 The AVR devices acquired by Microchip Technology in 2016 and hence, PICs and AVRs now belong to the same corporation.

motherboard's name from the menu "Tools→Board:" (e.g. Arduino Uno). If not automatically selected, when the board is plugged into the USB, the user may also select the serial device of the board from the menu "Tools→Port:" (e.g. Port 3).

The Arduino company first focused on the AVR devices and was built over the GNU GCC/G++ compiler. What this means is that it could be still used by the stunts supporters of C/C++ implementation for the AVR architectures. Despite the "getting started" process, the Arduino IDE incorporates built-in functions and ready-to-run examples that accelerate the development procedure.

Other corporations that focus on the development of software platforms, also offer built-in functions, preprocessor commands and ready-to-run examples that promise a quick jump-start of a μC-based application. However, those software platforms attempt to incorporate a plethora of microcontroller devices in order to satisfy the needs of different designers and increase the potential range of the market. This decision makes the software platform more complicated, especially for the novice designer. The notion of designing with μCs constitutes a complicated and multifaceted area of study and, hence, minimizing the options available to the novice users helps them gain a more immediate access to the application development procedure.

The latter strategy of Arduino company is applied to another issue that is related with the barriers to entry μC programming. That is, the IO operations with the outside world. The simplest operation that can be performed by a microcontroller unit, in order to get access to the outside world, is to output a digital signal to one of the device's pins. Thus, the simplest example that regularly a novice user performs is to connect a LED to one of the μC pins and make it blink. While this example might be considered simple enough, the hard part for the novice learner is to make a clear link between the firmware and the hardware.

To this end, the Arduino boards are designed with a silkscreen alongside the PCB's stackable headers, which describes its header pins. In addition, the Arduino-specific libraries that control the μC pins admit definitions that are identical to naming of those headers pins, and not to the naming of μC pins. This way the user observes directly the hardware and is able to use those identical symbols and names during the code development process. Otherwise, the code developer should think at which pin of μC device is the header pin connected, and then write code that refers to that identical μC pin. For instance, to blink the LED connected to pin 13 of Arduino Uno board one would normally had to get access to pin 5 of μC's port B (Figure 1.10). With the built-in functions of Arduino, the code developer can turn *ON* the LED (connected to pin 13) with the simple syntax ***digitalWrite(13, HIGH)***; or turn the LED *OFF* using the command ***digitalWrite(13, LOW)***.

According to the aforementioned information, the μC programming and application development with Arduino integrates software and hardware tools jointly and specifically designed, so as to provide a quick jump-start and flexibility in the implementation of a μC-based project. Nowadays, the Arduino has become a viral technology, as it has created a wave of the *ready-to-use* boards and *shareable* (over the internet) libraries. Moreover, many third-party partners provide Arduino-compatible hardware tools that expand the design options available in μC-based systems. For all these reasons, this book is oriented toward the Arduino IDE platform using popular and contemporary Arduino-compatible boards.

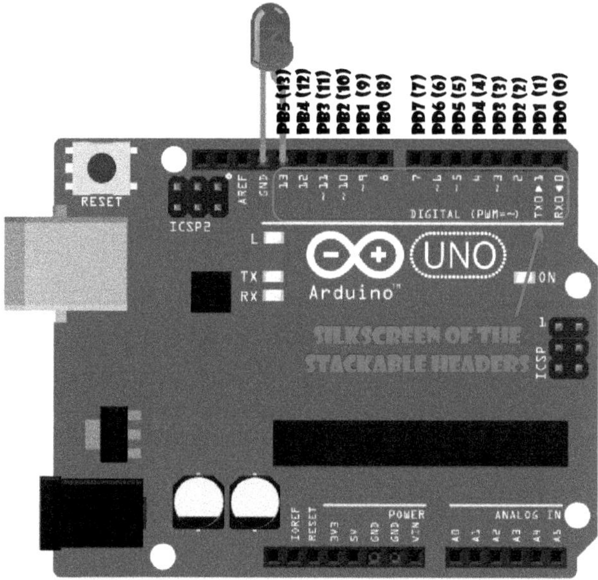

Figure 1.10 PCB silkscreen of the Arduino Uno stackable headers and the μC's digital IO pins.

In addition, the book attempts to explore creativity in μC product design with the utilization of contemporary sensor devices, as well as by the utilization of DIY culture in μC technology and the 3D printing technology of our age.

Where Is Creativity in Embedded Computing Devices Hidden?

To explore creativity in μC product design, we will consider the mean that paves the way for creativity and new solutions in one of the most popular embedded computing devices, nowadays. That is, the mobile computing devices, aka *smartphones*. The smartphones are gradually becoming powerful and complex embedded computing devices, while also giving rise to an opportunity for creativity and innovation.

Creativity in Mobile Computing Devices: Travel Light, Innovate Readily!

For all of us who have grown up in a period where *internet* was an unknown word and *gaming* a synonym to a bunch of kids kicking a ball in the neighborhood's street, the technological advancement came up as a revolution, which put extra weight on our back. Think of a person before leaving the house/apartment for a holiday trip and you will realize this extra weight is not just figure of speech. A few decades ago, one would normally check a bag full of portable gadgets for the trip (e.g. radio player, CD player along with some CDs, camera for shooting still pictures, camcorder for taking "high-quality" videos, extra memory for the cameras, alarm clock, handheld game console, as well as extra batteries and all the heavy chargers for its particular device). Nowadays, the same person would (in most of cases) put the smartphone in one pocket and make a last check for the phone's charger (perhaps in his/her other pocket).

By the way of introduction, the sense of portability has dramatically changed the last decades. Handheld devices have been merged together to make our life easier where the smartphone constitutes, in simple words, the *all-in-one* device. A user may utilize the smartphone for everyday activities (connect to the social media, send an email, be navigated while driving the car), entertainment (play a game, read a book, listen to the music), sophisticated operations (i.e. record sport activities, be guided inside a museum, monitor personal heath, control home appliances), and sometimes perform or respond to a call. This handheld device seems to be the smartest movement of the market the last decades, as it gave users the opportunity to carry all the favorite gadgets without the extra weight. If one had to select a single handheld device before leaving the house/apartment, the mobile phone would most probably be the first choice. Apparently, the feeling of secure (i.e. to be able to call for help whenever it is needed) is above all other feelings (pleasure, entertainment, and so forth).

Due to the advancement of computing systems and the gradual cost reduction of the incorporated (hardware) electronic components over the years, mobile phones have been remodeled into powerful handheld computers available to everyone. It is thought that more than 2 billion smartphones are in use, nowadays [49]. Accordingly, smartphones lead the market and society in many different ways, but also pave the way for creativity and new solutions. The spark for creativity and innovation came up with the Android open-source platform marketed by Google, while the boost established by the release of App Inventor for Android (which rendered the mobile application development accessible to the non-computer science majors) [50]. But where is the creativity hidden in these small but powerful computing devices?

To comply with the multifaceted and increasing needs of humans (e.g. navigation, gaming), the design of mobile devices has reached to the point where smartphones and embedded computer systems look quite alike. Design and development of an embedded computer system incorporates some IO units, which allow users to interface with the computing device, where in the case of mobile phones this role is undertaken by the touchscreen. Moreover, an embedded application is commonly arranged around interfaces, sensors, and actuators. Leaving aside the actuators, smartphones are commonly delivered with Bluetooth wireless technology, which constitutes a practical interface for exchanging data among nearby devices (and of course, it could be used for the control of an actuator at distance). In addition, modern smartphone devices incorporate several sensors that render feasible the implementation of creative applications. The types of sensors incorporated in modern mobile phones seem to hold the key for creativity and innovation, and to give credit where credit is due, an important share of them has been determined by the constantly growing network of gamers.

Mobile gaming has been considerably advanced from the time of Nokia's Snake game and is currently in agreement with motion-sensing operations. Motion-sensing games (originated by Nintendo [51] and currently being found in abundant smartphones apps) obtain data from *micro-electro-mechanical systems* (*MEMS*) sensors and allow users to pilot aircrafts, steer cars, and so forth. While some people see mobile sensors as a valuable tool for game apps, others value the potential utilization of a powerful embedded computer system with built-in MEMS sensors in research and development. Indoor navigation, gesture recognition, and virtual and augmented reality applications constitute examples that regularly apply to motion-sensing implementations. Such examples derive from data fusion of (i) the static and dynamic forces measured by 3-axis MEMS accelerometers, (ii) the angular velocity measured by 3-axis MEMS gyroscopes, (iii) the magnetic field measured by 3-axis MEMS magnetometers, and (iv) the barometric altitude determined by MEMS barometers.

Nowadays we have moved to the so called "combo" sensors that incorporate the functionality of two or more sensing elements (e.g. accelerometer and gyroscopes) into the same package, thereby minimizing the spatial requirements within the mobile device. Driven by the needs of smartphones, todays' "combo" solution are rearranged into a *system in package* (*SiP*) that incorporates an additional microcontroller device, addressed to deliver fused data from the built-in sensors. Hence, complementary to the minimization of the spatial requirements the smartphone can be relieved from complex processing operations, as well. This solution speeds up the mobile apps development process and therefore, it can be considered as being in agreement with today's maker culture in supporting creativity and innovation. Having recently announced the development of the world's smallest micro-computer (i.e. smaller than a grain of salt), IBM claims that such micro-computers (able to monitor, analyze, communicate, and even act on data) will be embedded in everyday devices within the next few years [52]. Thus, it would be wise to pay particular attention to the sensor devices when turning new and imaginative ideas (which incorporate embedded computers) into reality.

Communication with the Outside World: Sensors, Actuators, and Interfaces

Apart from the simple IO units (e.g. switches, LEDs), a microcontroller system is regularly arranged around sensors, actuators, and/or interfaces. According to the aforementioned analysis, the critical parts that give rise to an opportunity for creativity and innovation are, without doubt, the sensor devices. In the subsequent chapters of the book we will explore, in practice, how creative embedded computing may be implemented with the exploitation of the appropriate sensor devices.

Figure 1.11 illustrates a modern μC-based system which incorporates all the fundamental peripherals of a system, i.e. sensors, actuators, and interfaces. We address this example to also highlight the typical serial interfaces that are regularly used in microcontroller-based applications. That is, the UART, SPI, and I2C.

Apart from the four electric *actuators* (i.e. *DC motors*) of the drone, which are controlled either directly by a parallel interface (i.e. the μC's port pins) or indirectly through a DC motor driver board, the drone employs, at least, a *radio frequency* (*RF*) module. The latter is used to receive data from the remote controller (which also employs a similar or the same module) and it is regularly interfaced by the SPI protocol. The SPI is based on a synchronous type of serial communication. What this means is that the communication between the μC and the RF module is synchronized by a common clock line. It should be noted that each device on the SPI bus uses a different line to receive the incoming data, and a different line to transmit the outgoing information. If there is a need to connect more than one peripheral devices on the SPI bus, there is an additional line that is used to activate/deactivate

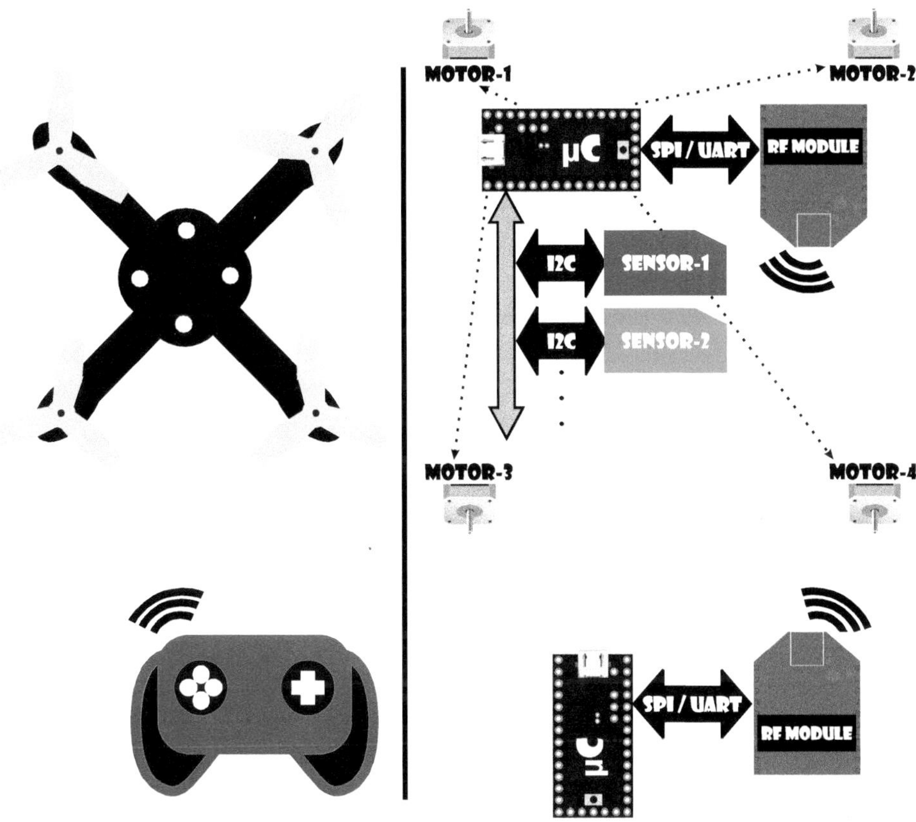

Figure 1.11 A drone is built around sensors, actuators, and interfaces.

each device (i.e. only one device should be activated in order to exchange information the system's microcontroller).

Some RF modules are interfaced by the UART protocol. The latter was the standard communication protocol between the μC and a personal computer, through the RS-232 serial port. This protocol is still very popular, nowadays, since many μC boards (such as, the Arduino Uno) incorporate a USB to UART converter in order provide backwards compatibility with older standard of serial communication. The UART applies to an asynchronous (and contrary to the SPI which uses *full-duplex*, the UART regularly addresses *half-duplex*[4]) communication. Since the μC and the peripheral device are synchronized by their own clock, they should be configured to the same clock rate.

The IC interface protocol is the most ordinary type of communication between the microcontroller and the system's sensor devices. The I2C applies also to a synchronous type of serial communication, but contrary to the SPI protocol, it applies to a half-duplex type of communication. This way, and along with the fact that the *data* line of the I2C

4 In full-duplex mode of communication data can flow simultaneously in both directions while in half-duplex; only one device at the time is able to transmit/receive data.

constitutes a bidirectional signal, this protocol uses only two signal lines (i.e. *clock* and *data*). The μC is the *master* device of the bus and each sensor device is regularly referred to as *slave* device. Each *slave* device on the I2C bus should be identified by a unique address, which is regularly a 7-bit address (i.e. $2^7 = 128$ different devices can be attached to the bus).

Conclusion

Embedded computer programming constitutes a multifaceted and interdisciplinary area of study, which falls within the disciplines of CS and EE. The programming of a regular computer encrypts the underlying operations with the hardware and, hence, the emphasis is placed on software practices (a learning approach appropriate for the discipline of CS). On the other hand, the programming and application development of an embedded computer entails more or less involvement with the hardware resources, and that involvement determines a likely position of the tutoring between the CS and EE disciplines. This chapter has recommended five types of embedded computers to help the instructor identify the educational possibilities in each category. Then, a *TPACK* analysis on the particularly interdisciplinary technology of microcontrollers was addressed.

μCs constitute a popular category of embedded computers nowadays. This technology has been experiencing, in the last decade, the widespread dissemination of DIY culture and has gradually shifted away from EE, hence, reducing the initial distance from CS discipline. It nowadays constitutes a low-cost and easily accessible technology that can be straightforwardly incorporated within a *makerspace*. Because of the *maker* movement in education, it would be wise to consider an upcoming turn to the educational research efforts related to microcomputer programming, at the expense of or complementary to the conventional computer programming courses. This attempt would raise questions such as: "What is the difference between programming a regular computer and a microcomputer? How could we arrange the content knowledge of a technical subject matter without too much focus on the specified technology?" The present chapter has addressed these issues and provided information that can be used as reference guide for further educational research in microcomputers, and embedded computers in general. Later in the chapter, the coined μCT term has explored the additional endeavor required to understand the capabilities of microcomputer technology as well as the thought processes involved in the solutions, carried out by microcomputer-based systems.

To understand the impact of μC technology on the maker industry, this chapter has also followed the advancement of μC programming and application development during the years of maturation. In consideration of the requisite time needed to learn, program, and develop a μC-based application, the author has identified the *long-* and *short-cycle* development eras. Then the chapter has explored how the modern μC tools provide quick jump-start and flexibility in the implementation of a μC-based project. Finally, the author has considered the recent trends in sensor devices and how they pave the way for creativity and new solutions in embedded computing devices.

2

Embedded Programming with Arduino

Microcontrollers have been experiencing the widespread dissemination of *do it yourself (DIY)* culture over the last decade. Today's wave of the *ready-to-use* boards and *shareable* – over the internet – libraries, renders feasible the rapid development of microcontroller-based applications. This chapter introduces the fundamentals of microcontroller programming with Arduino, with a sense of compatibility between the Arduino and embedded C programming. The chapter also makes a brief introduction to the μC interfacing techniques with the outside world.

Number Representation and Special-Function Codes

An application code that runs on a microcomputer system performs IO operations and data processing. Both types of computation require read/write requests from/to the device's memory. It could be said that all kind of processes carried out by a μC device, pass through its internal memory. First of all we need to understand numeral systems and special-function codes in order to realize how to arrange different types of information in μC's memory and how to represent data in the application code.

Table 2.1 depicts the three prevailing numeral systems that are regularly used in μC programming. The table represents the first 21 numbers in decimal system (starting from zero) and the way they are represented in 8-bit (i.e. byte) format. The prefix **"B"** denotes a *binary (bin)* representation, while the prefix **"0x"** denotes a *hexadecimal (hex)* representation of the number. When the number admits no prefix (or postfix) it is assumed to be represented in the – familiar for the humans – *decimal (dec)* system. It should be noted that the aforementioned representations are compatible with Arduino integrated development environment (IDE) and they may differ in other programming environments. At this point, it is worth making a reference to the way data are stored in μC's memory and the regular memory units that determine the μC architecture.

The most familiar unit that represents a group of bits is the *byte* (i.e. 8 bits). On the other hand, the *nibble* (i.e. 4 bits) is addressed for representing a single hexadecimal digit. Another popular unit is the *word,* which corresponds to a group of 16 bits.

Microcontroller Prototypes with Arduino and a 3D Printer: Learn, Program, Manufacture, First Edition.
Dimosthenis E. Bolanakis.

Table 2.1 Prevailing numeral systems in μC programming and Arduino representation.

Decimal (DEC)	Binary (BIN)	Hexadecimal (HEX)
0	**B**00000000	**0** ×00
1	**B**00000001	**0** ×01
2	**B**00000010	**0** ×02
3	**B**00000011	**0** ×03
4	**B**00000100	**0** ×04
5	**B**00000101	**0** ×05
6	**B**00000110	**0** ×06
7	**B**00000111	**0** ×07
8	**B**00001000	**0** ×08
9	**B**00001001	**0** ×09
10	**B**00001010	**0** ×0A
11	**B**00001011	**0** ×0B
12	**B**00001100	**0** ×0C
13	**B**00001101	**0** ×0D
14	**B**00001110	**0** ×0E
15	**B**00001111	**0** ×0F
16	**B**00010000	**0** ×10
17	**B**00010001	**0** ×11
18	**B**00010010	**0** ×12
19	**B**00010011	**0** ×13
20	**B**00010100	**0** ×15

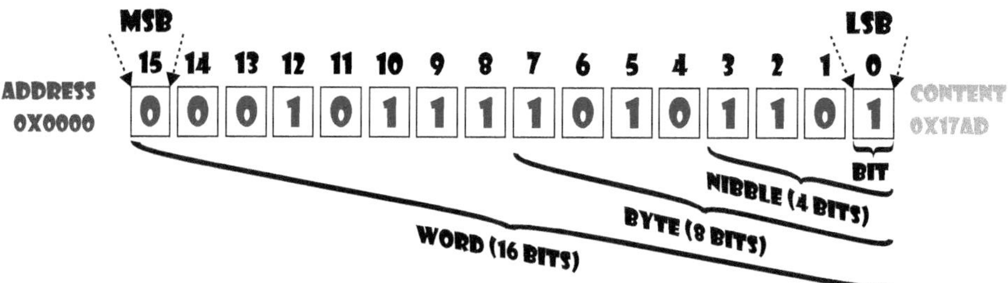

Figure 2.1 Example of a 16-bit register in μC's memory.

Figure 2.1 illustrates an example of a 16-bit register in μC's memory. The logical '0's and '1's held by the register form the hexadecimal value 0×17AD. It is worth noting that every single register in μC's memory feature a unique address (i.e. the address 0×0000 in this particular example). The 16th bit (i.e. bit 15) of the example holds the greater weight of the number and, hence, it is regularly referred to as *most significant bit* (*MSB*). On the other

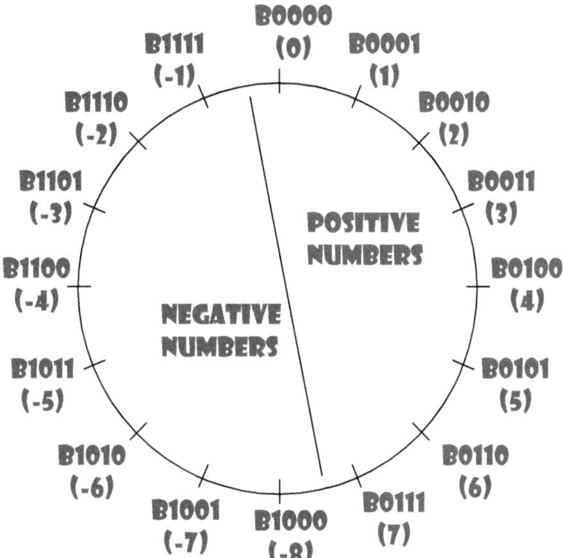

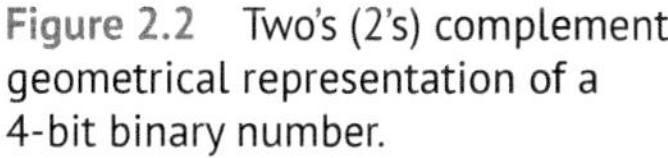
Figure 2.2 Two's (2's) complement geometrical representation of a 4-bit binary number.

hand, the 1st bit (i.e. bit 0) holds the least weight of the number and it is known as *least significant bit* (*LSB*). The same rule is valid for all the conventional numeral systems.[1] For instance, if we make an error to the *most significant digit* (*MSD*) of the decimal number 2000, which may represent a payment in Euros (€), then our error is equivalent to one or more thousand €. On the other hand, if we make an error to the *least significant digit* (*LSD*) of this number (i.e. 2000) then we may miss only a few €. Moreover, the MSB of a binary value may represent the sign of that number, in case the latter holds both positive and negative values. The most ordinary way to represent signed numbers in programming is the *two's (2's) complement* representation.

The *2's complement* constitutes a nonredundant representation system as there is only one representation for the number zero. Figure 2.2 illustrates a geometrical representation of a 4-bit signed number, in *2's complement* notation. A simple rule to switch between positive and negative binary numbers is to complement each bit of the original number (that is, the *one's complement* representation) and then add 1 to latter result. For example, we may switch between −7 and 7, in the binary system, as follows: $\sim(1001)+1 = 0110+1 = 0111$. And vice versa, to switch between 7 and −7 we do as follows: $\sim(0111)+1 = 1000+1 = 1001$.

Microcontrollers constitute digital microcomputer systems and, hence, the basic unit of information stored to their internal memory is the *binary digit (bit)*. In general, a bit stored to the memory of a digital computer is represented by two possible states; that is, the logical '1' and logical '0' (aka *true/false state*). Every register in μC's memory is arranged by a group of bits and that arrangement determines the μC architecture. For many years the market was orientated toward the development of 8-bit microcontrollers (e.g. the 8-bit AVR device employed by the popular Arduino Uno board). Nowadays, the μC market is focused on the

1 For more information the reader may refer to the *positional notation* of *base-b numeral systems.*

development of 32-bit devices (such as the 32-bit ARM Cortex-M4 processor of the familiar Teensy 3.2 board).

Despite the utilization of numbers, μC programming requires the utilization of particular coding schemes. A regular practice in embedded applications is the data transfer between a microcontroller and a personal computer. That particular practice requires the realization of the way *characters* are represented by a personal computer. Each character printed on the computer's screen corresponds to a 7-bit number. Table 2.2 presents all the printable and control (i.e. nonprintable) characters of the *American standard code for information interchange* (*ASCII*). For instance, the capital 'A' character corresponds to the hexadecimal number 0×41.

Another popular coding scheme that is commonly used in μC programming is the *binary-coded decimal* (*BCD*) representation. Table 2.3 presents a few random numbers in decimal, hexadecimal, binary, and the corresponding BCD encoding in nibble and byte representation. It is worth mentioning that the BCD belongs to the category of codes referred to as *weighted codes*, where (as in the conventional decimal, hexadecimal, and binary numeral systems) the weight of each digit depends on the position it holds in the number.

Arduino and C Common Language Reference

The *embedded programming* refers to the process of developing code for an embedded computer. The programming of an embedded computer is much alike to the conventional programming of a general-purpose personal computer. That is, the firmware execution in the μC device is in agreement with the so called *sequential programming* method. In sequential programming, the application code is executed as a series of steps, according to the textual order of statements in the source file. As mentioned earlier, the Arduino is built over the GNU GCC/G++ compiler and, hence, the Arduino code is quite similar to C/C++ code.

In contrast to the regular code of a personal computer, the application code designed to run by an embedded computer of no *operating system* (*OS*), is differentiated by two major factors. The first is related to the fact that the software code is responsible for the control of a dedicative hardware that is subjected to *real-time* response. The second refers to the control of the application-specific peripherals, which is addressed by the application system and aims in providing an interface with the outside world.

Hereafter, we make a brief overview to the fundamental aspects of sequential programming (common to the programming method of an *embedded system* as well as a *personal computer*). To provide compatibility between the supporters of C language programming and the supporters of Arduino programming, we provide Table 2.4 that incorporates the *directives*, *structures*, *operators*, *symbols*, etc., which are common between the two programming methods. Next, we explore the essentials of μC interfacing with the outside world using Arduino code. It is our intention to keep this overview as simple as possible in order to provide to the novice readers a quick jump-start in the application development with microcontrollers and Arduino. The readers will explore more advanced topics in μC-based design, in the subsequent and more practical chapters (addressed with a hands-on approach to learning).

Table 2.2 ASCII *printable* and *control* characters.

Low-order nibble	High-order nibble							
	(0×)0...	**(0×)1...**	**(0×)2...**	**(0×)3...**	**(0×)4...**	**(0×)5...**	**(0×)6...**	**(0×)7...**
...0	NUL	DLE	space	0	@	P	‘	p
...1	SOH	DC1	!	1	A	Q	a	q
...2	STX	DC2	“	2	B	R	b	r
...3	ETX	DC3	#	3	C	S	c	s
...4	EOT	DC4	$	4	D	T	d	t
...5	ENQ	NAK	%	5	E	U	e	u
...6	ACK	SYN	&	6	F	V	f	v
...7	BEL	ETB	‘	7	G	W	g	w
...8	BS	CAN	(	8	H	X	h	x
...9	HT	EM	)	9	I	Y	i	y
...A	LF	SUB	*	:	J	Z	j	z
...B	VT	ESC	+	;	K	[	k	{
...C	FF,	FS	,	<	L	\	l	|
...D	CR	GS	-	=	M	]	m	}
...E	SO	RS		>	N	^	n	~
...F	SI	US	/	?	O	_	o	DEL

Control (nonprintable) characters	
(0×00) Null	(0×10) Data link escape 1
(0×01) Start of heading	(0×11) Device control 1
(0×02) Start of text	(0×12) Device control 2
(0×03) End of text	(0×13) Device control 3
(0×04) End of transmission	(0×14) Device control 4
(0×05) Enquiry	(0×15) Negative acknowledgment
(0×06) Acknowledge	(0×16) Synchronous idle
(0×07) Bell	(0×17) End of transmitted block
(0×08) Backspace	(0×18) Cancel preceding message/block
(0×09) Horizontal tab	(0×19) End of medium
(0×0A) New line	(0×1A) Substitute for invalid character
(0×0B) Vertical tab	(0×1B) Escape
(0×0C) Form feed	(0×1C) File separator
(0×0D) Carriage return	(0×1D) Group separator
(0×0E) Shift out	(0×1E) Record separator
(0×0F) Shift in	(0×1F) Unit separator
	(0×7F) Delete

(*Continued*)

Table 2.2 (Continued)

Examples of escape sequences (\. . .)	
\n	New line (appends 0 × 0a)
\r	Carriage return (appends 0 × 0d)
\x0A	Any HEX value (appends 0 × 0a in this example)
\t	Horizontal tab (appends 0 × 09)
\”	Double quotation mark
\’	Single quotation mark
\\	Backslash

Table 2.3 BCD encoding (nibble and byte representation).

DEC	HEX	BIN	BCD (nibble representation)	BCD (byte representation)
12	**0** × C	**B** 1100	**B** 0001 0010	**B** 00000001 00000010
56	**0** × 38	**B** 0011 1000	**B** 0101 0110	**B** 00000101 00000110
1728	**0** × 6C0	**B** 0110 1100 0000	**B** 0001 0111 0010 1000	**B** 00000001 00000111 00000010 00001000

Working with Data (Variables, Constants, and Arrays)

Following the aforementioned brief overview on number representation, we may now explore the Arduino (and C language) features that are used to manipulate data in μC’s memory. According to Table 2.4, there are particular *datatypes*[2] that define the length of memory reserved by a variable:

i) ***bool*** reserves one byte of memory and may store two value only; that is, *true* and *false*. A possible syntax of *bool* datatype is as follows:

```
bool myVariable = true;
```

It is here noted that every programming line is terminated with a *semicolon (;)*.

ii) ***char*** reserves one byte of memory and it is recommended for storing ASCII characters. However, it may be used for storing signed numbers in the range of −128 to +127 (i.e. from -2^7 to 2^7-1). It should be noted that in Arduino (and embedded C) programming, single ASCII characters are placed within *single quotes (‘. . .’)*, while multiple ASCII characters are placed within *double quotes (“. . .”)*. A possible syntax of *char* datatype is as follows:

```
char myCharacter = 'A';
```

2 Arduino uses more *datatypes* but we make use of a limited set only, so as to achieve compatibility between Arduino and embedded C programming.

Table 2.4 Arduino/C common language reference.

<table>
<tr><th colspan="2">Preprocessor directives</th><th colspan="2">Assignment operator</th><th colspan="2">Other operators</th></tr>
<tr><td>/*. . .*/</td><td>Block comment</td><td>=</td><td>Assign to memory</td><td>{. . .}</td><td>Function/structure definition</td></tr>
<tr><td>//</td><td>Single line comment</td><td colspan="2">Arithmetic operators</td><td>[. . .]</td><td>Array subscript</td></tr>
<tr><td>#define</td><td></td><td>+</td><td>Addition</td><td>(. . .)</td><td>Function call</td></tr>
<tr><td>#include</td><td></td><td>–</td><td>Subtraction</td><td>&</td><td>Address-of</td></tr>
<tr><td>#ifdef</td><td></td><td>*</td><td>Multiplication</td><td>*</td><td>Indirection</td></tr>
<tr><td>#ifndef</td><td></td><td>/</td><td>Division</td><td>,</td><td>Separator</td></tr>
<tr><td>#else</td><td></td><td>%</td><td>Remainder</td><td>;</td><td>End of statement</td></tr>
<tr><td>#endif</td><td></td><td colspan="2">Bitwise operators</td><td>sizeof()</td><td>Returns number of bytes</td></tr>
<tr><td colspan="2">Control structures</td><td>~</td><td>Bitwise NOT</td><td colspan="2">Variable qualifiers</td></tr>
<tr><td>goto</td><td></td><td>|</td><td>Bitwise OR</td><td>const</td><td>Constant variable</td></tr>
<tr><td>if</td><td></td><td>&</td><td>Bitwise AND</td><td>volatile</td><td>RAM variable</td></tr>
<tr><td>else</td><td></td><td>^</td><td>Bitwise XOR</td><td>static</td><td>Preserves data in functions</td></tr>
<tr><td>switch. . .case</td><td></td><td><<</td><td>Bitwise shift (bit-shift) left</td><td colspan="2">Variable datatypes</td></tr>
<tr><td>break</td><td></td><td>>></td><td>Bit-shift right</td><td>bool</td><td>true/false (1 byte)</td></tr>
<tr><td>while</td><td></td><td colspan="2">Compound operators</td><td>char</td><td>Recommended for char</td></tr>
<tr><td>do. . .while</td><td></td><td>++</td><td>Assign unary increment</td><td>unsigned char</td><td>0–255 (1 byte)</td></tr>
<tr><td>for</td><td></td><td>−−</td><td>Assign unary decrement</td><td>short</td><td>2 bytes (signed)</td></tr>
<tr><td>continue</td><td></td><td>+=</td><td>Assign sum</td><td>unsigned short</td><td>at least 2 bytes</td></tr>
<tr><td>return</td><td></td><td>−=</td><td>Assign difference</td><td>long</td><td>4 bytes (signed); use L</td></tr>
<tr><td colspan="2">Comparison operators</td><td>*=</td><td>Assign product</td><td>unsigned long</td><td>4 bytes</td></tr>
<tr><td>==</td><td>equal to</td><td>/=</td><td>Assign quotient</td><td>int</td><td>2 or 4 bytes</td></tr>
<tr><td>!=</td><td>not equal to</td><td>%=</td><td>Assign remainder</td><td>unsigned int</td><td>2 of 4 bytes (signed)</td></tr>
<tr><td><</td><td>less than</td><td>|=</td><td>Assign bitwise OR</td><td>float</td><td>4 bytes floating-point</td></tr>
<tr><td><=</td><td>less than/equal to</td><td>&=</td><td>Assign bitwise AND</td><td>void</td><td>used in functions</td></tr>
<tr><td>></td><td>greater than</td><td>^=</td><td>Assign bitwise XOR</td><td colspan="2">Typecasting (conversion between datatypes)</td></tr>
<tr><td>>=</td><td>greater than/ equal to</td><td><<=</td><td>Assign bit-shift left</td><td>(char)variable</td><td>(unsigned char) variable</td></tr>
<tr><td colspan="2">Boolean operators</td><td>>>=</td><td>Assign bit-shift right</td><td>(short)variable</td><td>(unsigned short) variable</td></tr>
<tr><td>!</td><td>logical NOT</td><td></td><td></td><td>(long)variable</td><td>(unsigned long)variable</td></tr>
<tr><td>||</td><td>logical OR</td><td></td><td></td><td>(int)variable</td><td>(unsigned int)variable</td></tr>
<tr><td>&&</td><td>logical AND</td><td></td><td></td><td>(float)variable</td><td></td></tr>
</table>

The above syntax is equivalent to the following examples given in hexadecimal, binary, and decimal numeral system, respectively:

```
char myCharacter1 = 0 X 41; //hexadecimal
char myCharacter2 = B01000001; //binary
char myCharacter3 = 65; //decimal
```

It is here noted that *double slash (//)* prefix introduces a single line comment and everything written after this preprocessor directive will not be compiled into machine code.

```
char myArray1[] = "Hello World!\n";
char myArray2[] = {'H','e','l','l','o','
','W','o','r','l','d','!','\n','\0'};
char myArray3[14];
```

The above three code lines illustrate how to declare an array of characters in μC's memory. The first two declarations obtain the exact same outcome. In detail, the former syntax encloses the overall characters within *double quotes*, while the latter separates each character within *single quotes*, using the *comma (,)* operator. The latter method requires enclosing the array elements within *curly braces ({. . .})* and it is essential for initializing an array of numbers. The array declaration requires using the *square bracket ([. . .]) operators* after the *datatype* and the indicative array *name* (e.g. myArrray1). Square brackets are also used to access a particular array element, where the foremost element is stored to the zero position, e.g. myArray1[0] (aka *subscript*). Optionally, we may define the size of the array (as accomplished in the third example) but if the array is originally initialized with elements, then the compiler automatically defines its size. It should be noted that during the declaration of a sequence of characters, the compiler reserves one additional byte for the terminating *null (\0)* character (as illustrated in the second example). A sequence of characters that is terminated by the *null* character, is regularly referred to as *string*. Because there is no printable ASCII character of zero value, the *null* character at the end of an array provides particular flexibility when working with *strings* (such as, the concatenation of two individual arrays into a single *string*).

```
const char myArray1[] =
{0×48,0×65,0×6C,0×6C,0×6F,0×20,0×57,0×6F,0×72,
0×6C,0×64,0×21,0×0a,0};
```

The above example uses the *const* variable qualifier in order to configure the array elements as "read-only" data (i.e. the content of the array cannot be modified during the execution of the code). The content of the above array is the same as the previous "Hello World!\n" example, but it uses hexadecimal numbers instead of characters. The '*\n' escape sequence* appends 0 × 0a byte in μC's memory (see Table 2.2). It is worth noting that the *escape sequences* in programming consist of character sequences that are translated into other characters when invoked.

iii) ***unsigned char*** is the same as the *char* datatype but it encodes positive numbers only, that is from 0 to 255 (i.e. 2^8-1).
iv) ***short*** datatype is used for assigning a 16-bit signed value in μC's memory, that is from −32 768 (-2^{15}) to +32 767 ($2^{15}-1$). A possible syntax is as follows:

```
short myVariable = 0×8000; // the 0 × 8000 signed number
is equivalent to the decimal -32768.
```

v) ***unsigned short*** is the same as the *short* datatype but it encodes positive numbers only (of at least 16 bits), that is from 0 to 65 535 (2^{16}-1).

vi) ***long*** datatype is used for assigning a 32-bit signed value in μC's memory, that is from −2 147 483 648 (-2^{31}) to +2 147 483 647 (2^{31}-1). A possible syntax is as follows:

```
long myVariable = 0×FFFFFFFF; // the 0 × FFFFFFFF signed
number is equivalent to the decimal -1.
```

vii) ***unsigned long*** is the same as the *long* datatype but it encodes positive numbers only, that is from 0 to 4 294 967 295 (2^{32}-1). A possible syntax is as follows:

```
long myVariable = 0×FFFFFFFF; // this 0 × FFFFFFFF
unsigned number is equivalent to the decimal 4294967295.
```

viii) ***int*** datatype constitutes a 16-bit (i.e. *short*) or 32-bit (i.e. *long*) signed value, where the length depends on the incorporated μC device.

ix) ***unsigned int*** is the same as the *int* datatype but it encodes positive numbers only (of 16 bits or 32 bits according to the incorporated μC device).

x) ***float*** datatype defers from all the above types that apply to the spectrum of integers, that is, {−n, . . ., −2, −1, 0, 1, 2, . . ., +n}. The *float* datatype refers to numbers that feature a *fractional* part (i.e. carry a decimal point). A float variable can be as large as 3.4028235E + 38 and as low as −3.4028235E+38. A possible syntax is as follows:

```
float myVariable = 1.728;
```

xi) ***void*** is used only in the declaration of *functions* and denotes that the function returns no information and/or admits no arguments. The *void* keyword can be omitted if other *datatypes* are unnecessary. It is here noted that functions will be explored next in this chapter. A possible syntax of *void* is as follows:

```
void loop() {
// ...
}
```

Due to the diverse microcontroller architectures, the length of memory reserved by each data type may differ from devices to device. To this end, it is wised to explore the *sizeof()* operator before programming a μC board, in order to evaluate the number of bytes reserved by each *datatype*. This operator returns the number of bytes reserved by a variable or array. Next, we give some practical examples with Arduino Uno board, in terms of the utilization the various *datatypes*, as well as conversion between different *datatypes* (the latter conversion method is regularly referred to as *typecasting*).

Arduino UART Interface to the Outside World (Printing Data)

To get started with the following example, we need an Arduino Uno board along with a universal series bus (USB) (*type A* to *type B*) cable, and a personal computer incorporating USB port. Arduino IDE can be downloaded for free at www.arduino.cc/en/Main/Software

Figure 2.3 Arduino Uno board connected to the USB port of a laptop (using *type A* to *B* cable).

(it is noted that the examples of this book apply to the Arduino 1.8.9 version). As mentioned earlier, all Arduino boards incorporate a USB to universal asynchronous receiver/transmitter (UART) serial port that can be used for exchanging information with a personal computer. Moreover, the Arduino IDE employs built-in Arduino-specific functions that support an easy control of device's serial port. The physical connection of an Arduino Uno board to the USB port of a laptop is presented in Figure 2.3.

Arduino Ex.2-1

The first example code that will be uploaded to the Arduino Uno board is given in Figure 2.4. To upload this code we first select *'Arduino/Genuino Uno'* from the IDE menu *'Tools→Board:'*, as well as the serial COM port where the board is plugged into, from the menu *'Tools→Port:'*, e.g. *'COM11 (Arduino/Genuino Uno)'*. Then, we select the menu *'Sketch→Upload'* (or we just press the *'Upload'* button from the central screen). As soon as the application code is uploaded into the μC's memory (a process that is confirmed by the message *'Done Uploading'*, which appears on the bottom of IDE main screen), we open a built-in application of Arduino from the menu *'Tools→ Serial Monitor'*. The latter console (Figure 2.5a) prints the incoming data that travel over the USB port (from the Arduino Uno board). The *Serial Monitor* can also be used for sending data to an Arduino board. The *Serial Monitor* is an ASCII terminal console and, hence, it can only portray printable characters on its screen. However, it is sometimes crucial to inspect nonprintable characters on a terminal console. For this purpose, we make use of the *freeware* Termite console (Figure 2.5b), which provides the ability of printing the hex value of each received printable/nonprintable character.

Before the explanation of the main code, it is wised to make a reference to the main Arduino-specific functions that are used in every Arduino code, that is, the *setup()* and *loop()* functions. The former is used for all the initializations of the application code, while

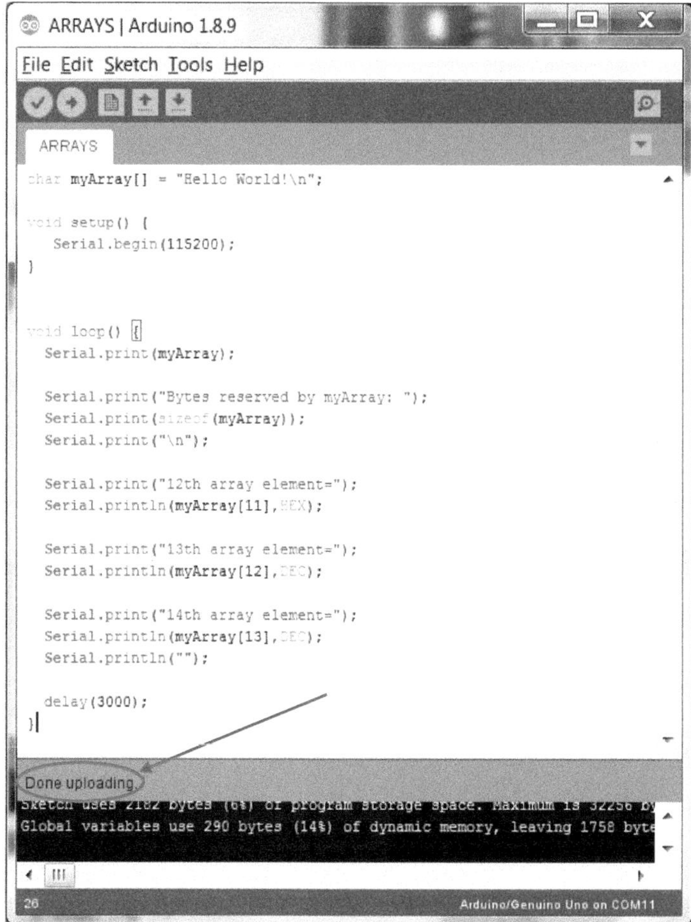

Figure 2.4 Arduino Ex.2-1a: output data from the μC to a computer through UART. *Source:* Arduino Software.

the latter implements an endless loop that incorporates and iterates the main code. It is here noted that functions in Arduino (and embedded C) programming are indicated by the *parentheses ()* operator.

Other functions used by this particular example are the *Serial.begin()*, *Serial.print()*, *Serial.println()*, and *delay()*. The first three functions are part of the *serial* library, which is used for the communication between the Arduino board and a personal computer (or the serial communication between two or more Arduino boards).

The *Serial.begin()* admits one or two arguments, where the first argument determines the data rate of the serial transmission, aka *baud rate*. In this particular syntax the data rate is set to 115 200 baud. Since the μC and the host computer feature their own individual synchronization clock, the exact same *baud rate* should be selected for the program that runs on the personal computer (as selected for the *Serial Monitor* and *Termite* consoles of

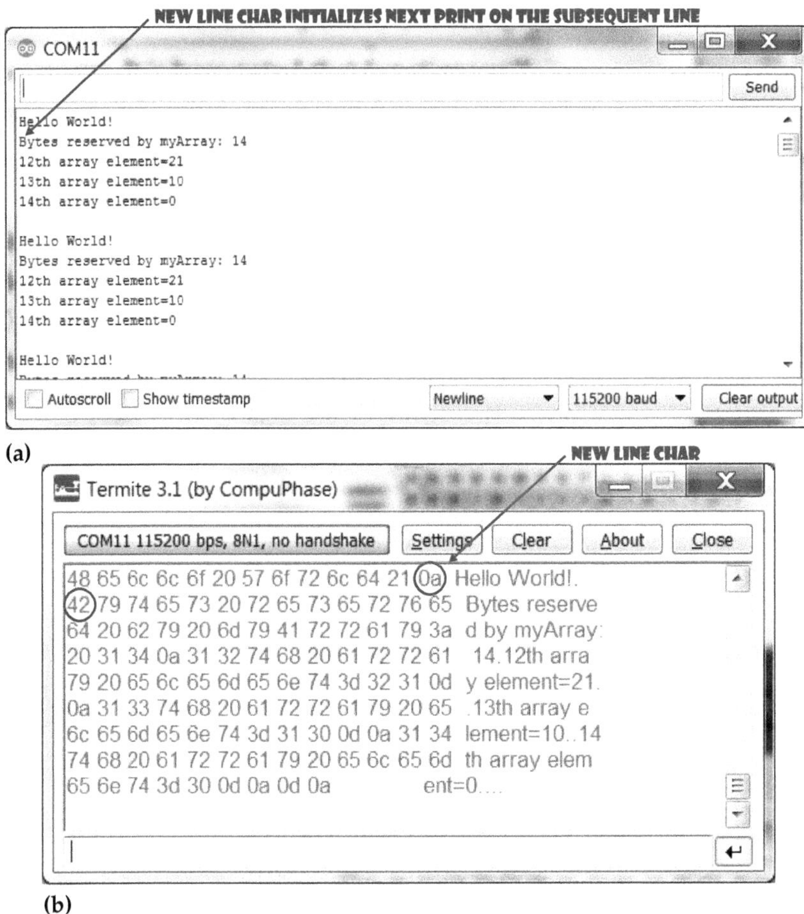

Figure 2.5 RS 232 terminal consoles: (a) Arduino serial monitor. (b) Termite. *Source:* Arduino Software.

Figure 2.5). The second (and optional) argument determines number of *data*, *parity,* and *stop* bits of every single transmitted frame. If not used, the Arduino board is configured by the default argument; that is, *SERIAL_8N1* (i.e. every transmitted frame incorporates eight *data* bits, one *parity* bit, and one *stop* bit).

Because the *Serial.begin()* configures the serial transmission of the microcontroller device, it is enclosed within the *setup()* function. The rest of the functions are part of the main application code and, hence, they are enclosed (and endlessly repeated) within the *loop()* function.

The *Serial.print()* and *Serial.println()* functions print data to the serial port as ASCII text. The difference between the two functions is that the latter appends a *carriage return (\r)* as well as a *new line (\n)* character after the user-defined text. Having defined the array "Hello World!" at the very first line of the source code (Figure 2.4), those two functions obtain the array elements and print them to the serial port. The fact that the array is declared outside the *setup()* and *loop()* functions, makes that variable *global*. Contrary to the *local* variables that are declared within a function and accessed only by its inside statements, *global* variables can be seen by every function/statement of the application code.

The first line of the main code (i.e. the first code line of the *loop()* function) prints the text "Hello World!" Because the latest array element employs the *new line (\n)* character, the next print takes place at the subsequent line of the terminal console, as depicted by Figure 2.5a. To verify the transmission of the nonprintable *new line (\n)* character to the host computer, we may observe its hex value by the Termite console. That is, the 0×0a value, which appears at the first line of Figure 2.5b.

The *Serial.print()* and *Serial.println()* functions can also admit a single or text of characters (enclosed within single or double quotes, respectively). Thereby, the second code line of *loop()* function prints the text "Bytes reserved by myArray:<space>". The latest array element of this text is the *space (' ')* character, which corresponds to the hex value 0×20 (this number can be confirmed by the printed byte at the 1st column / 4th line of the Termite console, in Figure 2.5b).

Despite characters, *Serial.print()* and *Serial.println()* functions can also admit numbers, where each digit of the number is printed with an ASCII character. The *Serial.print()* function of the third code line of *loop()* admits the *sizeof()* operator, which is used for returning the number of bytes reserved by elements of *myArray*. This code line prints the number 14 (which results from the print of the ASCII characters '1' and '4', as can be seen by the corresponding hex values 0×31 and 0×34 of the 2nd and 3rd column / 4th line of the Termite console, in Figure 2.5b). The string "Hello World!" consists of 12 printable characters, as well as the *new line (0 × 0a)* and *null (0 × 00)* characters, where the latter character is automatically appended by the compiler in order to initialize a *string*. Then the *Serial.print()* function of the fourth code line of the *loop()* prints merely the *new line* character so as to initialize the next print at the subsequent line of the terminal console.

If there is a need to evaluate the printed control characters without using a console like the Termite (which prints hex values of the received ASCII characters), we can manually print the *decimal*, *hexadecimal*, or *binary* value of each ASCII character, as performed by the *Serial.println()* functions of the application code. In detail, the *Serial.println(myArray[11],HEX);* code line prints the hexadecimal value of the 12th array element (that is, the '!' char of the array). It is here noted that the *square bracket operators ([. . .])* of this syntax access a particular array element, where numbering starts from zero. (The first array element is the *myArray[0]* which, in this particular example, employs the *'H'* character.) The subsequent *Serial.println()* functions of the code print the decimal values of *new line* and *null* characters (i.e. 10 and 0, respectively).

An important issue in embedded programming is the memory handing, as the capacity of μC's memory is much lesser than the physical memory of a personal computer. While the memory handing in conventional computer programming might be treated as featuring unlimited storage resources, embedded programming requires considering the capacity of memory and in particular, the volatile (data) memory. For instance, the Atmega328p microcontroller (employed by the Arduino Uno board) features 32 kilobyte (kB) of nonvolatile Flash memory (where the bootloader code reserves 0.5 kB) and 2 kB of volatile *static RAM (SRAM)*. The latter memory constitutes a dynamic memory, the content of which can be modified during the execution of the code (but does not hold its content without power supply).[3]

The example code presented herein stores all the incorporated ASCII texts to the data (SRAM) memory. Since there is no need to change the content of those texts, we could force the compiler to store them into the program (Flash) memory, along with the rest of the machine code. This implementation is performed with the utilization of the Arduino-specific *F()* macro. Enclosing a *string* within the *F()* macro, as depicted by Figure 2.6b, we initialize a Flash-based *string*. The result can be evaluated by the information provided at the bottom area of the Arduino IDE screen, after the compilation of the source code. The first compilation without the utilization of *F()* macro (Figure 2.6a), reserves 290 *bytes (B)* of data memory (i.e. 14% of SRAM). The second compilation with the utilization of *F()* macro (Figure 2.6b), reserves 202 B of data memory (i.e. 9% of SRAM). Since that the Flash memory is much larger from the SRAM, both implementation make no difference in terms reserved recourses of Flash (6% or reserved Flash memory in both cases, that is, 2182 reserved bytes without the *F()* macro, and 2236 reserved bytes with the use of the *F()* macro).

At this point, it would be wise to make a reference to an Arduino-specific utility, named *PROGMEM*. This utility is part of the *pgmspace.h* library and constitutes a variable modifier, which forces the compiler to store information into the Flash memory. This keyword could be used for assigning the elements of an array (e.g. *myArray*) in Flash memory. However, the *PROGMEM* modifier requires using the *datatypes* defined in *pgmspace.h* library, as well as special functions (also defined in *pgmspace.h*) for obtaining Flash data. The utilization of *PROGMEM* modifier requires an influx of new information, and hence, its study shifts away from the introductory character the example code presented herein.

The last Arduino-specific function of the example code is the *delay()* function. This function admits an *unsigned long* value related to *milliseconds (mss)* and, as soon executed, it delays the program for the amount of time defined by its argument.

Arduino Ex.2–2

The example code of Figure 2.7 explores further the *datatypes* and *typecasting* process in Arduino programming. The first code line (Figure 2.7a) initializes the "Hello World!\n" array that was used in the previous example, but using *decimal* values instead of *characters*. The subsequent six lines initialize the global variables *ucharVAR*, *shortVAR*, *ushortVAR*, *longVAR*, *ulongVAR*, and *floatVAR* of datatype *unsigned char*, *short*, *unsigned short*, *long*, *unsigned long*, and *float* respectively. In addition, the *shortVAR* is assigned to the hexadecimal value 0×8000 (code line 3), the *longVAR* is assigned to the hexadecimal value 0×FFFFFFFF (code line 5), and the *floatVAR* is assigned to the floating point value

3 Some μC devices incorporate an additional nonvolatile memory, which allows changing its content during the code execution. This memory (regularly referred to as EEPROM in μC's datasheets) is particularly useful in applications, such as a home alarm system, where the user might desire to dynamically change the alarm password. Hence, there is no need to revise the overall firmware code, as the alarm password in not stored in the Flash memory. It should be noted that the μC's EEPROM constitutes a peripheral IO subsystem, handled by particular functions. The Arduino-specific functions can be found in the EEPROM library.

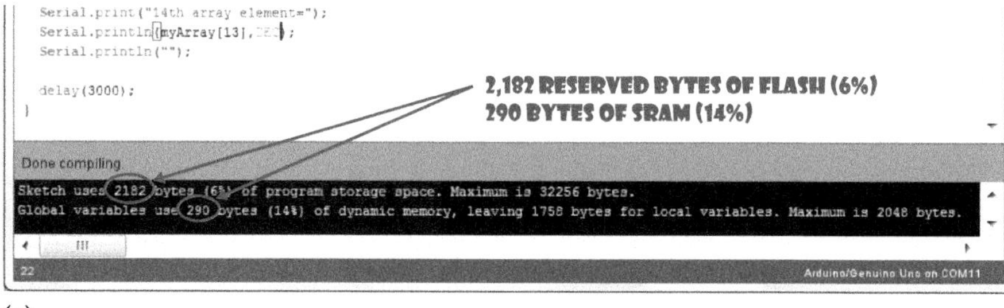

(a)

ARRAYS | Arduino 1.8.9

File Edit Sketch Tools Help

ARRAYS

```
char myArray[] = "Hello World!\n";

void setup() {
  Serial.begin(115200);
}

void loop() {
  Serial.print(myArray);

  Serial.print(F("Bytes reserved by myArray: "));
  Serial.print(sizeof(myArray));
  Serial.print("\n");

  Serial.print(F("12th array element="));
  Serial.println(myArray[11],HEX);

  Serial.print(F("13th array element="));
  Serial.println(myArray[12],DEC);

  Serial.print(F("14th array element="));
  Serial.println(myArray[13],DEC);
  Serial.println("");

  delay(3000);
}
```

FLASH-MEMORY BASED STRINGS

2,236 RESERVED BYTES OF FLASH (6%)
202 BYTES OF SRAM (9%)

Done compiling.

Sketch uses 2236 bytes (6%) of program storage space. Maximum is 32256 bytes.
Global variables use 202 bytes (9%) of dynamic memory, leaving 1846 bytes for local variables. Maximum is 2048 bytes.

Arduino/Genuino Uno on COM11

(b)

Figure 2.6 Arduino Ex.2-1b: (a) SRAM-memory and (b) Flash-memory strings. *Source:* Arduino Software.

3.141 592 (code line 7). Next, the *setup()* function performs the proper configuration of the Arduino Uno serial port, as explored by the previous example code.

The first seven code lines inside the *loop()* function print (to the Arduino Uno serial port) the reserved bytes of the *bool*, *char*, *short*, *long*, *int*, and *float* datatypes; that is 1, 1, 2, 4, 2, and 4 bytes, respectively. The result of the printing process is depicted by the first seven lines of *Serial Monitor* screen (Figure 2.7b).

The subsequent three code lines print the decimal value of the *shortVAR* variable and since the variable admits signed numbers of *short* datatype, the printed value is equal to −32 768 (line 8 of Figure 2.7b). The next two lines first perform a *typecasting* conversion

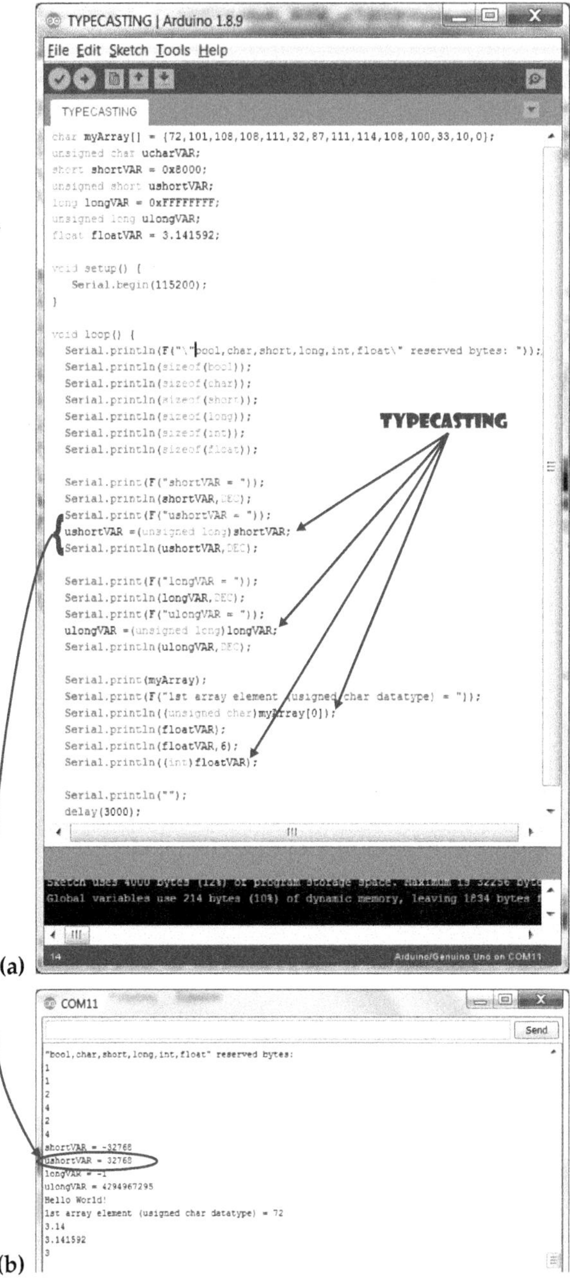

Figure 2.7 Arduino Ex.2-2: *datatypes* length and *typecasting*. (a) source code, (b) printed results. *Source:* Arduino Software.

to the *shortVAR* (i.e. convert the *short* variable to *unsigned short*), and then assign the result to the *ushortVAR*, and finally print the latter decimal value in Arduino serial port. The printed result is equal to the positive number 32 768 (line 9 of Figure 2.7b).

To realize the printed result of the signed *shortVAR*, and the unsigned *ushortVART* variable, we may refer to the *2's* complementary geometrical representation that was explored earlier in Figure 2.2. Since the *short* datatype reserves two bytes of memory, it admits hexadecimal values from 0×0000 to 0×FFFF. If this range represents unsigned values then the variable admits decimal values from 0 to 65 535, and hence, the 0×8000 is equal to the positive decimal number 32 768. On the other hand, if the same range represents signed values, then the variable admits decimal values from −32 768 to +32 767 and, hence the 0×8000 is equal to the negative decimal number −32 768.

The subsequent five lines inside the loop perform the exact same operation for the *long* datatype. They first print the decimal and signed value of the *longVAR* variable, and then the *typecasting* conversion to the *longVAR* converts and prints the corresponding unsigned value. In detail, the signed 0×FFFFFFFF value corresponds to the decimal −1, while the corresponding unsigned decimal value corresponds to the positive 4 294 967 295 (i.e. 2^{32}-1). The corresponding printed results appear on the 10th and 11th line of the *Serial Monitor* screen (Figure 2.7b).

The subsequent three code lines of the *loop()* function print the "Hello World!\n" string to the serial port, and then they perform on the fly *typecasting* conversion to the 'H' character of this string (which is the first array element) and print the corresponding *unsigned char* value (i.e. the *decimal* number 72).

The last five lines of the *loop()* function perform the following tasks: (i) print the *floatVAR* with two fractional digits, which is the default configuration of the *Serial.println()* function when used for printing floating point numbers, (ii) print the *floatVAR* value with six fractional digits precision, (iii) perform *on the fly* typecasting to the *floatVAR* and print the *integer*[4] result to the serial port, and (iv) transmit *carriage return* and *new line* characters to the serial port and initiate a delay of three *seconds (s)*. The printed results appear on the last three lines of the *Serial Monitor* terminal (Figure 2.7b).

Program Flow of Control (Arithmetic and Bitwise Operations)

The *instruction execution* in sequential programming is performed in *sequential order*. However, in programming, it is often required to alternate the normal *flow of control* and make, for example, a choice of execution among two statements, repeat a few statements, and so on. We have already explored the common *control structures* in Arduino (and embedded C) programming (Table 2.4) and, next, we explore the various implementations that apply to the program *flow of control* (aka *control flow*). To facilitate the understanding of *control flow* we herein address *flowcharts* and *pseudo-codes*.

To keep things simple, we herein explore only the elementary flowchart symbols, which are presented by Table 2.5. The *arrows* illustrate the direction of the code. The *terminal* symbols are used for opening and closing the code, respectively. The *connector* symbols can be used for breaking down large flowcharts into smaller ones. The *in/out* symbol illustrates incoming data to the μC device from the outside world, or outgoing data (from the μC device to the outside world). The *process* symbol declares that the μC's processor (i.e. CPU) performs an internal operation. The *predefined process* symbol illustrates a function call or

4 It should be noted that the conversion of a *float* to *integer* results in a truncation of the integer part of the floating point value.

Table 2.5 Elementary flowchart symbols.

Symbol	Description
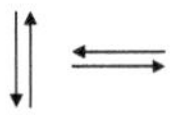 (arrows)	The ***arrows*** indicate the direction of the code.
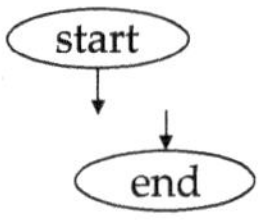 start / end	The ***terminal*** (start and end) symbols are used for opening/closing the code.
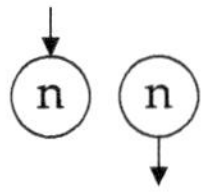 n n	The ***connector*** admits *character(s)*, which could be a *numbers (n)*, and it is used for breaking down large flowcharts into smaller portions.
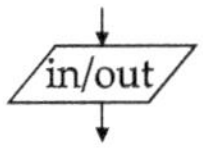 in/out	The ***in/out*** symbol illustrates incoming/outgoing data to/from the μC device.
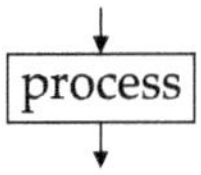 process	The ***process*** symbol indicates an internal operation performed by the μC's processor.
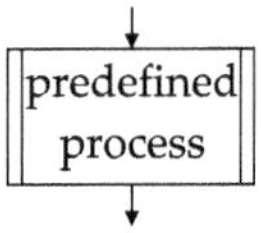 predefined process	The ***predefined process*** symbol indicates calls to functions or interrupts.
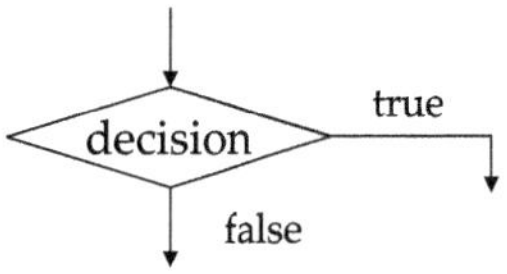 decision (true / false)	The ***decision*** symbol determines the direction of the code according to a decision.

a service of interrupt. The decision symbol determines the direction of the code according to the outcome of a particular decision.

The control flow statements in Arduino (and embedded C) programming are presented in Figure 2.8. We may identify two fundamental groups: (i) statements that implement an *unconditional*, *conditional*, or *multiway* branch to the code (Figure 2.8a–j), and (ii) statements that implement an *infinite* or *finite* loop, where the *finite* loop may feature a *predefined* or *nonpredefined* number of iterations (Figure 2.8k–q).

i) ***goto***

Figure 2.8a,b present the syntax of the *goto* statement, which implements an *unconditional branch* to the sequence of the program execution. The *goto* keyword admits a *label* and when executed, it transfers the control flow to the position where the *label* is traced (i.e. performs a jump from either *down to top* or *top to down*). In the *down to top* branch example of Figure 2.8b, the execution of the *goto* statement transfers the control flow to the label *start*. In Arduino (and embedded C) programming *labels* are syntax with a *colon (:)*. It should be noted that the utilization of the *goto* structure is not recommended in high-level programming. There are other structures that can be used to implement *unconditional branch*, which guarantee a more readable code.

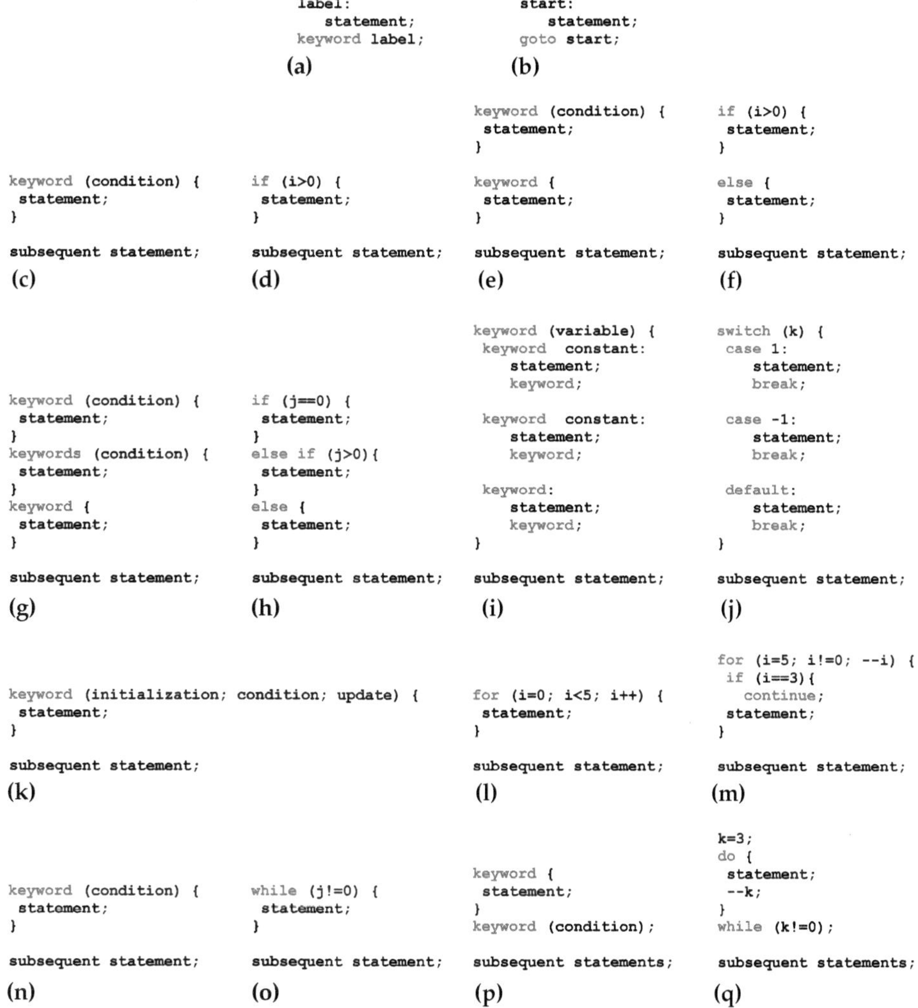

Figure 2.8 Control flow statements in Arduino (and embedded C) programming. (a) general syntax of the *goto* statement, (b) example of the *goto* statement, (c) general syntax of the *if* statement, (d) example of the *if* statement, (e) general syntax of the *if/else* statement, (f) example of the *if*/else statement, (g) general syntax of the *if/else if* statement, (h) example of the *if/else if* statement, (i) general syntax of the *switch. . .case* statement, (j) example of the *switch. . .case* statement, (k) general syntax of the *for* statement, (l) example of the *for* statement, (m) example of the *for* loop along with the *continue* statement, (n) general syntax of the *while* statement, (o) example of the *while* statement, (p) general syntax of the *do. . .while* statement, (q) example of the *do. . .while* statement.

ii) ***if(), else, else if()***

The *if()* statement is used to provide a *conditional branch* to the sequence of the program execution. The *if()* statement admits a condition (i.e. a Boolean expression), which is inserted into parentheses (Figure 2.8c). The condition is placed next to the *if* keyword and when its value is considered to be *true*, the body of the statement is executed. The *body* of the *if()* statement is enclosed within *curly braces ({. . .})* and its execution is skipped in case the evaluated condition is considered to be *false*. In the example of Figure 2.8d, the body of the statement will be executed in case the value of *i* variable is greater than zero. Otherwise, the control flow moves to the subsequent statement (i.e. the execution of the *if()* statement body is skipped).

The *else* statement can be optionally used to separate the *true* from the *false* condition (Figure 2.8e). Figure 2.8f illustrates an *if/else* implementation equivalent to the aforementioned *if()* example (Figure 2.8d). In detail, if the content of *i* variable is greater than zero (i.e. *true* condition) then the body of the *if()* statement is executed. Otherwise, if the content of *i* variable is less than or equal to zero (i.e. *false* condition), the body of the *else* statement is executed. It is noted here that the *else* statement does not admit a condition (i.e. it is executed for all the other valid conditions).

The *else if* expression can be used for implementing a *multiway conditional branch* (Figure 2.8g) to the sequence of the program execution. The example of Figure 2.8h evaluates three individual conditions. First, if the *j* variable is equal to zero, then the body of *if()* statement is executed. Second, if the *j* variable is greater than zero, then the body of *else if()* statement is executed. Third, if the *j* variable is less than zero, then the body of *else* statement is executed. It is noted here that the condition of *if()*, *else if()* statements is regularly syntax with comparison operators (as presented earlier by Table 2.4).

iii) ***switch()...case***

The *switch()...case* expression is also used for implementing a *multiway conditional branch* (Figure 2.8i) to the sequence of the program execution. The syntax is initiated with the *switch()* keyword and next to the keyword is placed, within *parentheses*, the evaluated condition. Next, the body of the statement is placed within *curly braces*. Contrary to the *if()*, *else if()* statements, the *switch()...case* condition admits a variable of *int* or *char* datatype. Inside of the statement's body, the *case* keyword determines the execution of one or more statements, provided that the variable's content is equal to the *constant* value placed immediately after the *case* keyword. It is here noted that a *colon (:)* should be inserted immediately after the *constant* value. After the execution of the statement(s) within a *case*, the keyword *break* is used to transfer the control flow at the end of the *switch()...case* statement. Optionally, an additional case can be used at the end of the *switch()..case*, which is executed when none of the earlier statements is evaluated to be *true*. This additional case is defined by the *default* keyword, along with a *colon (:)* immediately after that keyword.

The example of Figure 2.8j evaluates the content of the *k* variable. If its value is equal to 1 then the *statement* of the first *case* is executed. Then, the *break* keyword terminates the execution of the *switch()...case*. In case the content of *k* variable is equal to *minus one (−1)* then the *statement* of the second *case* is executed (and then, the *break* keyword terminates the execution of the *switch()...case*). For all the other potential values of *k* variable, the *default* statement is executed and the program execution moves to the subsequent statement.

iv) ***for()***

The *for()* statement is regularly addressed for implementing a *finite loop* of *predefined iterations*. The *for()* statement admits three expressions within the parentheses, which are syntax immediately after the *for()* keyword (Figure 2.8k). The *initialization* expression is executed only once as soon as the *for()* execution begins and sets, for example, an initial value to a variable (aka *loop counter*). The *condition* is evaluated on each iteration (e.g. tests the value of the *loop counter*) and determines if the *for()* execution will be repeated or terminated. As implied by its name, the *update* expression is addressed for updating the content of the loop variable every time the *for()* statement is executed.

The example of Figure 2.8l initializes the content of *i* variable to zero as soon as the execution of the *for()* statement begins. Then, the body of the *for()* statement (i.e. the statement within the *curly braces*) is executed. Next, the content of *i* variable is evaluated and if its content is less than five ($i < 5$), the *loop counter* is unary incremented ($i{+}{+}$) and the loop execution is repeated. The expression $i{+}{+}$ is equivalent to the expression $i = i + 1$. The *unary increment (++)* operator can be placed *before (++i)* of *after (i++)* the variable, forming a *prefix* or *postfix increment*, respectively. However, the result is the same in this *loop* example,[5] which runs five consecutive time (i.e. from $i = 0$ to $i = 4$).

The example of Figure 2.8m initializes a similar *for()* loop, but with *counting down* task instead of *counting up*, which is also repeated five times. However, the *if()* statement inside the loop evaluates the content of the *loop counter* and when its value reaches number three ($i = 3$), it executes the *continue* statement. The latter execution skips the current iteration of the *loop* and continues to the subsequent one. Hence, the statements inside the *loop* are executed four times instead of five. The flowchart of this particular example is given in Figure 2.9.

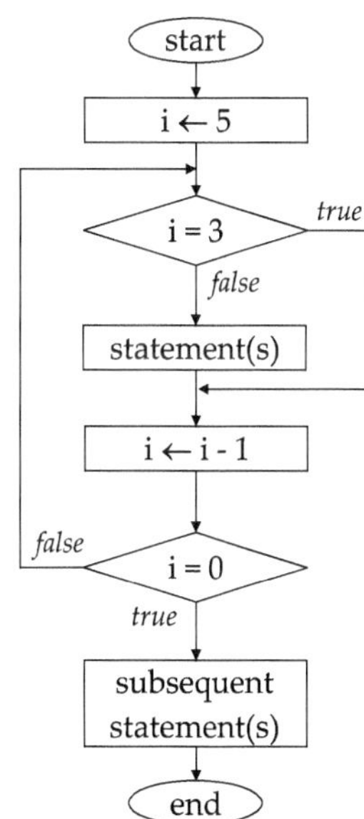

Figure 2.9 Flowchart of the *for()* loop example given in Figure 2.14m.

It is worth noting that the *update* expression of the *for()* statement admits other operators,[6] as well, while a syntax with empty expressions, i.e. *for(; ;)*, implements an *infinite* loop, which executes its body endlessly.

v) ***while(), do. . .while()***

The *while()* and *do. . .while()* statements are commonly addressed for implementing *finite loops* of *nonpredefined iterations*. A possible example in μC applications could be a *loop* that waits until the user presses a button on the hardware system. However, the *while()* and *do. . .while()* statements can be used to implement *loops* of *predefined iterations*, as well.

Figure 2.8n,o illustrate the syntax of the *while()* statement and a pseudo-code example, respectively. The syntax and a pseudo-code example of the *do. . .while()* statement are presented in Figure 2.8p,q, respectively. The latter

5 An example like **a = b++;** is different from **a = ++b;** as the former loads the content of *b* variable to *a* variable and then increments *b* variable, while the latter first increments *b* variable, and then loads the outcome to *a* variable.

6 An example like **for(i = 1;i <= 128;i* = 2) {. . .}** runs for the values $i = 1, 2, 4, 8, 16, 32, 64$, and 128.

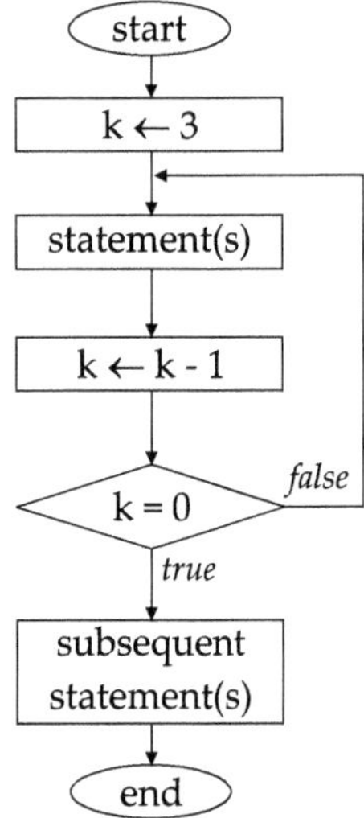

Figure 2.10 Flowchart of the *do. . .while()* loop example given in Figure 2.8q.

example illustrates a *loop* of *predefined iterations* with the *do. . .while()* statement. Both statements evaluate a condition inside the parentheses and if this condition is found to be *true*, they repeat the execution of the *loop*. Otherwise, they send the control flow outside the *loop*. The difference between the two statements is that the *while()* first evaluates the condition and then executes its body (i.e. the statements enclosed within *curly braces*), while the *do. . .while()* first executes its body and then decides the repetition of the *loop*.

Accordingly, the *while()* example (Figure 2.8o) evaluates the content of *j* variable and repeats the execution of the *loop* if the variable value is other than zero. On the hand, the *do. . .while()* example (Figure 2.8q) executes the *loop* and then evaluates the content of *k* variable. If the latter value is other than zero, the *loop* execution is repeated. Because the *k* variable is set to the value three ($k = 3$) immediately before the *do. . .while()* statement, and since the variable is unary decreased inside the *loop* ($--k$), the *do. . .while()* loop is executed three successive times (that is, a *loop* of *predefined iterations*). The flowchart of the latter example is given in Figure 2.10.

Arduino UART Interface (Flow of Control and Arithmetic/Bitwise Examples)

Hereafter, we explore the *flow of control* statements along with some *bitwise* and *arithmetic* operations that are essential in μC programming. As before, we make use of the Arduino Uno UART interface so as to print data on the terminal console running on a personal computer.

Arduino Ex.2-3

Figure 2.11a presents the source code, which exploits a *for()* statement in order to successively transmit all printable ASCII characters to the Arduino Uno serial port. Inside the main code (i.e. the *loop()* function) we first declare the loop counter, that is, the *i* variable of *int* datatype. Then, we initialize the *for()* loop to run for the counter values 0×20–0×7E, which constitute the hexadecimal values of all printable characters, i.e. from *space ('. . .')* till *tilde ('~')* character (Table 2.2). The body of the *for()* statement incorporates the Arduino *Serial.print* function, which performs on the fly *typecasting* to the loop counter and converts its value to *char* datatype (so as to transmit the current loop counter value to the serial port). As soon as the *for()* statement is terminated, the code appends a *new line (\n)* character to the serial port and then delays the code for three seconds before the next (overall) execution of the *for()* loop. The printed result on the *Serial Monitor* terminal console is presented in Figure 2.11b, where every single printed line corresponds to a single run of the *for()* loop (i.e. for the values 0×020–0×7E).

Because every *for()* loop demands the declaration of a variable, a good practice is to initialize the loop counter inside the *for()* statement, as presented in Figure 2.11c. This coding

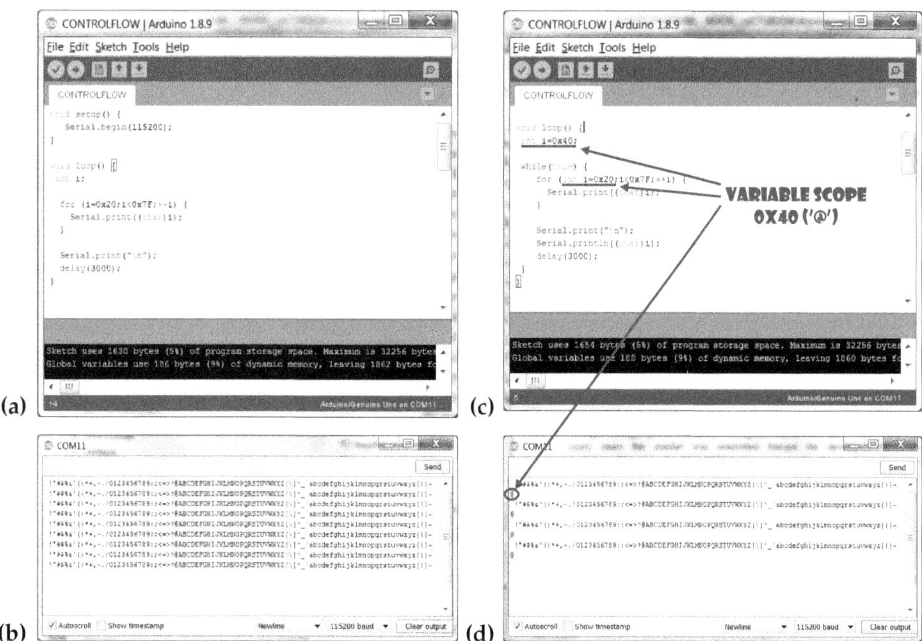

Figure 2.11 (a) Arduino Ex.2–3: *for()* & *while()* statements and variable *scope*. (a) *for* loop example code, (b) *for* loop printed results, (c) *while* loop example code, (d) *while* loop printed results. *Source:* Arduino Software.

style is peculiar to a property in Arduino (and embedded C) programming, called variable *scope*, which determines the region where the variable has its existence. In this particular example, the *i* counter can only be accessed inside the *for()* statement. To make that clear we have declared another variable, with the exact same name at the beginning of the *loop()* function, and configured that variable to an initial value ($i = 0 \times 040$). Below this declaration we have incorporated a *while()* statement that encompasses the previous code. The condition of the *while()* statement is permanently set to *true* and, hence, we have implemented an *infinite* loop, which encompasses the *for()* statement along with the following three code lines. The purpose of the *while()* statement is to prevent the *loop()* function to reconfigure the content of the *i* variable above the *while()* statement.

The printed result of the revised source code is given in Figure 2.11d. The *for()* statement prints each ASCII *character* equivalent of the *i* counter (i.e. 0×20–0×7E). As soon as the *for()* loop is terminated, the code prints the *i* variable, which has been declared and configured once, inside the Arduino *loop()* function. The latter printed result is equal to the ASCII character '@', which corresponds to the hexadecimal value 0×40.

Arduino Ex.2–4

Figure 2.12 presents a revision of the previous source code, which prints only the alphabet letters (instead of all printable characters). In detail, the code of Figure 2.12a prints the upper case letters of the alphabet, while the code of Figure 2.12c prints the upper and lower

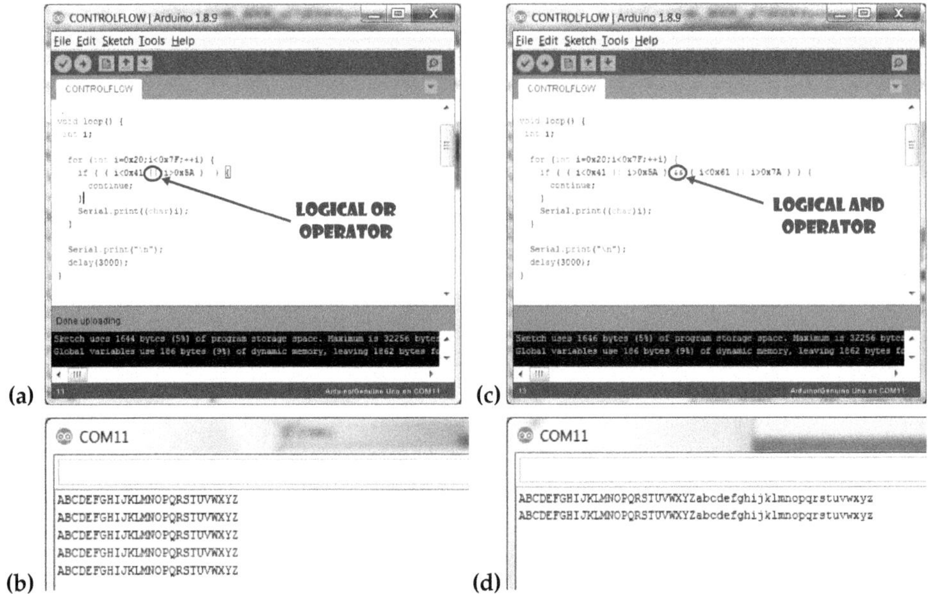

Figure 2.12 Arduino Ex.2-4: *for()* statement and *Boolean* operators (upper/lower case letters). (a) *logical OR* example code, (b) *logical OR* printed results, (c) *logical AND* example code, (d) *logical AND* printed results. *Source:* Arduino Software.

case letters. The printed results are illustrated in Figure 2.12b,d, respectively. To accomplish those tasks, the source code make use *logical OR* and *logical AND* Boolean operators.

With the utilization of the *logical OR* (||) operator, the *if()* statement inside the *for()* loop explores if the hex value of the loop counter is outside the range 0×41–0×5A (Figure 2.12a). In case the condition is true, the *continue* statement is executed and the current loop iteration is skipped (i.e. the *for()* loop resumes to the subsequent counter value). Since the hex values 0×41–0×5A corresponds to the ASCII characters 'A'–'Z', this example prints the upper case alphabet letters (Figure 2.12b). With the additional utilization of the *logical AND (&&)* operator, the example of Figure 2.12c prints the ASCII characters within the range 0×41–0×5A ('A'–'Z'), as well as within the range 0×61–0×7A ('a'–'z').

Arduino Ex.2-5

Figure 2.13 presents the flowchart and the source code (Figure 2.13a,b, respectively) for inputting data to the μC from a personal computer. The code first initializes an array of 20 elements of *char* datatype (i.e. *myArray*), and then the Arduino *Serial.begin()* within the *setup()* function configures the serial transmission of the microcontroller device at 115 200 baud rate.

The first code line inside the *loop()* configures a *pointer* of name *ptr*. In Arduino (and embedded C) programming, the pointer declaration is possible with the use of the *indirection (*)* operator. In programming, the pointer is a variable that points to a specific memory location and it is commonly addressed for array indexing techniques. The task of forcing a

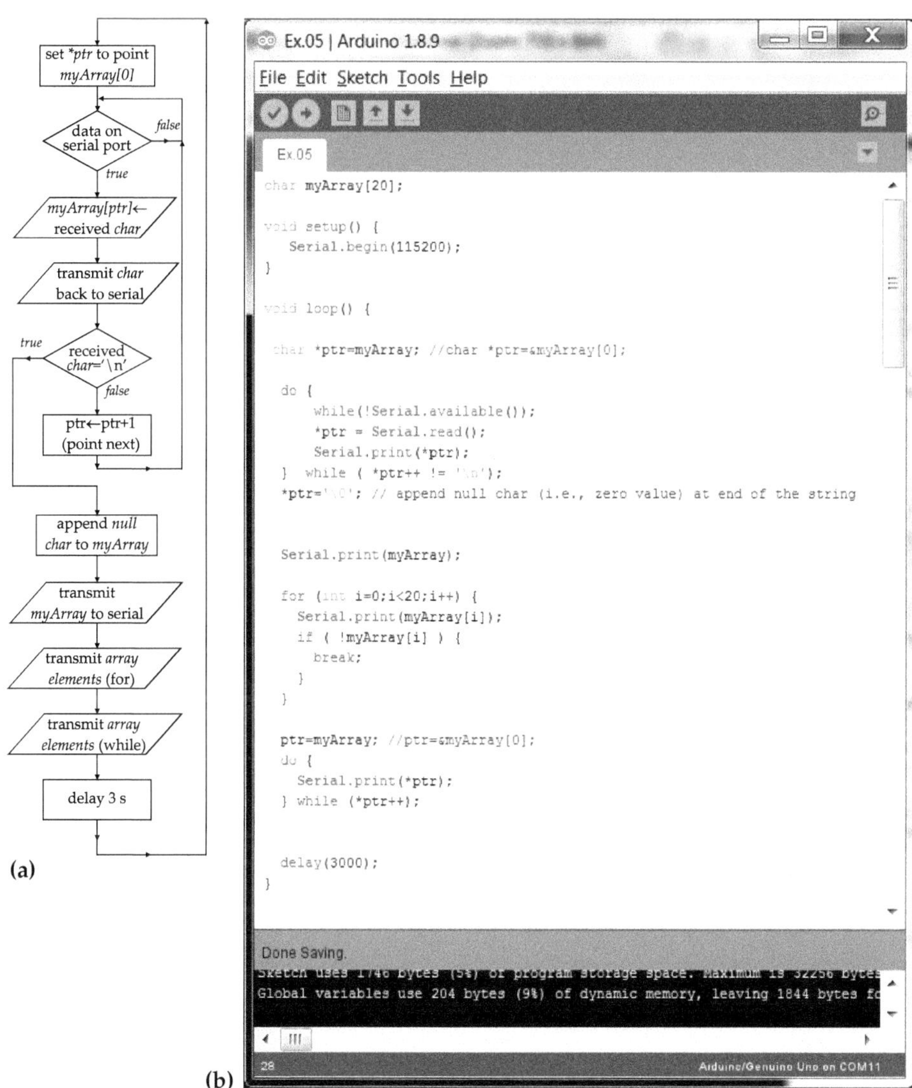

Figure 2.13 Arduino Ex.2-5: input data to the μC from a computer through UART (pointers). (a) flowchart, (b) source code. *Source:* Arduino Software.

pointer to point to a particular memory address is achieved with the use the *address-of (&)* operator. For example, the syntax char **ptr=&myArray[0];* declares a pointer and the assigns the latter to point to the first array element. If the pointer has already been initialized in the program, we leave out the *indirection (*)* operator. For instance, the syntax *ptr=&myArray[2];* assigns the pointer to point to the third array element. If we optionally leave out the array *subscript* (i.e. *ptr=myArray*), then pointer points to the first array element.

Subsequent to the pointer initialization, a *do. . .while()* statement obtains data from the serial port till the *new line ('\n')* character is fetched. At first, a *while()* statement repeats the execution of the Arduino *Serial.available()* function until a byte appears on the μC's serial port. The latter task is accomplished with the use of the *logical NOT (!)* operator at the front of the *Serial.available()* function. As soon a *character* arrives to the serial port, the Arduino *Serial.read()* function obtains and assigns that character to the memory location pointed by the **ptr* variable. The subsequent *Serial.print()* function loops back (to the serial port) the received character. The condition of the *do. . .while()* statement evaluates if the content of the **ptr* variable (which hold the most recent received character) equals to the *new line ('\n')* character, and then increases the pointer to point to the next array element. If the condition is found to be *false*, the *do. . .while()* statement is repeated in order to obtain, from the serial port, the next available character. Otherwise, the *do. . .while()* statement is terminated and the subsequent code line appends a *null ('\0')* at the final position of *myArray* (hence, initializing a *string* array that incorporates the ASCII characters received by the μC's serial port).

It should be noted that the *do. . .while()* statement, herein, highlights the role of *finite loops* of *nonpredefined iterations*. The loop termination of this example is an option of the user, who determines the end of the generated *string* array upon pressing the <Enter> key on the computer's keypad.

As soon as the string array is defined (i.e. after the user presses the <Enter> key), the code prints the generated *string* three successive times. The first one is accomplished with the used of the Arduino *Serial.print()* function, while the other two prints are accomplished by the *for()* and *do. . .while()* statements, respectively. The two latter prints are addressed for highlighting the effectiveness of the *null ('\0')* terminated character of a *string*, the effectiveness of *pointers*, as well as the effectiveness of *finite loops* of *nonpredefined iterations*.

In consideration of the *for()* statement, the loop is configured to run for 20 successive times, as *myArray* is initialized with 20 elements. Inside the loop, the array elements are referenced by a *subscript*. Hence, the foremost code line prints each individual array element, starting from *myArray[0]*. While the array is configured to hold 20 elements, the size of the generated *string* is determined by the user. Hence, the *if()* statement inside the loop evaluates if the value of the current array element equals to zero. If the evaluated condition is *true*, the loop has reached the last element of the user-defined *string* (i.e. the *null* character) and, hence, the *break* keyword is executed. With the exploitation of the *null ('\0')* terminated character of a *string*, the *break* execution transfers the control flow outside the *for()* loop.

The effectiveness of *pointers* along with a *do. . .while()* statement when working with *strings*, is clearly illustrated in the latter print task of *myArray*. At first, we initialize the pointer to point to the first array element, and then the *Serial.print()* function prints the current array element (pointed by **ptr*). Then, the condition of the *do. . .while()* statement first evaluates the content of **ptr* and then performs an *unary increment* to the pointer. If the evaluated condition equals to zero (i.e. the *null* character of the *string* has been obtained), the *do..while()* statement is terminated and the *delay()* function is executed. Otherwise, the *do..while()* loop is repeated and the subsequent array element is fetched and printed to the serial port.

Table 2.6 illustrates the possible syntax of *unary increment (++)* operator when used with *pointers*. The corresponding tasks can be performed with the use of the *unary decrement (−−)* operator, as well. The printed results of Arduino Ex.2–5 along with some revised tests are illustrated in Figure 2.14.

Table 2.6 Unary increment operator applied to pointers.

Syntax	Description
*ptr++	The expression is treated as *(p++). In the Arduino Ex.2–5, the *do. . .while(*ptr++)* would first evaluate if the content of the current array element (pointed by **ptr*) is zero, and then it would increment the pointer to point to the next array element.
*++ptr	The expression is treated as *(++p). In the Arduino Ex.2–5, the *do. . .while(*++ptr)* would first increment the pointer to point to the next array element, and then it would evaluate if the content of the current array element (pointed by the **ptr*) is zero.
++*ptr	The expression is treated as ++(*p). In the Arduino Ex.2–5, the *do. . .while(++*ptr)* would first increment the value of the current array element pointed by the **ptr*, and then it would evaluate if the latter value is zero.

A simple way to transmit ASCII character to the μC's serial port is through a terminal console. Each *character* or *character set* is transmitted by writing the data through the computer's keyboard to the proper field of the terminal console, and then pressing <Enter> key (as depicted by Figure 2.14a). Particular attention should be taken into account when reading data from the serial port of an Arduino board. For instance, when the termite console is initially connected to the serial port where the Arduino Uno board is connected (and executes the source code of Arduino Ex.2–5) four bytes appear on the terminal screen, that is, the 0×F0, 0×F0, 0×F0, 0×F0. Those bytes are most likely sent by the Arduino bootloader code, which runs upon the board connection to the computer's USB port.

Thereby, if we enter the characters "test" through the terminal console and press <Enter>, we observe that the aforementioned characters appear at the beginning of our string (Figure 2.14b). This particular error appears only the first time we enter a *character set* from the keyboard. If we reenter the characters "test" and press <Enter>, we observe the print of an errorless string (Figure 2.14c). Another way to obtain an errorless print from the very begging is to press the *reset* button on the Arduino Uno board before entering the opening string (Figure 2.14d).

The printed examples of Figure 2.14d,e,f present the three possible combinations of the *unary increment (++)* operator when used with *pointers*. All the combinations apply to the last *do. . .while()* statement of Arduino Ex.2–5 source code. The output result of the original version of Ex.2–5 prints the *"test\n"* string along with the terminal *null* '*\n*' character (Figure 2.14d). The first revision of Ex.2–5 code prints the *"test\n"* string without the *null* '*\n*' character (Figure 2.14e), as the *do. . .while(*++ptr)* first increments the pointer to point to the next array element and then evaluates if the content of the current array element is zero.

The second revision of Ex.2–5 code prints the 't' character (of the previously generated "test" string), and then increments the value of the current array element pointed by the *ptr (i.e. *myArray[0]*), and afterward evaluates if the latter value is zero. Thereby, *myArray[0]* content is unary incremented and, hence, it obtains values from 0×75 (which is the

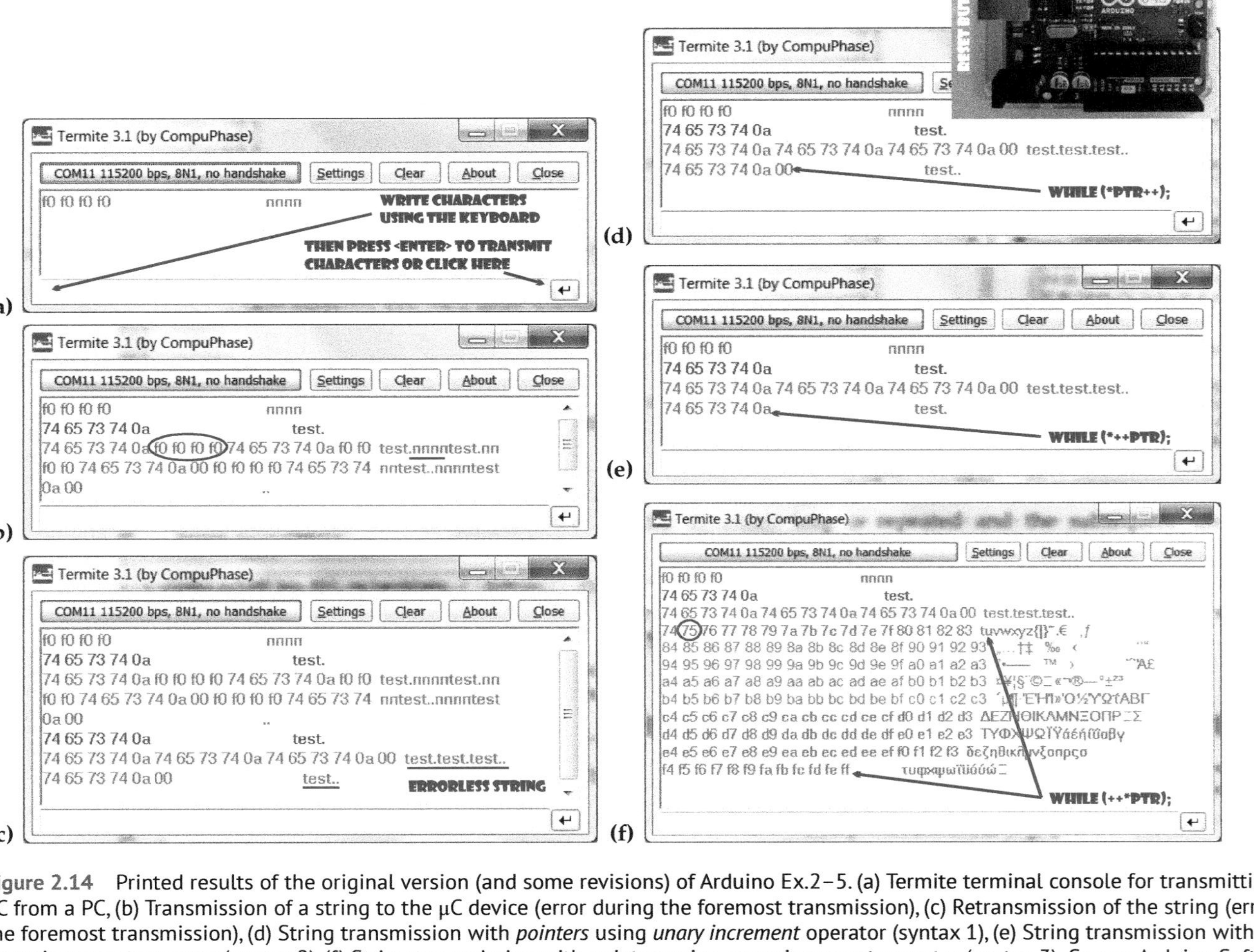

Figure 2.14 Printed results of the original version (and some revisions) of Arduino Ex.2–5. (a) Termite terminal console for transmitting characters to the μC from a PC, (b) Transmission of a string to the μC device (error during the foremost transmission), (c) Retransmission of the string (errorless results after the foremost transmission), (d) String transmission with *pointers* using *unary increment* operator (syntax 1), (e) String transmission with *pointers* using *unary increment* operator (syntax 2), (f) String transmission with *pointers* using *unary increment* operator (syntax 3). *Source:* Arduino Software.

hex value of the *character* 'u' that follows *character* 't') till 0×FF (which is the highest value a *char* datatype can receive). An unary increment to *myArray[0]* when the latter element holds 0×FF value, causes a reset of its content (aka *overflow*[7]). Thereby, the *do. . .while()* statement verifies a zero value and transfers the control flow outside the loop.

Arduino Ex.2-6

Figure 2.15a,b,c presents, respectively, the flowchart, source code, and printed results of an arithmetic operation that transforms a numeric string to binary value. The gray part of the flowchart is the exact same as in the previous example, which was used for the formation of a string array from *characters* sent by the user. The new part of code assumes that the user transmits ASCII characters that correspond to numbers (i.e. from '0' to '9'). The new code part (i.e. inside the red rectangle of Figure 2.15b) transforms the numeric string to binary value, which is assigned to an *unsigned char* variable named *number*. Because the *unsigned char* datatype can hold a byte value only, the code works for numeric strings "0"–"255".

The new code part inside the red rectangle (Figure 2.15b) works as follows. At first the *number* variable is cleared (i.e. set to zero value), and the *ptr* pointer is set to point to the first array element (i.e. *myArray[0]*). The condition inside the *while()* statement explores if the current value pointed by the *ptr* is among the ASCII numbers '0'–'9' (i.e. 0×30–0×39). If the condition is not met the loop is terminated. Since the numeric string is sent by the user as soon as he/she presses the <Enter> key on the personal computer, the code is terminated as soon as the *new line* character is obtained from *myArray*.

The code inside the *while()* initially transforms each ASCII character to binary number. This task is performed by subtracting the hexadecimal value 0×30 by each *character*. For instance, if the user enters the ASCII nine ('9' = 0×39), then the subtraction 0×039–0×30 = 0×09 converts the ASCII nine ('9') to the binary nine (0×09 = 0b00001001 = 9). As soon as the conversion finishes, the code addresses the *positional notation* of the base-10 numeral system to reform the numeric sting to binary value.

Figure 2.16 illustrates how the numeric string "128" is converted to the binary value 128 by the successive iterations of the *while()* statement. Figure 2.16b illustrates the storage of the ASCII numbers '1', '2', and '8' to the first three elements of *myArray*. According to the *positional notation* of base-10 numeral system, the worth of the number 128 is defined as follows: $1\times10^2 + 2\times10^1 + 8\times10^0$ (=100+20+8). Thereby, Figure 2.16c illustrates the three successive iterations (of the *while()* statement), which implement the aforementioned calculations (while the ASCII to binary conversion of each individual array element is given in Figure 2.16d).

The binary value of the aforementioned conversion is stored to the *number* variable, and then the variable is sent to the μC's serial port using the Arduino functions *Serial.println* and *Serial.write*. The former function retransforms the binary value to ASCII characters and sends them to the serial port, while the latter transmits directly the byte value.

---

7 The corresponding process of decreasing the content of a variable when the latter holds zero value is referred to as *underflow*. In case of a *char* datatype, the variable will receive the value 0×FF when decreased from zero value.

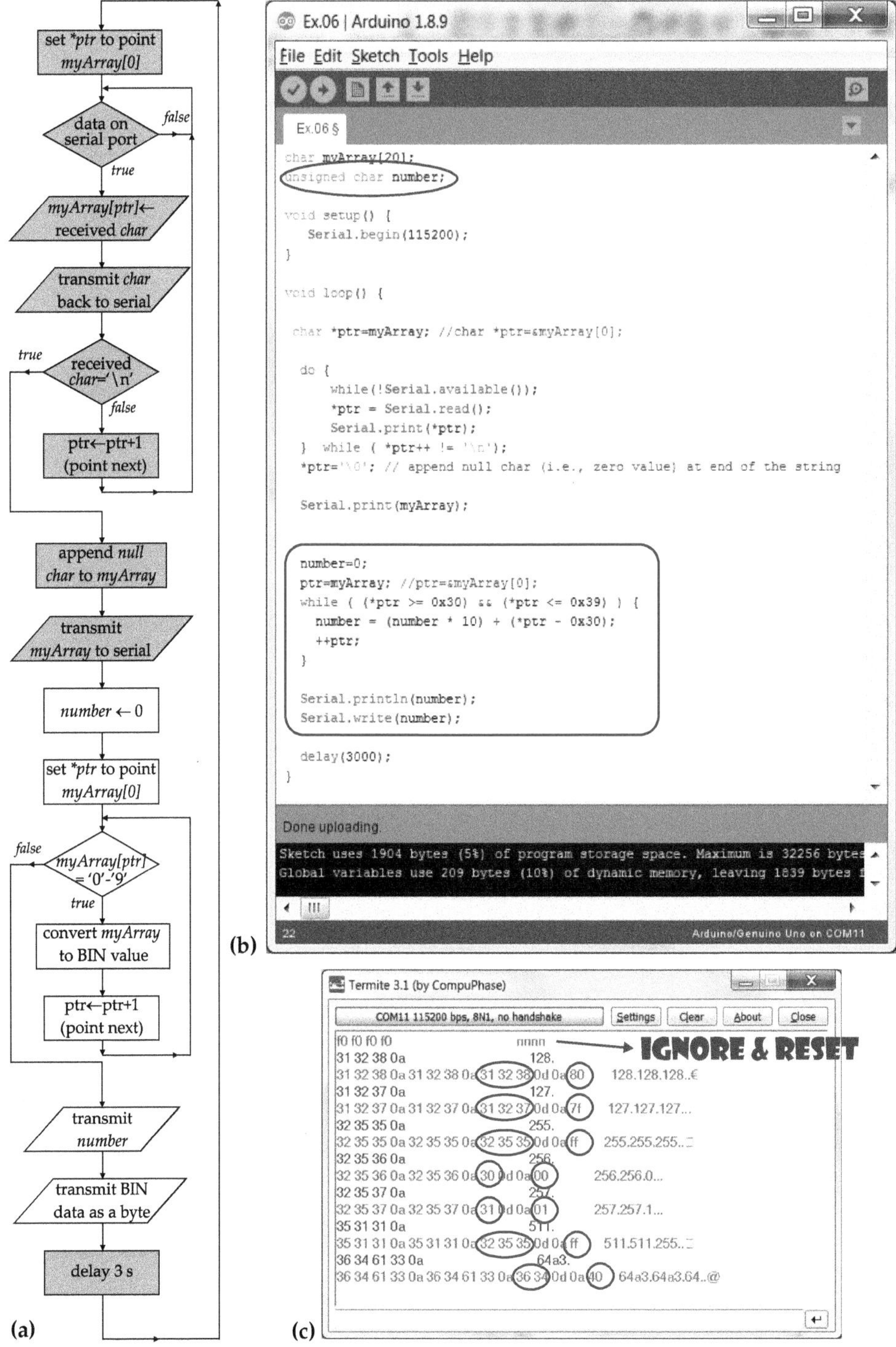

Figure 2.15 Arduino Ex.2-6a: arithmetic operations (numeric string to binary value). (a) flowchart, (b) source code, (c) printed results. *Source:* Arduino Software.

(a)

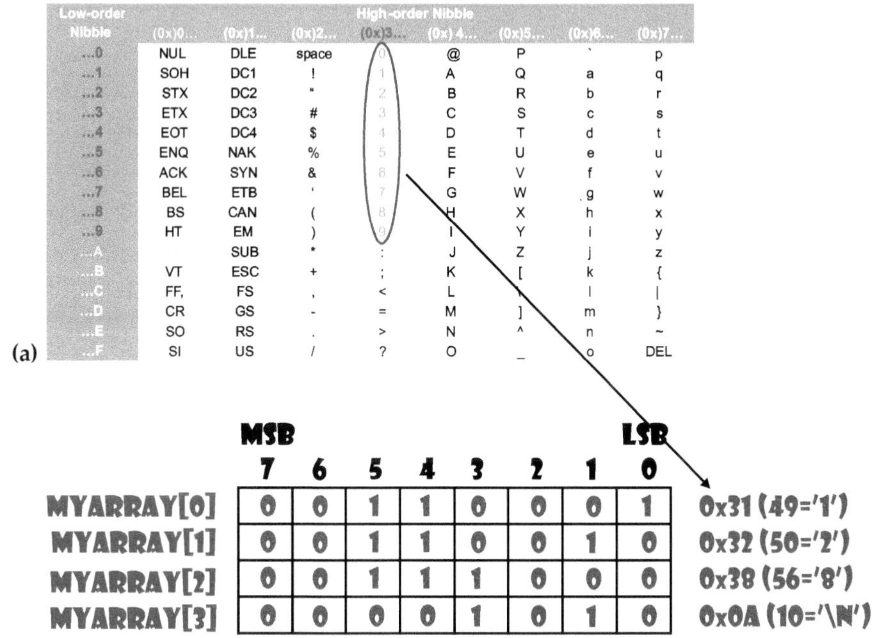

(b)

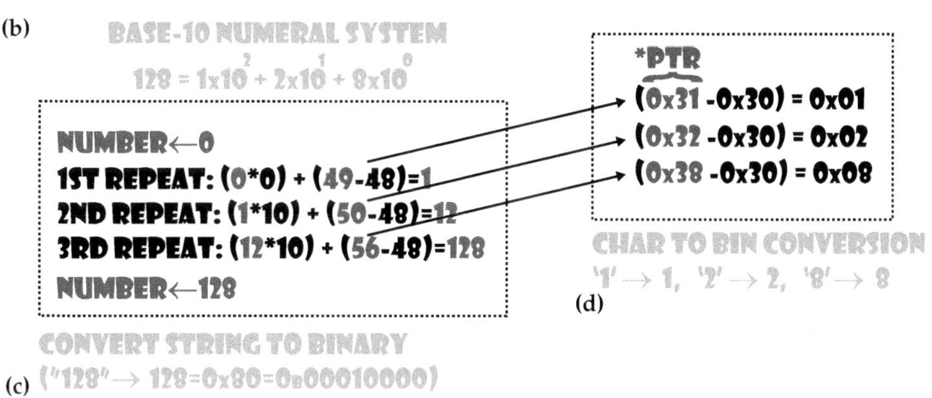

(c)

Figure 2.16 Transforming a numeric string to binary value ("128"→128). (a) the ASCII numbers '1'–'9', (b) storage of the ASCII numbers '1', '2', and '8' to three memory locations, (c) process of converting a string to binary number through a loop.

The printed results when transmitting the numeric sting "128", is presented in the red circles of Figure 2.15c. The Termite console also presents the printed results of the numeric stings "127", "255", "256", "257", "511", and "64". The numeric sting "256", "257", "511" are higher than a byte value (i.e. 0 × 100, 0 × 101, 0 × 1FF). Hence, only the least significant byte appears on the Termite console in the corresponding transitions (i.e. 0 × 00, 0 × 01, and 0 × FF, respectively). It is worth noting that if the user sends *characters* others than '0'–'9', then the code is terminated. In the very last example, the user sent the numeric sting "64a" but the conversion is performed only to the numeric sting "64".

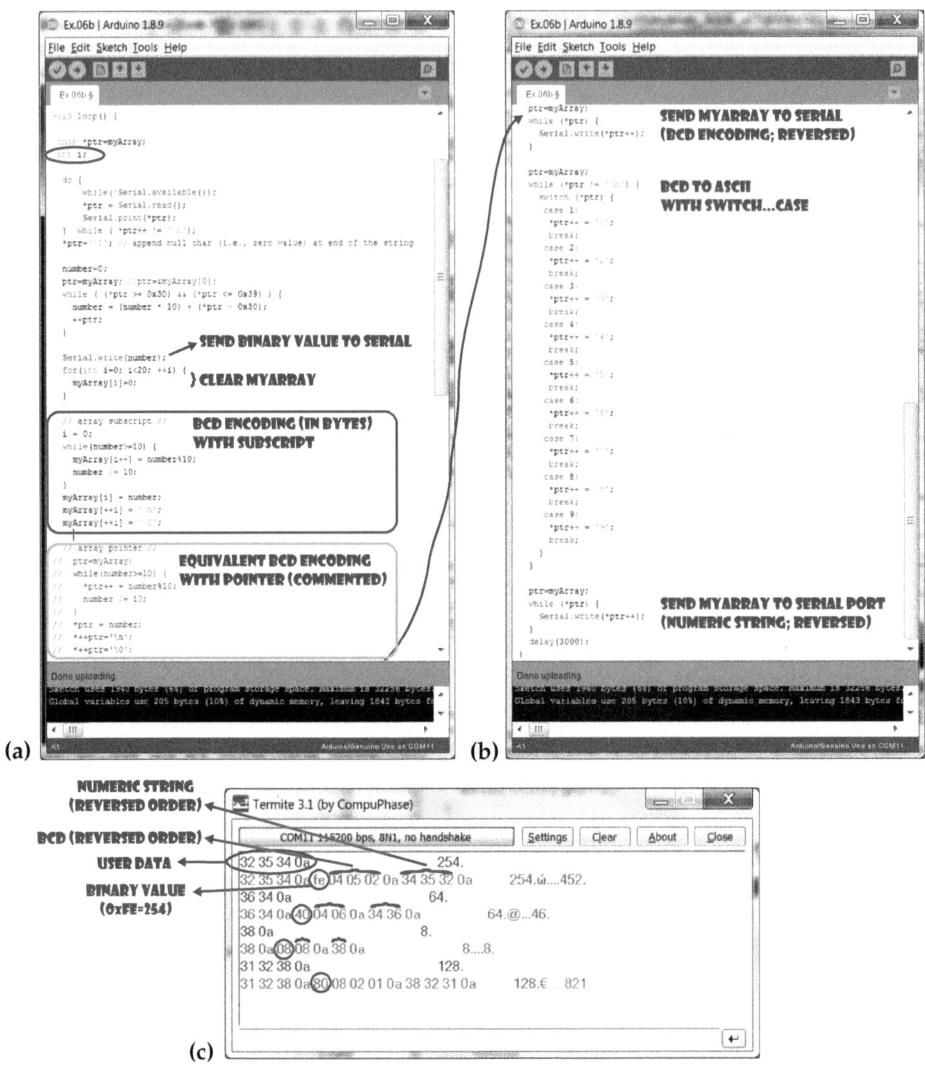

Figure 2.17 Arduino Ex.2-6b: arithmetic operations (binary value to numeric string). (a) source code (part 1 of 2), (b) source code (part 2 of 2), (c) printed results. *Source:* Arduino Software.

Figure 2.17a,b presents an upgraded version of the aforementioned example code, which additionally transforms the binary value (stored to the *number* variable) to numeric string. It should be noted that the example code is designed to work for positive numbers only. The arithmetic operations that convert the *number* variable to BCD are performed inside the *while()* statement (which is enclosed within the red rectangle). Applying to the *positional notation* of base-10 numeral system, the *while()* statement successively divides the *number* variable to the number 10, until the division's *quotient (quo)* is less than 10. On each division the *remainder (rem)* is assigned to the corresponding element of *myArray*, starting with *myArray[0]*. Thereby, the outcome is illustrated in reversed order. The *while()*

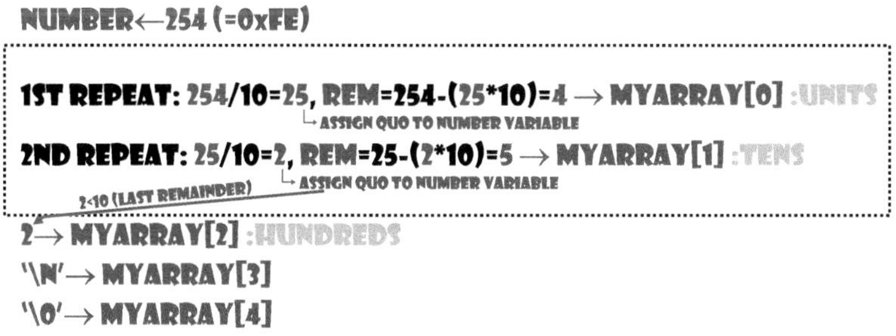

NUMBER←254 (=0xFE)

1ST REPEAT: 254/10=25, REM=254-(25*10)=4 → MYARRAY[0] :UNITS
└→ ASSIGN QUO TO NUMBER VARIABLE
2ND REPEAT: 25/10=2, REM=25-(2*10)=5 → MYARRAY[1] :TENS
└→ ASSIGN QUO TO NUMBER VARIABLE
2<10 (LAST REMAINDER)
2→ MYARRAY[2] :HUNDREDS
'\N'→ MYARRAY[3]
'\0'→ MYARRAY[4]

Figure 2.18 Transforming a binary value to BCD in reversed order (254→"452").

statement addresses *array subscript* (i.e. the *i* variable) to perform the BCD conversion. The corresponding code using *array pointer* is enclosed within the green rectangle[8] of Figure 2.17a. It is here noted that the *quotient* and *remainder* are generated with the utilization of the compound operators *assign quotient (/=)* and *assign remainder (%=)*, respectively, which have been earlier presented by the chapter (see Table 2.4).

A conversion example of the decimal value 254 is illustrated in Figure 2.18. The *while()* statement in this particular case runs two successive time. During the first run the worth of the *number* variable is divided by number 10 and the remainder is assigned to the first array element (i.e. *myArray[0]*). During the second run the previous *quotient* (which is stored back to the *number* variable) is divided again by 10 and the remainder is assigned to the subsequent array element (i.e. *myArray[1]*). The new *quotient* is less than 10 and, hence, the *while()* loop is terminated, and the final remainder is assigned to the next array element (i.e. *myArray[2]*) outside the loop. Finally, the *new line ('\n')* and *null ('\0')* characters are appended to the subsequent array elements (i.e. *myArray[3]* and *myArray[4]*, respectively).

The rest of code first prints the BCD encoding and then addressed a *switch()...case* statement in order to reform BCD encoding of *myArray* to numeric string (Figure 2.17b). With the utilization of an array pointer, the *switch()...case* statement is enclosed within a *while()* loop, which runs from *myArray[0]* element until the element holding the *new line ('\n')* character. The *switch()...case* statement reforms each byte holding a binary value from 0 to 9, to the equivalent ASCII character '0'–'9'. Afterward, the code transmits to the μC's serial port the equivalent numeric string (in reversed order) of the number originally assigned to the *number* variable. The printed results of the user-defined numbers 254, 64, 8, and 128 are presented in Figure 2.17c.

Arduino Ex.2–7

The example of Figure 2.19 constitutes a revision of the previous example that attempts to highlight the value of bitwise operations. The application code, depicted by Figure 2.19a,b, evaluates a single-byte number (i.e. 0–255) and prints to the serial port the corresponding positive (0 to +127) and negative (−128 to −1) value of that number, in terms of the *2's complement* representation system.

8 To test the equivalent code uncomment the code lines within the green rectangle (*array pointer*) and comment the lines within the red rectangle (*array subscript*).

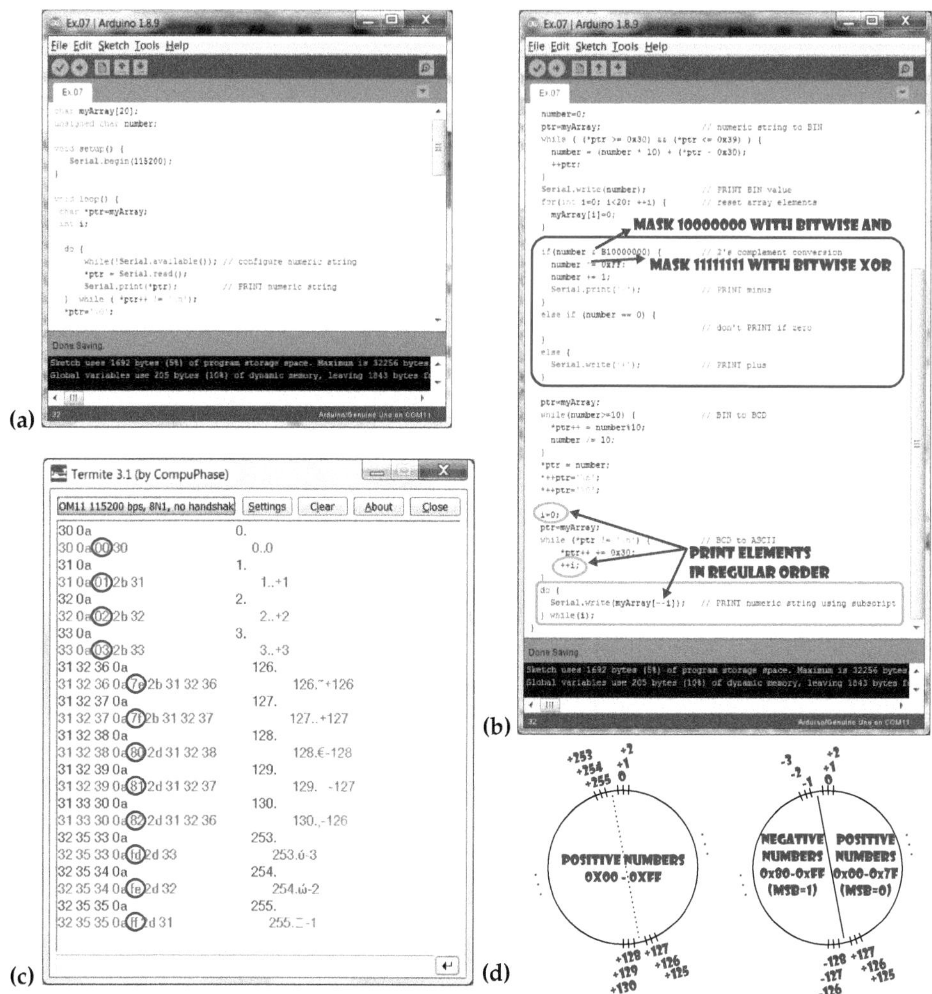

Figure 2.19 Arduino Ex.2-7: bitwise operations (unsigned byte to signed byte conversion). (a) source code (part 1 of 2) (b) source code (part 2 of 2) (c) printed results (d) geometrical representation of unsigned and 2's complement signed numbers (byte-size long). *Source:* Arduino Software.

As before, the code configures a numeric string from numbers received by the μC's serial port which are stored to *myArray[]*, while also looping back to the serial port each *character* until the *new line ('\n')* is received (Figure 2.19a). Next, the code reforms the numeric string to a single-byte (binary) number, it prints that value to the serial port, and then it clears all elements of *myArray[]* (Figure 2.19b).

The new code part within the red rectangle (Figure 2.19b) evaluates if the *number* variable (i.e. the user-defined number in *binary* arrangement) is either of positive or of negative value. The assessment is accomplished by executing *bitwise AND (&)* of *number* variable to the binary value *B10000000*. In programming, the sequence of bits that has an effect on a *bitwise* operation so as to extract particular information, is known as *mask* or

Table 2.7 Bitwise AND (&), OR (|), and XOR (^) of two bits (Truth table).

X	Y	AND (X&Y)	OR (X|Y)	XOR (X^Y)
0	0	0	0	0
0	1	0	1	1
1	0	0	1	1
1	1	1	1	0

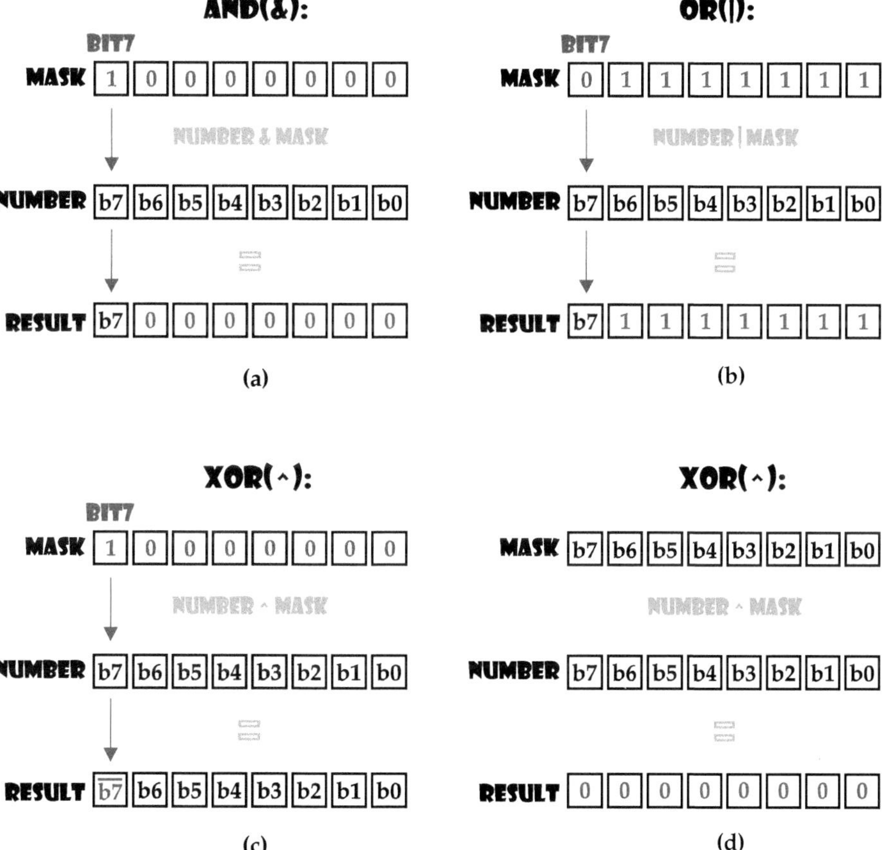

Figure 2.20 Bitwise AND, OR, XOR with bitmasks. (a) bitwise AND (b) bitwise OR (c) bitwise XOR (toward inverting bits) (c) bitwise XOR (toward evaluating the equality of variables).

bitmask. Herein, the *bitwise AND* operation is used to clear all but the MSB of the *number* variable, which represents the sign bit. Table 2.7 presents the *truth table* of the *bitwise* AND, OR, and XOR, which are commonly used with *bitmasks*. Figure 2.20 illustrates the process of (i) forcing to logical '0' all but bit7 (b7) of a variable with *bitwise AND* (Figure 2.20a), (ii) forcing to logical '1' all but b7 of a variable with *bitwise OR* (Figure 2.20b), (iii) inversing the state of b7 of a variable with *bitwise XOR* (Figure 2.20c), and (iv) evaluating the equality of two bytes with *bitwise XOR* (Figure 2.20d). It is here noted that the bitwise XOR on two bits of same value (i.e. 1 ^ 1 or 0 ^ 0) generates zero value.

If the *bitwise AND* operation within the ***if()**. . .else if(). . .else* expression is evaluated *true* (i.e. the *bitwise AND* generates *nonzero* value), the *2's complement* of the *number* variable is calculated and the *minus ('–')* character is transmitted to the serial port. If the latter assessment is found *false*, the *if(). . .**else if()**. . .else* expression evaluates *number* variable. Accordingly, if the *number* variable is of zero value, a none ASCII character is transmitted to the serial port. Finally, if the *number* variable if of positive value, the *if(). . .else if(). . .**else*** expression transmits the *plus ('+')* character.

The two subsequent *while()* statements of the example code convert the binary value of *number* variable to BCD and ASCII characters, respectively. Contrary to the previous example that utilized a *switch. . .case()* statement, the BCD to ASCII conversion is performed by adding the hexadecimal value 0×30 to each element of *myArray[]*. Moreover, the second *while()* statement exploits the *i* variable as *counter* in order to determine the number of the array elements that were reserved for the storage of the numeric string. Hence, the following *do. . .while()* statement prints the array elements (which have been initially arranged in reversed order) in regular (i.e. nonreversed) order.

Figure 2.19c depicts the results for when the user enters the consecutive numeric strings "0", "1", "2", "3", "126", "127", "128", "129", "130", "253", "254", and "255". The code makes the appropriate conversions and prints the corresponding signed values: "0", "+1", "+2", "+3", "+126", "+127", "−128", "−127", "−126", "−3", "−2", and "−1". All numbers that have been circled represent the binary value of the original numeric string. Figure 2.19d depicts the unsigned numbers (left circle) of a byte value and the corresponding signed numbers in 2's complement representation (right circle).

The storage of the string "128", which has been converted to BCD is illustrated Figure 2.21a, and the *while()* and *do. . .(while)* iterations of that case are given in Figure 2.21b. In detail, the elements of *myArray[]* hold the number 0×08, 0×02, and 0×01, which is the BCD representation of the "128" numeric string in reversed order (Figure 2.21a). Before entering the *while()* loop, the *i* counter is reset to zero and **ptr* pointer is configured to point to *myArray[0]* element. The first time the *while()* loop is executed, *myArray[0]* element (pointed by **ptr*) is added to the HEX value 0×30 so as to be converted to the corresponding ASCII number (that is, the 0×08 is converted to the ASCII '8'). The second time the *while()* loop is executed, *myArray[1]* element (pointed by **ptr*) is converted from HEX value 0×02 to the ASCII '2'. The third time the *while()* loop is executed, *myArray[2]* element (pointed by **ptr*) is converted from HEX value 0×01 to the ASCII '1'. After that, the *while()* statement verifies that **ptr* points to the array element that holds the *new line ('\n')* character and, hence, the loop is terminated.

On each repetition of the *while()* loop, the *i* variable has been unary incremented and has finally received the value three (*i* = *3*). Thereby, the *i* counter is configured to point to the last array element (i.e. *myArray[3]='\n'*). The subsequent *do. . .while()* loop utilizes the *i* counter in order to print the reversed elements of *myArray[]*, in regular order. Since the *unary decrement (−−)* operator precedes the *i* variable (i.e. −−*i*), the code first decreases *i* counter by one and then prints the array element pointed by the *i* subscript. Thereby, the

	7 (MSB)	6	5	4	3	2	1	0 (LSB)	
MYARRAY[0]	0	0	0	0	1	0	0	0	0x08
MYARRAY[1]	0	0	0	0	0	0	1	0	0x02
MYARRAY[2]	0	0	0	0	0	0	0	1	0x01
MYARRAY[3]	0	0	0	0	1	0	1	0	0x0A (='\N')

(a)

(b)

Figure 2.21 Converting the array elements from BCD to ASCII and printing them in regular order. (a) storage of the byte-size numbers 1, 2, 8 in memory, in reverse order, (b) execution of the *while* loop for converting numbers to ASCII characters and execution of the *do. . .while* loop for rearranging the order of characters in memory.

code prints first the element *myArray[i=2]*, then the element *myArray[i=1]*, and finally the element *myArray[i=0]*. After the last print, the *do. . .while()* statement verifies that $i = 0$ and terminates the loop execution.

It is worth making a reference to the *bit-shift left (<<)* and *bit-shift right (>>)* operators, which are particularly useful for multiplications/divisions of a binary number to a power of two (i.e. number*2^X, or number/2^X). It should be noted that this multiplication/division method applies to the *positional notation* of the base-2 numeral system. As in the base-10 numeral system where every digit features a value 10 times more than the digit found immediately next to it (on the right side), in base-2 numeral system every bit on the left equals 2 times more than the preceding bit.

In the previous example code we have inserted four code lines in the position depicted by the red rectangle of Figure 2.22a. The new lines shift the *number* variable three positions to the left (*number<< = 3*) and print the latter result to the serial port. Then, they shift the latter outcome three positions to the right (*number>> = 3*) and print again the new result. The printed results are presented in Figure 2.22b, where the highlighted information in red color represents the multiplication by 2^3, while the highlighted information in green color represents the division by 2^3.

Figure 2.23 illustrates the process of shifting a byte-sized variable three positions to the left (Figure 2.23a) and three positions to the right (Figure 2.23b). The outcome when shifting the number 0 × 03 three positions to the left (i.e. multiply by 2^3) is given in Figure 2.23c, while the outcome when shifting the number 0 × 18 three positions to the right (i.e. divide by 2^3) is given in Figure 2.23d.

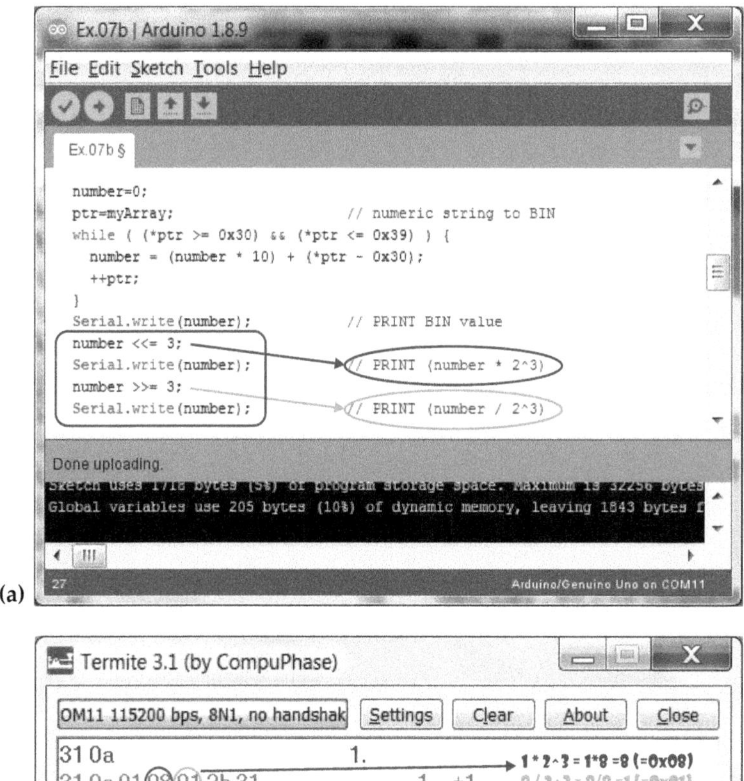

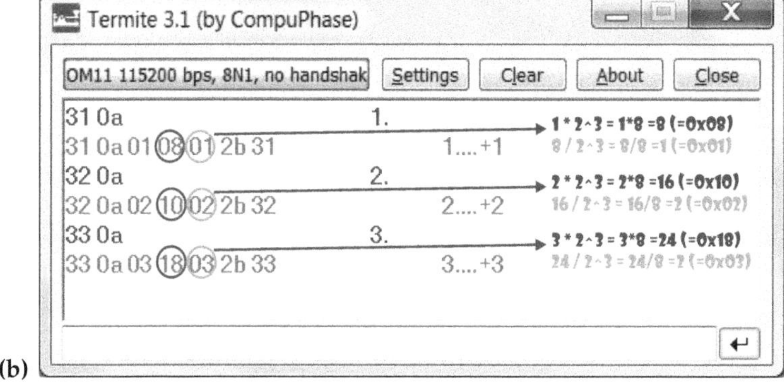

Figure 2.22 Bit-shift *left (<<)* and *right (>>)* of a binary value (multiply and divide by 2^x). (a) source code, (b) printed results. *Source:* Arduino Software.

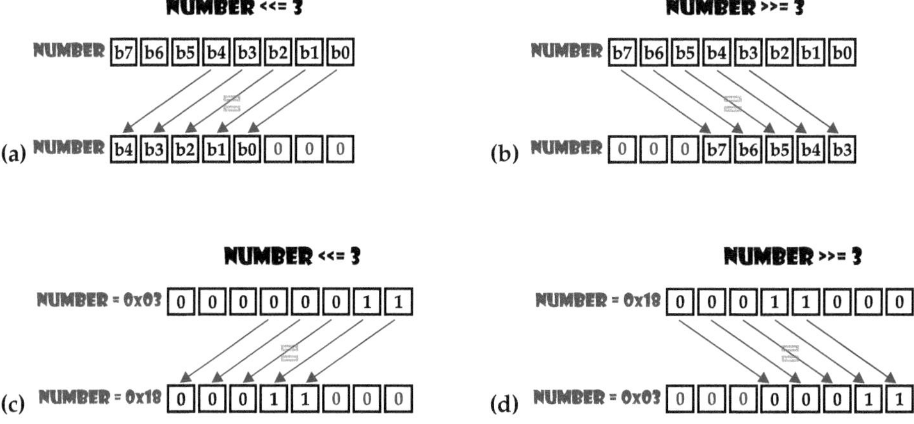

Figure 2.23 Three positions *bit-shift left* and *right* (i.e. multiply and divided by 2^3). (a) general description of *bit-shift left* execution, (b) general description of *bit-shift right* execution, (c) example of *bit-shift left* execution, (d) example of *bit-shift right* execution.

Code Decomposition (Functions and Directives)

Functions in programming are commonly used for the code decomposition into smaller segments, which can be reused in other example codes, as well. The user-defined functions make available a more readable and easily upgradable code. The designer may proceed to either the *top-down* or *bottom-up* programming approach, when developing an application code with user-defined functions. In the former case, the designer first develops the main code and then the code of each sub-process (i.e. each function code). In the latter case, the designer first implements the code of each function and then ties them up to the main code.

The function declaration in Arduino (and C) programming is given in Figure 2.24a. Optionally, the function may return a value, the *datatype* of which is declared before the (user-defined) function name. Moreover, the function may optionally admit one or more arguments, which are placed within parentheses after the function name. For functions that do not return a value or admit none arguments, the *void* keyword is used for declaring that particular feature (Figure 2.24b). The function body is placed after the arguments and within *curly braces ({. . .})*. The *pseudo-code* example of Figure 2.24c initializes a function (of name *myFunction*), which returns a value of *int* datatype and admits two arguments, also of *int* datatype. The function is configured to return the product of the two arguments (i.e. *arg1* and *arg2*).

Arduino Ex.2-8

The example code presented hereafter applies to a previous example code (that is, the Ex.2–7 given in Figure 2.19) and implements the same tasks with functions. The code of the mains tasks has been enclosed within functions in a separate file, named *__CH2_functions.h*. While declaration of the user-defined functions can be enclosed within the source code, it is regularly applied in a separate file, aka *header* file (i.e. a file of an *.h* extension). A different freeware tool has been addressed for the writing of that file; that is, the Notepad++, which constitutes a *source code editor* (Figure 2.25a).

The *__CH2_functions.h* file incorporates five distinct functions:

i) ***void Serial_ReadNumericString (char *ptr):*** This function returns no value and admits a *pointer* type argument. *Serial_ReadNumericString()* incorporates the *do. . .while()* statement that configures the user-defined numeric string until the *new line ('\n')* arrives to the μC's serial port. It is here noted that a function that accepts a *pointer* type argument can also accept an array.
ii) ***unsigned char NumericString_toBin (char *ptr):*** This function returns an *unsigned char* type value, while it also admits a *pointer* type argument. *NumericString_toBin* function incorporates the *while()* statement, which converts the user-defined numeric string into binary (byte-size) value.
iii) ***unsigned char Negative_toPositive (unsigned char binaryValue):*** This function returns, and also admits an *unsigned char* type value. The *Negative_toPositive()* evaluates if the previously generated binary variable is of negative value and if the evaluation is confirmed, it reforms that value into the corresponding positive number (in consideration of the *2's complement* representation system).

```
return_type function_name (optional arguments) {
 function body;
}
```

(a)

```
void myFunction (void) {
 statement;
}
```

(b)

```
int myFunction (int arg1, int arg2,) {
  optional statements;
 return arg1*arg2;
}
```

(c)

Figure 2.24 Function declaration in Arduino (and Embedded C) programming. (a) general syntax of a *function* declaration in C language statement, (b) example of a *function* declaration of neither arguments nor returning value, (c) example of a *function* declaration admitting arguments as well as returning value.

iv) ***void Binary_toBCD (char *ptr, unsigned char binaryValue):*** This function returns no value and admits two argument. The former is a *pointer* type argument, while the latter an *unsigned char* type value. The *Binary_toBCD()* converts the second argument, which represents a binary value, into BCD representation. The equivalent BCD outcome is assigned to an array pointed by the first argument.
v) ***int BCD_toASCII (char *ptr):*** This function returns an *int* type value and admits a *pointer* type argument. The *BCD_toASCII ()* converts the previously generated array of BCD representation into ASCII characters.

Figure 2.19b depicts the main code, which invokes the aforementioned functions. The separate file can be included in the main code with the preprocessing directive *#include*. The Arduino (and C language) directives regularly begin with the *hush (#)* symbol and provide information, to the preprocessor or the compiler, on how to handle the source code. For instance, the block comment (/*. . .*/) symbols direct the compiler to ignore all the information embedded within the two symbols.

There are two different ways to include a *header* file in Arduino IDE. The first way is to include the full disk drive path (where the header file is located) insides *double quotes* (*". . ."*), as presented in the foremost (commented) line of Figure 2.25b. The second way is to write, immediately next to the *#define* directive, the file name without quotes. The latter way requires the storage of the *header* file inside the *libraries* folder, which is found in the root folder of the installed the Arduino IDE. The *header* file is regularly inserted inside a folder. For instance, the *__CH2_functions.h* is herein assigned to the path "C:\arduino-1.8.9\libraries__CH2_functions".

The main code declares the *char* type global array of 20 elements (i.e. *myArray[20]*) to hold the user-defined numeric string, as well as the *unsigned char* type global variable (i.e. *number*) to hold the equivalent binary value of the numeric string. The main code within the *loop()* function, performs the following: (i) Invokes the *Serial_ReadNumericString()* function, which admits *myArray* as argument and stores at that part of memory, the user-defined numeric string; (ii) Invokes the *NumericString_toBin* function, which admits *myArray* as argument and converts the user-defined numeric string (found at that array) to binary value, and returns the conversion outcome to the *number* variable; (iii) Assesses the

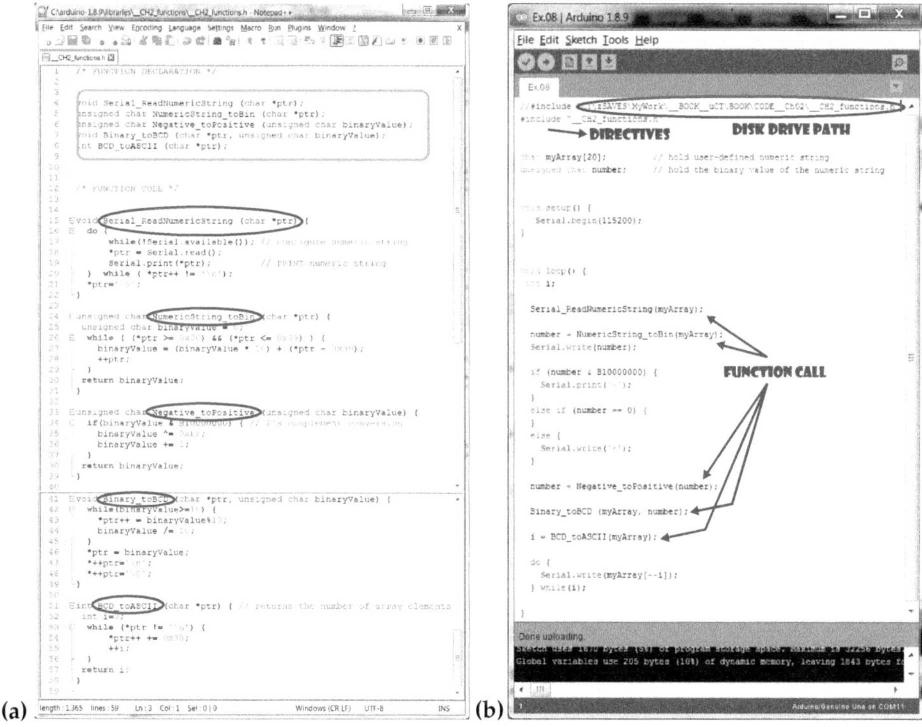

Figure 2.25 Arduino Ex.2-8: code decomposition with functions (as applied to Arduino Ex.2-7). (a) header file, (b) source code invoking the header file. *Source:* Arduino Software.

content of *number*, and in case of a negative value, the code prints to the serial port the *minus ('–')* symbol (while in case of positive value the code prints the *plus ('+')* character, and for zero value the code prints nothing); (iv) Invokes the *Negative_toPositive()* function, which admits the *number* variable as argument and if the latter value is negative, it calculates the corresponding positive value (in terms of the 2's complement representation) and returns the outcome to the *number* variable; (v) Invokes the *Binary_toBCD()* function, which admits two arguments, *myArray* and *number*, and converts the binary value of the latter argument to BCD representation, where each BCD byte is assigned to *myArray*. The *Binary_toBCD()* function also returns the number of the array elements to the *i* variable; and (vi) Finally, the *do. . .while()* statement first decreases the content of the *i* variable by one, and then prints each element of *myArray* pointed by *i* subscript.

The printed results of this example code are the exact same as in the Ex.2-7 (which were previously presented in Figure 2.19c).

Another way to decompose and/or implement a more readable code is to define *macros*. The *macro* constitutes a fragment of code that is initialized with the preprocessor directive *#define* before the main code, and when invoked, the macro code is expanded at the specified position. This feature makes the *macro* execution more rapid than the function call, but attention should be paid to the amount of program memory reserved by the macro code every time it is invoked by the main program.

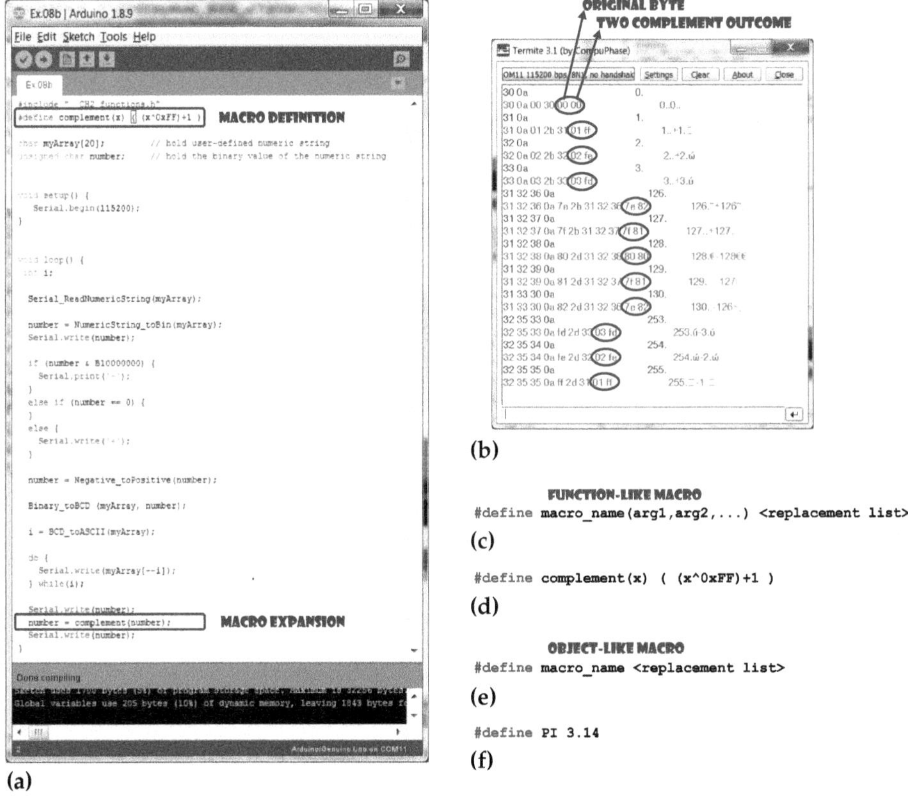

Figure 2.26 Arduino Ex.2–8b: code decomposition with functions and macros. (a) source code invoking macros, (b) printed results, (c) general syntax of a *function-like macro*, (d) example of a *function-like macro,* (e) general syntax of an *object-like macro,* (f) example of an *object-like macro.* *Source:* Arduino Software.

Figure 2.26a presents a revision of Ex.2–8 that defines a *macro* for the 2's complement conversion (of the employed argument). The last three code lines print to the serial port the *number* variable, and then expand the *macro* code that converts the *number* to the equivalent positive or negative value (in 2's complement representation), and print again to the serial port the latter outcome. The printed results are given in Figure 2.26b. This approach is regularly referred to as *function-like macro* and may admit one or more arguments (Figure 2.26c,d). *Macros* that do not admit arguments are regularly referred to as *object-like macros*, like the PI definition example of Figure 2.26e,f.

Conclusion

The chapter has applied to Arduino software and hardware tools as they provide a quick jump-start and flexibility in the implementation of microcontroller-based projects and, hence, they have become a viral technology nowadays. The chapter has provided a brief overview to the fundamental aspects of sequential programming, in consideration of a relative compatibility between the Arduino and embedded C programming methods, while example codes designed to run on the Arduino Uno board, have also been explored. Hereafter are some unsolved problems for the reader.

Problem 2-1 (Data Output from the μC Device: Datatypes and Bytes Reserved by the hw)

Explore via **Ex.2–2** the number of bytes reserved by each *datatype* (i.e. short, int, long, and float) in other than the 8-bit Arduino Uno board.

Problem 2-2 (Data Output from the μC Device: Logical Operators in Control Flow)

Revise **Ex.2–4** in order to repeatedly transmit to the terminal console the ASCII characters representing numbers '0', '1', '2',. . ., '9', and then transmit the even ('0', '2', '4', '8') followed by the odd ('1', '3', '5', '7', '9') numbers.

Problem 2-3 (Data Input to the μC Device: Arithmetic and Bitwise Operations)

Revise **Ex.2–5** so that the μC device waits the user to insert a password from the host PC. The string that stores the password should encompass at least one ASCII character representing a number (i.e. '0', '1', '2',. . ., '9'). If the condition is met, the μC sends back to the host PC the message *"Valid entry"* and the code is terminated, otherwise, it transmits the message *"Invalid entry, please reenter password"* and code is repeated (until a valid password is verified).

Tips: To terminate the code, insert an infinite execution of a loop incorporating no instructions within its body.

Problem 2-4 (Code Decomposition)

Write the code part of an alarm system that waits the user to insert the password in order to deactivate the alarm. Provided that a 4-digit password is stored to the μC's memory, evaluate the password given from the host PC and print a message of your choice (in case the password is correct or not). Make use of code decomposition techniques during the firmware development process.

3

Hardware Interface with the Outside World

The main difference between programming a regular computer and a microcomputer arises from the fact that the latter system requires significant effort on practices related to the hardware interface with the outside world. In detail, every microcontroller-based system features unique specifications that are determined by the employed *input/output (IO)* units, which may radically differ from system to system. The design process commonly begins from the selection of the microcontroller motherboard along with the proper IO units (e.g. switches, keypads, sensor) that attempt to satisfy the system specifications. Thereafter, the microcontroller programming can be separated into two main coding practices: (i) the handing of the system's incorporated IO units, and (ii) the data processing. While the data processing could be characterized quite alike to the corresponding programming procedure of a conventional computer, the handing of the IO units demonstrates the hardware interface practices of microcontroller programming.

This chapter applies to the fundamental hardware interface practices in microcontroller applications, starting from the familiarization and utilization of the *ready-to-use* Arduino libraries, and gradually moving to more advanced matters, such as interrupting the regular program execution, building custom libraries for Arduino, etc. The chapter intends to affect the perception of the reader on the hardware interface practices, and establish a robust background on such issues that are of particular importance when building hardware projects around microcontrollers.

Digital Pin Interface

In order to realize the fundamental aspects of sequential programming with Arduino, the previous chapter performed an introduction to the μC interfacing techniques with the outside world, using the universal serial bus (USB) to universal asynchronous receiver/transmitter (UART) serial port found in an Arduino Uno board. Hereafter, we explore the regular parallel interfaces that can be exploited by an ordinary μC device. The code examples apply to the popular and classic Arduino Uno board.

Microcontroller Prototypes with Arduino and a 3D Printer: Learn, Program, Manufacture, First Edition.
Dimosthenis E. Bolanakis.

Arduino Ex.3.1

To simplest example to begin with when uploading code to a microcontroller motherboard, is the blinking light-emitter diode (LED). The Arduino and Arduino-compatible boards are regularly delivered with a LED that allows the user to output the simplest form of data to the outside world, that is, to implement a pin (output) interface to the outside world.

Figure 3.1a depicts the position where the LED is found on the Arduino Uno board. The LED is driven by the PIN13. Figure 3.1b represents the flowchart, which makes the LED blink every second. At first, PIN13 is configured output, and then the logical value on that pin is alternated between the states *HIGH* and *LOW*. The latter process is repeated forever within an endless loop. The corresponding code is depicted by Figure 3.1c. The Arduino *pinMode()* function is used to configure PIN13 as output, while the Arduino *digitalWrite()* function is used to write a logical ‘1’ or ‘0’ to that pin, and hence change the state of the LED pin to *HIGH* or *LOW* (i.e. to turn the LED *ON* or *OFF*). The Arduino *delay()* function waits one second before changing again the state of the LED.

Figure 3.1d,e assign the names *LED* and *LED_delay* to the numbers 13 and 1000, respectively, which represent the PIN number that controls the LED, and the delay (in ms), which

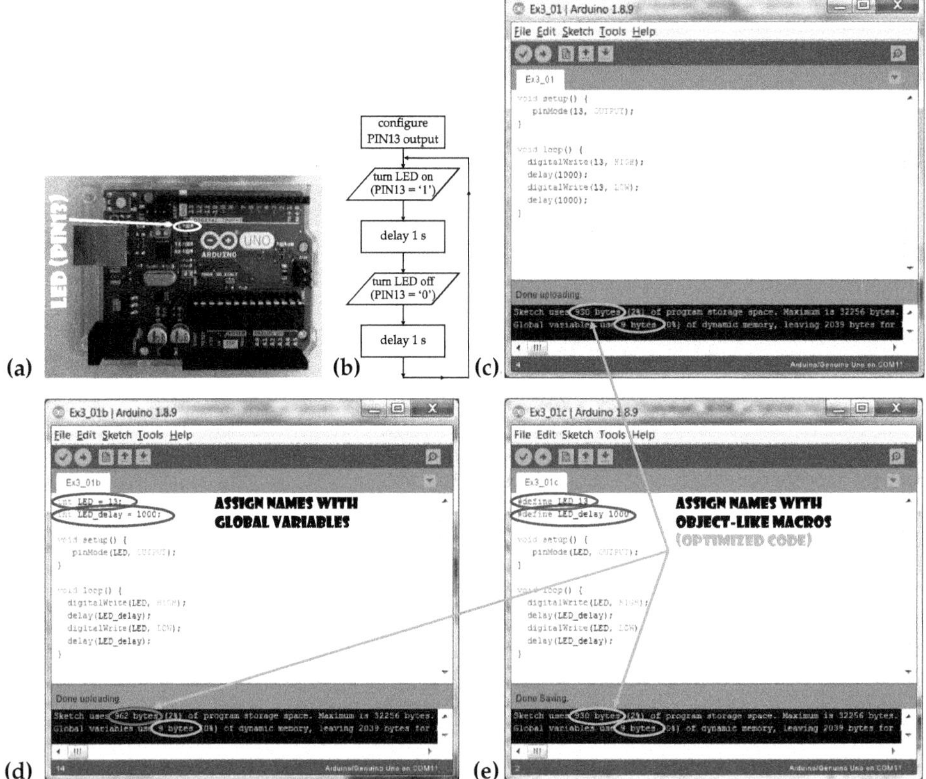

Figure 3.1 Pin interface with the outside world: OUTPUT DATA. (a) LED on Arduino Uno board, (b) flowchart of the example code, (c) example firmware code, (d) updated firmware using global variables, (e) updated firmware using object-like macros. *Source:* Arduino Software.

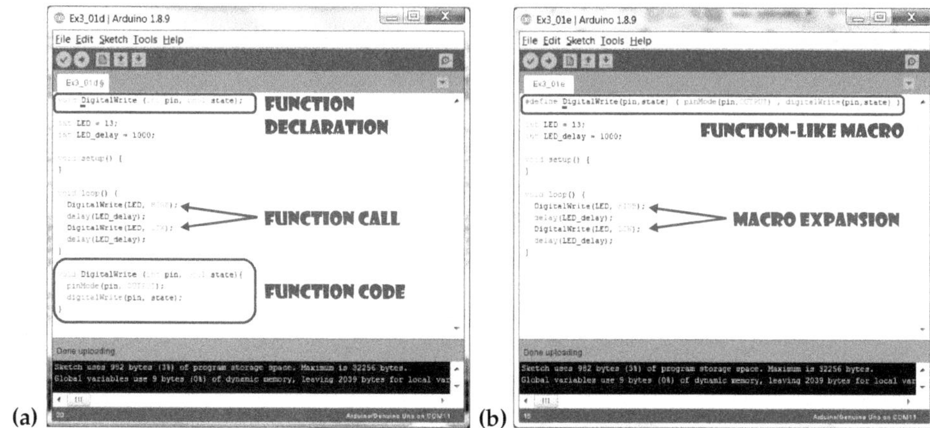

Figure 3.2 Blinking LED with user-defined *function* and *macro*. (a) firmware code implemented with *functions*, (b) firmware code implemented with *function-like macros*. *Source:* Arduino Software.

is used between the alternations of the LED state. The assignment is accomplished by two different ways, that is, with the use of (i) global variables (Figure 3.1d) and (ii) object-like macros (Figure 3.1e). It is worth noting that the second type of assignment is the optimum way, as it reserves less space in program memory than the first one (i.e. 930 bytes instead of 962 bytes of Flash memory).

Figure 3.2a,b present a revision of the blinking LED example with custom-defined *function* and *function-like macro*, respectively. In both cases, the user-defined *function* and *macro* encompass the configuration of PIN13 as output. Thereby, the Arduino *setup()* function is left empty. Because the microcontroller pins are regularly bidirectional (i.e. they can be configured either as output or input pins), this revision provides the flexibility of changing the direction of the pin inside the function/macro. However, the function call and/or macro expansion is overloaded with the extra work (and additional delay) that is needed to configure the pin direction before changing/reading the logical state of the pin.

Arduino Ex.3.2

Every μC device consists of digital[1] IO pins that are grouped in ports. The maximum number of pins per port is determined by the μC architecture. The Arduino Uno board is of 8-bit architecture (i.e. the memory registers are of 8-bit arrangement), and hence, each port consists of up to 8 pins maximum. The Arduino Uno board consists of three ports, that is, port B, C, and D, where the first two consist of six pins each (i.e. PB0–PB5 and PC0–PC5), and the third consists of eight pins (PD0–PD7). The IO (digital) pinout arrangement of Arduino Uno board is illustrated by Figure 3.3.

1 The digital IO pins of a μC device are those pins, which when configured outputs, are able to provide a logical signal (i.e. '1' or '0') upon the user request through the application code. In Arduino Uno board, the logical '1' corresponds to 5 volts (V), while the '0' corresponds to 0 V. When the digital IO pins are configured inputs, they are able to read a logical signal, which is applied by the external environment (e.g. through a switch), and automatically assign that logical value to the μC's data memory.

Figure 3.3 Pin arrangement in an Arduino Uno board (port B, port C, and port D).

A faster manipulation of the microcontroller's digital IO pins is through the port registers. The μC device of the Arduino Uno motherboard is the Atmega328p chip and each port of the device is controlled by three different registers, that is DDRx, PORTx, and PINx (where x denotes the indicative name of the port, e.g. DDRB, PORTB, and PINB). Because all the digital IO pins of Atmega328p μC device are bidirectional, the DDRx register is used to control the direction of each port pin. The assignment of a logical '1' to a bit of DDRx register, forces the corresponding pin to be configured output, while the logical '0' forces the pin to be configured input. If a pin is configured output, the setting of a bit of PORTx register forces the corresponding pin to *HIGH* state (i.e. 5 V appear on that particular pin of Arduino Uno board). Otherwise, the resetting of a bit of PORTx register forces the corresponding pin to *LOW* state (i.e. 0 V appear on that particular pin of Arduino Uno board).

According to the aforementioned information, DDRx and PORTx registers are writable registers. On the other hand, the PINx register is used to read a logical '1' or '0' from the bit that corresponds to an input port pin. Hence, PINx constitutes a read-only register. All the aforementioned registers are defined with those specific names in the Arduino integrated development environment (IDE), and hence, they can be used by the application code in order to control the ports of Atmega328p device.

Figure 3.4a presents the application code of the blinking LED through the direct manipulation of the port B registers (where the LED is driven by PB5 pin, that is, PIN13 on Arduino Uno board). Inside the Arduino *setup()* function, the DDRB register is assigned to the binary value B00100000, which configures PB5 pin as output (while the

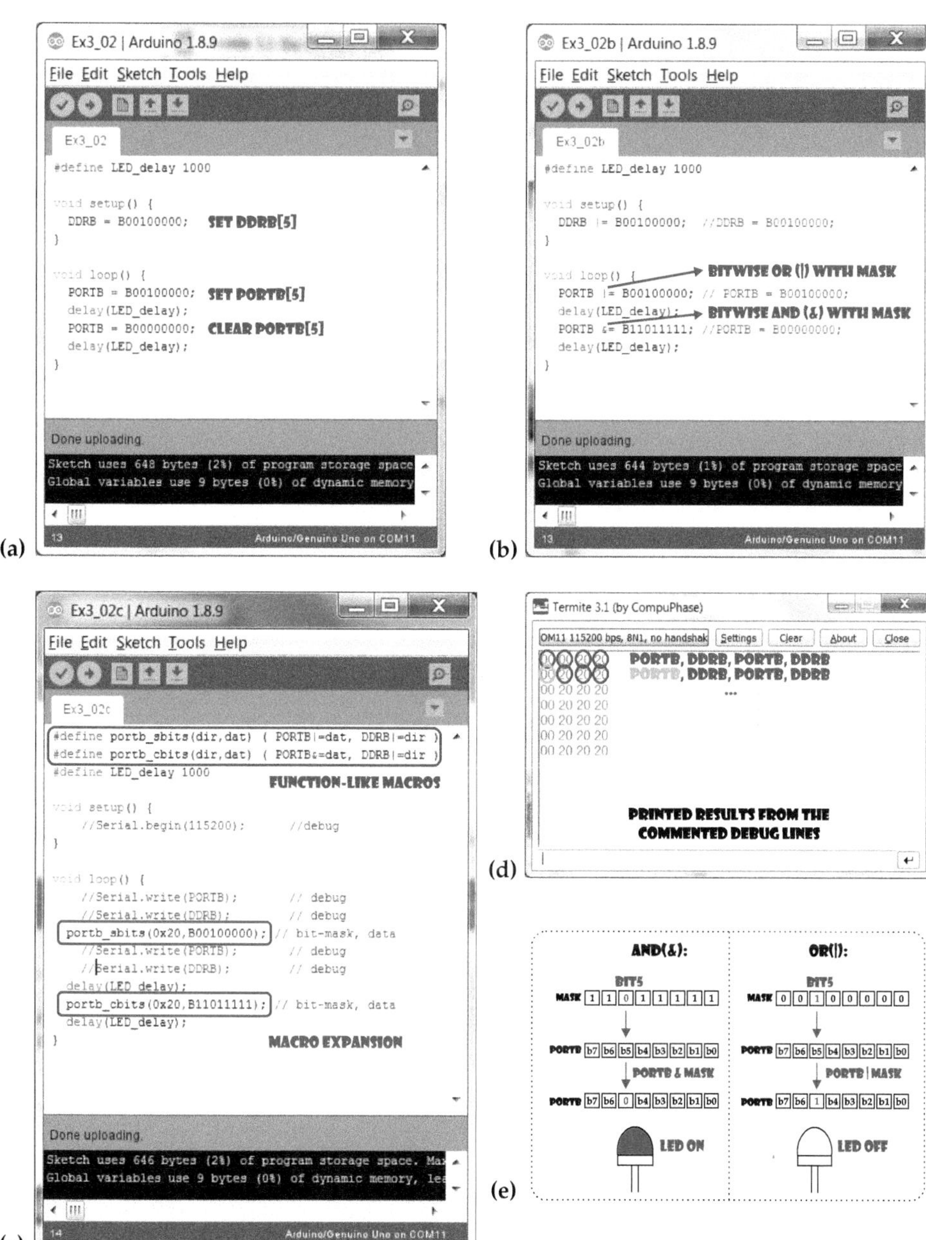

Figure 3.4 Blinking LED through the direct manipulation of the port registers. (a) firmware example code with direct port manipulation, (b) updated code with bitwise operations, (c) updated code with function-like macros, (d) printed results on Arduino *serial monitor* tool, (e) the execution of bitwise AND/OR operations for turning the LED ON/OFF. *Source:* Arduino Software.

rest port B pins are configured inputs). The first code line inside the Arduino *loop()* function sets bit 5 of PORTB register to logical '1' and forces the LED to turn ON (while the rest bits of PORTB are cleared). The subsequent *delay()* function retains the latter state of the LED for one second. Afterward, the subsequent code line clears bit 5 of PORTB register (as well as the rest bits of PORTB), and hence the LED is turned OFF. The subsequent *delay()* function retains the OFF state of the LED for 1s, and then, the *loop* is repeated.

A good practice when controlling some (but not all) pins of a port, is to make use of a bitwise operation along with a proper bitmask. Figure 3.4b illustrates how to achieve the same result using the bitwise OR and AND to the DDRB and PORTB registers. In detail, the *setup()* function configures PB5 pin as output while leaving the direction of the rest pins unchanged. This operation is performed with a bitwise OR to the content of DDRB register along with the binary mask B00100000. Moreover, the main *loop()* blinks the LED on PB5 pin without changing the value of the rest bits of PORTB register. Again, the bitwise OR operation to the bitmask B00100000 sets PB5 and turns the LED ON, while the bitwise AND to the mask B11011111, clears PB5 and turns the LED OFF.

In the code example of Figure 3.4c we have created two function-like *macros* that perform the same result with bitmasks while they also configure the direction of PB5 pin on each macro expansion. If we uncomment the code lines that exploit the Arduino serial port, we may observe the printed results of DDRB and PORTB registers.

Figure 3.4d presents the content of PORTB and DDRB registers. At first the content of both registers is of zero value, which is the initial value of those registers. Then, both registers receive the value 0x020 (i.e. B00100000), which configures the direction of pin5 on port B and sets its value to logical '1'. Then, DDRB5 is reconfigured output (i.e. DDRB = 0x20) and PB5 bit is cleared (i.e. PORTB = 0x00) so as to turn the LED OFF. Afterward, the overall process is repeated. The bitwise AND, OR operations on PORTB register are depicted by Figure 3.4e. It should be noted that if we assign a value to PORTB register, that value will be stored to the register but only the pins of port B that are set as outputs will be affected (and will admit values in accordance with the content of the PORTB register).

Figure 3.5a depicts the connection of an oscilloscope on the PB5 pin of an Arduino Uno board, in order to measure the voltage that makes the LED blink. The reason for this setup is to explore the extra delay that is added to the pulse of PB5 pin, regarding the different coding approach. To make a clear sense of that issue, we have removed the *delay(LED_DELAY)* code lines from the code examples.

Figure 3.5b illustrates the PB5 pulse of the EX2_02 code example. At first, we observe that the oscilloscope is set to 2V/division, and hence, the PB5 pulse is alternated between the levels 5V and 0V, while the *time period*[2] *(T)* of the wave is equal to 375 ns. The fact that the *Ton* duration is less than the *Toff* relies on the machine instructions that make the LED blink. In detail, one machine instruction sets PB5 to logical '1', then another instruction clears PB5 to '0' and, afterward, the subsequent machine instruction sends the control flow

2 It is noted that the *time period* or *period* of a wave is mathematically expressed as $1/F$, where F the *frequency* of the wave expressed in *Hertz (Hz)*.

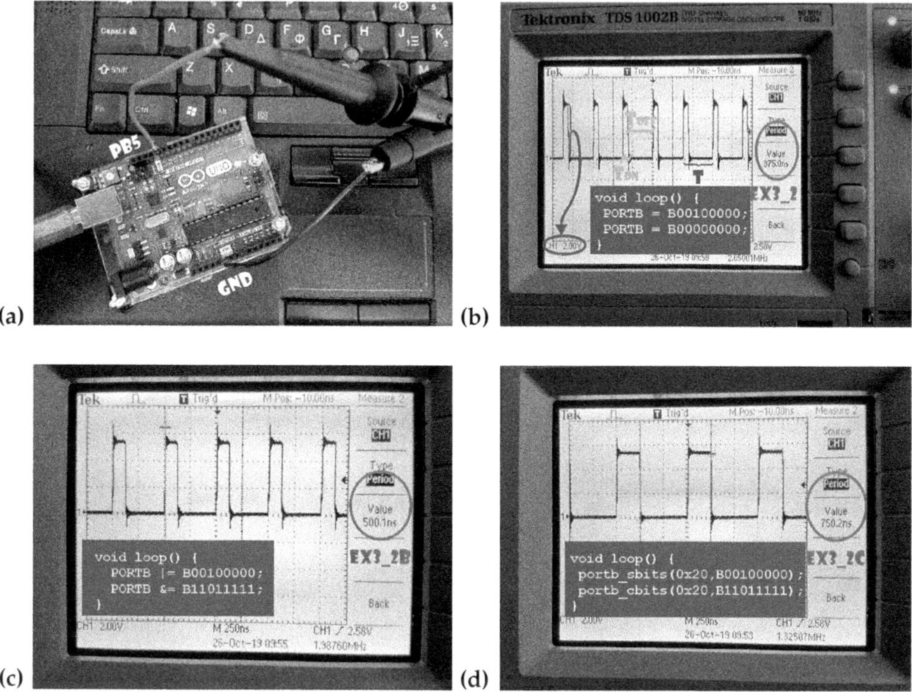

Figure 3.5 Waveforms on PB5 pin (blinking LED without the one second delay). (a) connection of the oscilloscope to pin PB5 of the Arduino Uno to acquire the blinking period, (b) pulse of the PB5 when applying direct manipulation of the port pins (375ns), (c) pulse of the PB5 when applying bitwise operations to the port pins (500ns), (d) pulse of the PB5 when using user-defined macros to control the port pins (750ns).

to the beginning of the loop in order to repeat the process. The latter machine instruction adds an extra delay to the duration of the low-level pulse of PB5 pin.

Figure 3.5c illustrates the PB5 pulse of the EX2_02b code example, which makes use of bitwise operations with bitmasks in order to make the LED blink. The bitwise operations require more instructions at the machine level, and hence, they increase the period of the blinking LED to approximately 500 ns. The *period T* is increased even more in Figure 3.5d, which illustrates the implementation of EX2_02c code example. The latter example configures the direction of PB5 pin before setting the logical level on the output pin, and hence, the additional machine instructions increase the period of the blinking LED to approximately 750 ns.

Arduino Ex.3.3

An important issue in μC programming, when working with port pins, is the latency that is generated by the code instructions. Figure 3.6a presents an example code that allows us to control PB5 (PIN13) and PB4 (PIN12) either *sequentially* or *concurrently*. To control the two pins sequentially we should uncomment the code lines inside the blue frame and comment the code lines inside the red frame, while the concurrent control of the two pins occurs by

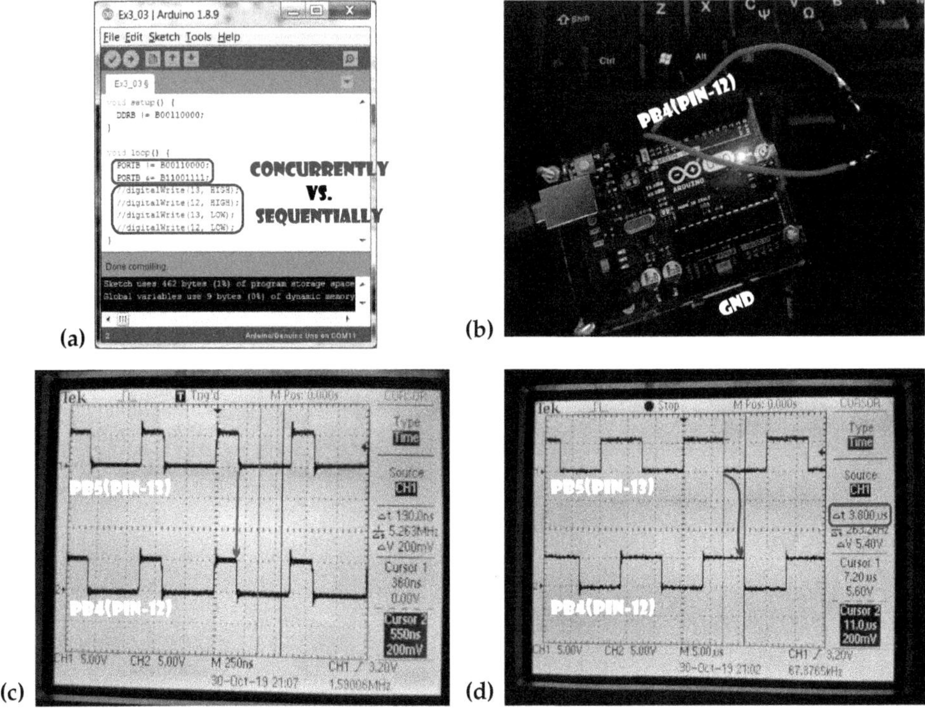

Figure 3.6 *Concurrent* vs. *sequential* manipulation of the μC's port pins. (a) example firmware, (b) connection of the oscilloscope to pins PB4 and PB5 of the Arduino Uno, (c) delay when manipulating the port pins concurrently (ns), (d) delay when manipulating the port pins sequentially (μs). *Source:* Arduino Software.

the code as is. Figure 3.6b illustrates the setup for acquiring the waveforms of the two pins of Arduino Uno board through an oscilloscope.

The results from the concurrent and sequential control of the two pins are respectively presented in Figures 3.6c and 3.6d. The direct manipulation of the microcontroller's IO pins through the PORTB register results in the direct response of the logical value on PB5 and PB4 pins (Figure 3.6c). On the other hand, the manipulation of each pin individually through the *digitalWrite()* functions, causes a latency of 3.8 μs in between the response of the two pins (Figure 3.6d).

Arduino Ex.3.4

Subsequent to the examination of the data output process through the μC's digital IO pins, we herein explore the data input procedure through the IO pins. The simplest way to insert data to μC device is through a switch. In the example presented hereafter we make use of the push-button.[3] The latter type of switch holds the ON state for as long as the user keeps pressing the switch; otherwise the switch returns to the OFF state.

3 Despite the *push-button* switch, there is another type of switch called *latch switch*, which maintains its state (i.e. ON or OFF) after being set to ON or OFF.

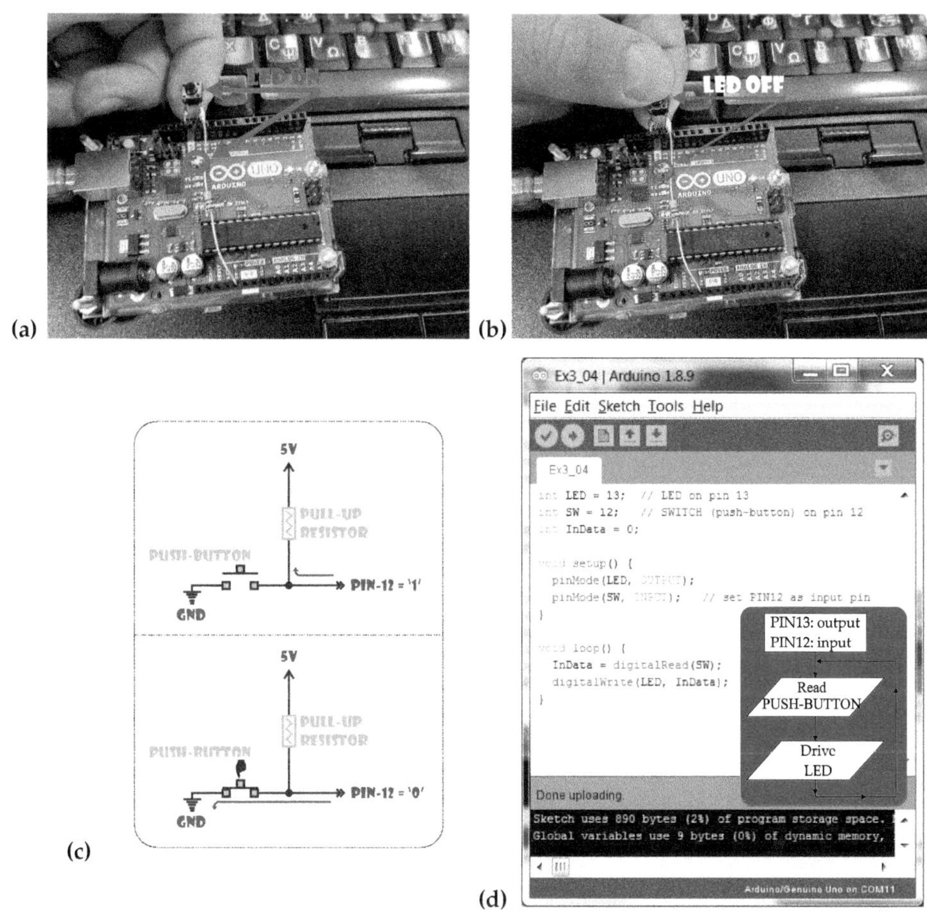

Figure 3.7 Pin interface with the outside world: INPUT DATA. (a) release push-button to turn the LED ON, (b) press the push-button to turn the LED OFF, (c) connection diagram of the push-button to PIN12, (d) example firmware and flowchart. *Source:* Arduino Software.

Figure 3.7a,b presents the setup and functionality of the example code. In detail, when the push-button is not pressed at all, the LED is turned ON (Figure 3.7a). If the user presses the button, the LED is turned OFF and remains at that particular state for as long the user keeps pressing the button (Figure 3.7b).

The schematic diagram that represents the push-button connection is given in Figure 3.7c. The one edge of the button is connected to the *ground (GND)*, while the other one is connected to the μC's input pin (i.e. PIN12) as well as to resistor of 1 *kilohm (KΩ)* value. The other edge of the so-called *pull-up* resistor is connected to the 5 V source of the Arduino board and, hence, it ensures a known state to the input signal when the button is open (that is, a logical '1'). When the button is pushed by the user, it creates a closed circuit with the ground and, hence, it forces PIN12 to the state of logical '0'.

The source code along with the flowchart (inside of the red frame) is given in Figure 3.7d. The global variables at the very beginning define the indicative names LED and switch (SW), for the pins 13 and 12, respectively. The InData variable is used to hold the logical value (i.e. either logical '1' or '0') determined by the state (either *opened* or *closed*) of the switch. The *setup()* function configures PIN13 and PIN12 as output and input pins, respectively. The *loop()* function first assigns the current logical value of the switch to the InData variable, and then, the content of the latter variable is assigned to PIN13, which drives the LED. The *loop()* function is repeated endlessly. It is here noted that the *digitalRead()* Arduino function reads the logical value from the corresponding memory location linked to the input pin, and returns either a HIGH or LOW value.

Arduino Ex.3.5

In Figure 3.8 we explore a critical phenomenon that appears when reading the state of a switch and we discuss a possible method to resolve that unwanted situation. The contacts of a switch may generate spurious open/close transitions upon pressing/releasing the switch. Those transitions often become perceivable by the rapid instruction execution of the μC's CPU. The result is that a single switch press (or release) by the user, may be identified by the μC as a process of multiple presses and releases of the switch and, hence, generate an incorrectly response of the source code. A software method to resolve the so-called *switch bounce* phenomenon, is through the utilization of a 50 ms delay immediately after the very first transition of the switch state (i.e. either an *open-to-close* or *close-to-open* transition). The delay ensures that the switch state becomes stable after the passage of that particular time.

Figure 3.8a presents the setup of the user pressing the push-button switch, while Figure 3.8b presents the response of the input signal upon the switch press (at t1 time). The upper frame of Figure 3.8b depicts the expected outcome, while the lower frame illustrates the actual response on the input signal with the inspected switch bounce phenomenon.

The flowchart and code example of Figure 3.8c and Figure 3.8d, respectively, clarify how to use the debounce delay immediately after the very first transition of the switch state. The main loop of the code example is executed as follows: (i) the first *while()* statement waits until the switch is pressed by the user, (ii) the first *delay()* function performs a switch debounce as soon as the user presses the switch, (iii) the second *while()* statement along with the utilization of the *logical NOT (!)* operator waits until the switch is released by the user, (iv) the second *delay()* function performs a switch debounce as soon as the user releases the switch, and (v) the *bitwise XOR (^)* operation on the content of PORTB along with the appropriate bitmask causes the state of the LED to be alternated (i.e. if state is HIGH it is changed to LOW, and vice versa), and the overall process is repeated from the beginning of the loop.

The code example of Figure 3.9a performs the exact same operation but the reading of the switch state is now accomplished through the direct manipulation of PINB register of port B. As illustrated by the upper frame of Figure 3.9b, the *logical AND (&)* operation to the content of PINB register with the bitmask B00010000 causes a reset to all bits of PINB except bit 4, which holds the logical value that appears on pin PB4 (i.e. PIN12). The lower frame of Figure 3.9b illustrates the *logical XOR (^)* operation to the content of PORTB registers, which complements bit 4, and hence, it causes the LED driven by PB4 to toggle (i.e. to switch from one state to another).

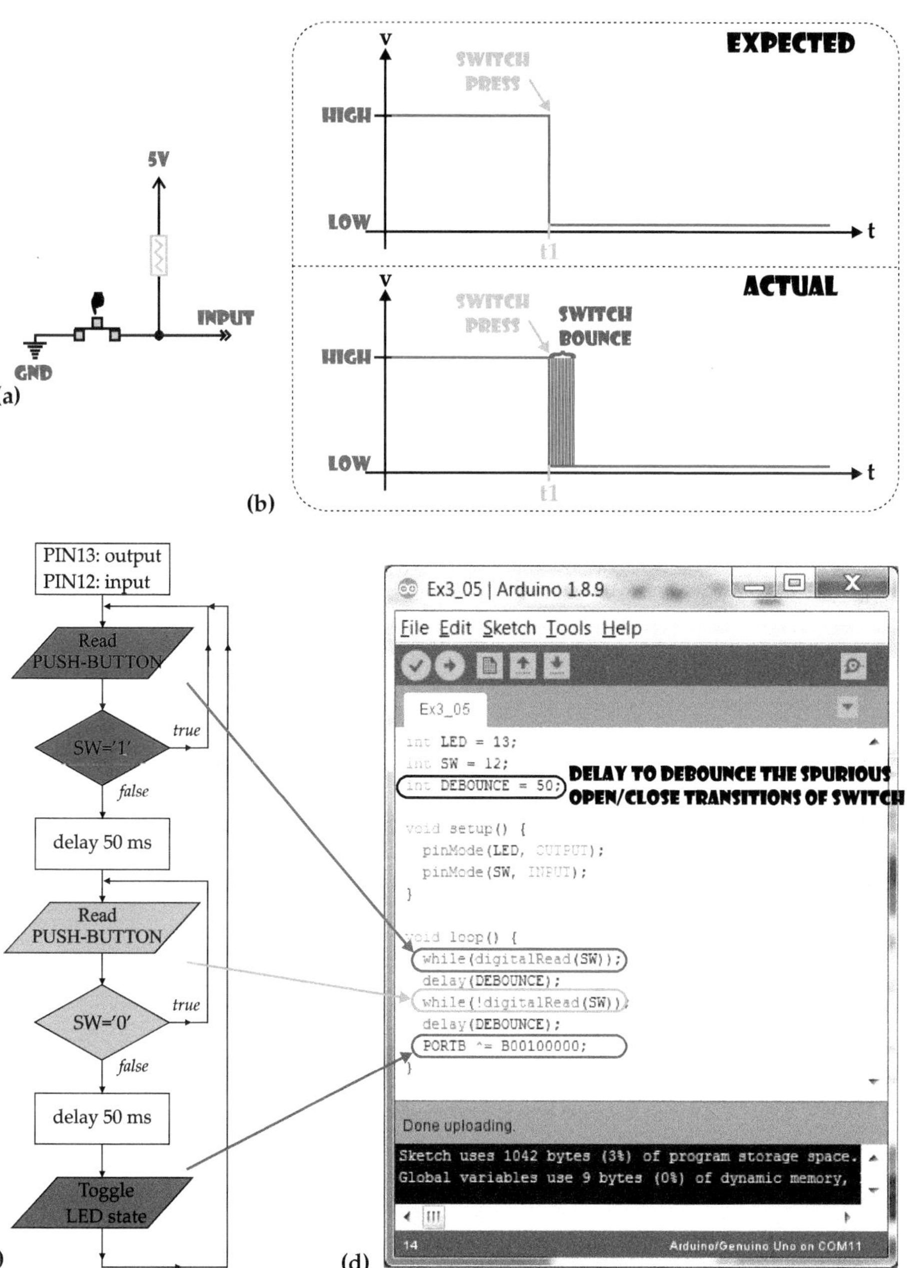

Figure 3.8 Debounce the spurious switch (open/close) transitions. (a) connection diagram of a push-button with pull-up resistor, (b) *expected* and *actual* signal waveform when pressing the push-button, (c) flowchart of the application example, (d) firmware of the application example. *Source:* Arduino Software.

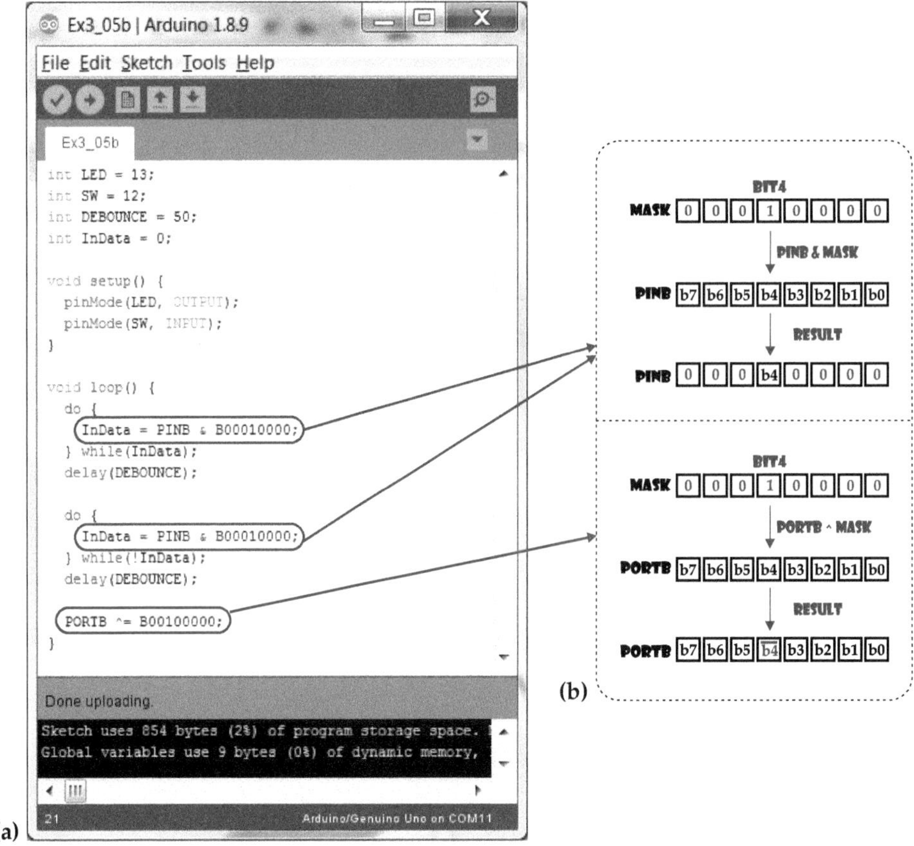

Figure 3.9 Reading the switch state through the direct manipulation of PINB register. (a) example firmware, (b) bitwise operations on the register of PORTB. *Source:* Arduino Software.

Analog Pin Interface

Because microcontroller applications often require the acquisition of an analog signal, the μC devices regularly incorporate one or more *analog-to-digital converters (ADCs)*. Some μCs may incorporate *digital-to-analog converters (DACs)*, as well, which are able to generate an analog signal through the μC pins but, unfortunately, an Arduino Uno board does not embed any DAC subsystem. An indirect way to provide an analog output interface from one of the μC pins is to produce a digital output pulse with an average value determined by the application code. The latter method is called *pulse width modulation* (*PWM*) and can be used to implement applications, such as dimmer (i.e. a light fixture of varying brightness), or driving motors at various speeds, etc.

Figure 3.10a illustrates an analog signal that admits any value between the LOW and HIGH state of a digital signal (and could represent the output of DAC system). A digital signal that resembles a “square” wave of the exact same period *T* but of different *duty cycle* (*DC*) is illustrated in Figure 3.10b–d. The term DC refers to the t_{ON}/T percentage, and

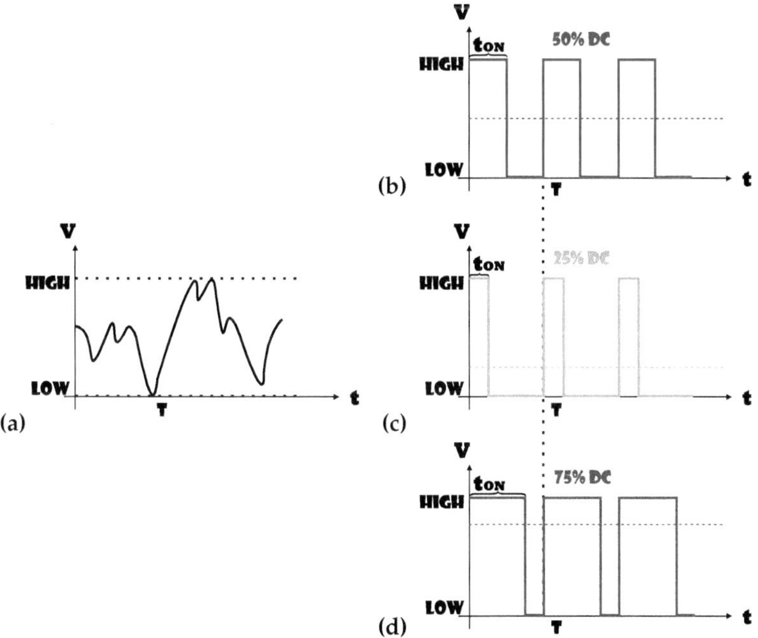

Figure 3.10 Analog signal and PWM arrangement in an Arduino Uno board. (a) example of an analog signal, (b) PWM example of 50%DC, (c) PWM example of 25%DC, (d) PWM example of 75%DC.

hence, the greater the ON time (i.e. the pulse width) the greater the DC value. Figure 3.10b–d represents how to modulate the pulse width of a digital signal in order to adjust the average voltage at the 50%, 25%, and 75% of the HIGH state, respectively.

Figure 3.11 depicts the pin arrangement of the ADC and PWM channels (that is, six channel of each subsystem) found on the Arduino Uno board. It is worth noting that the ADC subsystem is of 10-bit resolution, i.e. the acquired voltage is mapped into an integer value of range equal to 2^{10} (from 0 to 1023). The maximum input voltage that corresponds to the top ADC value (i.e. 1023) is determined by the *reference* voltage of the ADC subsystem (the default reference[4] voltage of Arduino Uno board is 5 V).

Arduino Ex.3.6

The example presented hereafter explores how to read the voltage of an open/close switch condition through an analog pin. Figure 3.12a presents the push-button setup (that was previously analyzed) along with a cable that provides a short circuit of the input pin which reads the switch state to the analog A0 pin of an Arduino Uno board.

---

4 It is noted that the reference voltage of the ADC subsystem can be modified by the Arduino *analogReference()* function and if set to EXTERNAL, the desired reference voltage should be externally applied to the AREF pin of the board.

Figure 3.11 ADC and PWM pin arrangement in an Arduino Uno board.

The source code of the example is given in Figure 3.12b. The example uses the serial port to send the value acquired by the ADC subsystem, and hence, the *setup()* function configures the desired baud rate of the port. The first instruction inside the *loop()* function, that is, the Arduino *analogRead()* function, acquires the analog value that appears on PINA0 and assigns that number to the *ADCvalue* integer variable. The subsequent arithmetic operation converts the ADC binary value into voltage by multiplying the binary number to the *reference* voltage (i.e. 5 V) of the ADC subsystem and dividing the result to the maximum ADC binary value (i.e. 1023). The outcome is stored to the float variable named *Voltage*. Then, the *SerialPrintln()* function transmits the *ADCvalue* to the serial port, while the printed result is depicted by Figure 3.12c. Because the switch state is alternated between HIGH and LOW (i.e. between 5 V and 0 V) we observe that the printed values are the max and min ADC binary values (i.e. 1023 and 0).

In order to print the *Voltage* variable instead of the *ADCvalue* variable, we should comment the first *SerialPrintln()* and uncomment the subsequent one (Figure 3.12b). The printed results of that particular case are given in Figure 3.12d, where the possible printed values are the 5 V and 0 V. Figure 3.12e depicts another built-in application of the Arduino IDE, which is called *Serial Plotter* and can be initiated from the menu *'Tools → Serial*

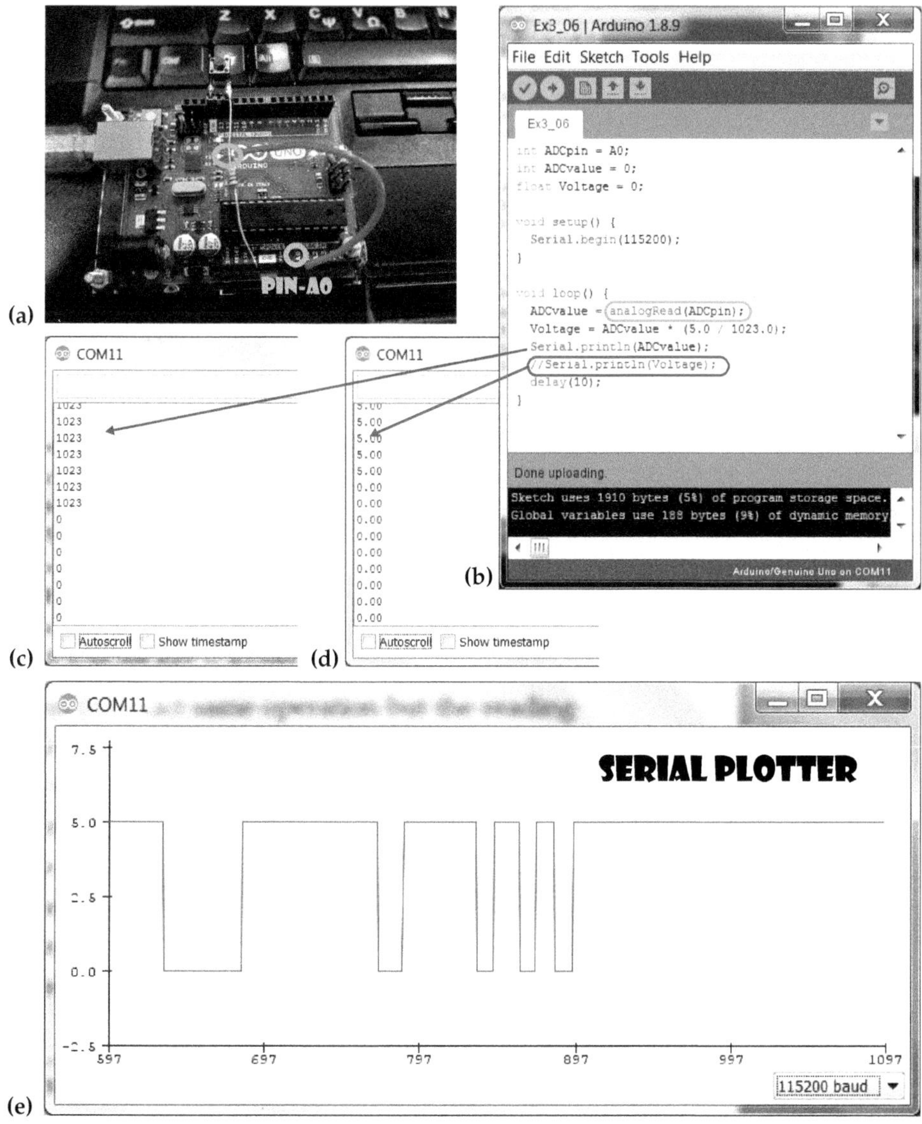

Figure 3.12 Reading the switch state through the analog input pin A0. (a) hardware setup, (b) example firmware, (c) printed results (ADC value in decimal representation), (d) printed results (ADC value in volts), (e) plot of the ADC value in volts with the Arduino *serial plotter*. *Source:* Arduino Software.

Plotter'. The *Serial Plotter* tool is useful for visualizing data that arrive on the serial port of the Arduino board. The example of Figure 3.12e represents the state of the push button in real time (5V for an open-switch condition and 0V for a close-switch condition).

The example of Figure 3.13 presents a more realistic representation of an analog signal, through the acquisition of the 3V3 power pin found on Arduino Uno board. To do so, we connect of the 3V3 power to the A0 analog pin (Figure 3.13a) and then upload the

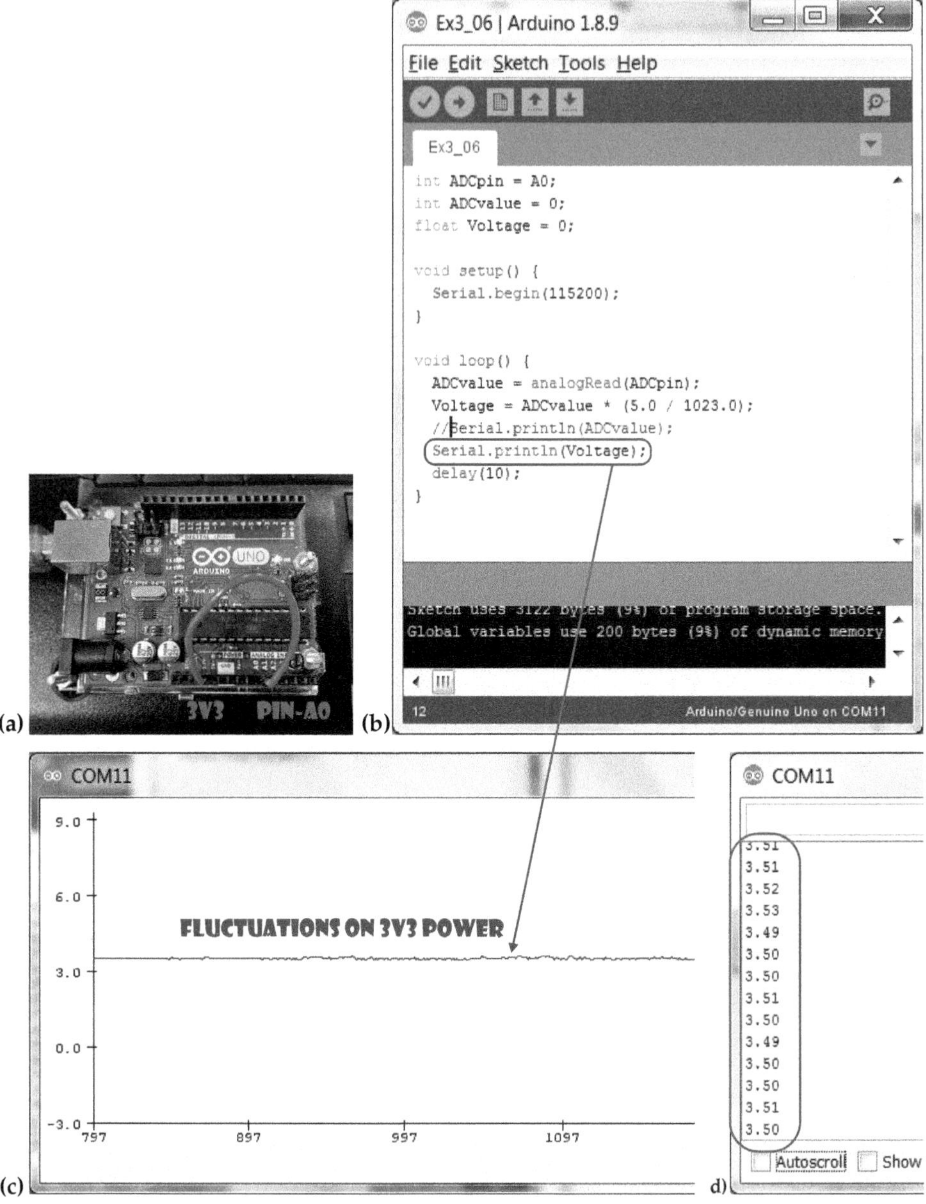

Figure 3.13 Reading the 3V3 power on an Arduino Uno board through the analog A0 pin. (a) hardware setup, (b) example firmware, (c) plot of acquired signal, (d) printed results of acquired signal. *Source:* Arduino Software.

aforementioned source code (Figure 3.13b). The printed results of this example are illustrated by the *Serial Plotter* tool of Figure 3.13c, as well as by the *Serial Monitor* tool of Figure 3.13d. The resolution of the ADC subsystem is capable of acquiring the fluctuations that appear on the output of the 3V3 voltage regulator (employed by the Arduino Uno board). We observe that the output of the 3V3 regulator fluctuates among the values 3.49–3.53 V (Figure 3.13d).

Arduino Ex.3.7

The example of Figure 3.14 presents how to generate a PWM wave from the proper pin of an Arduino Uno board. The example generates a PWM wave on PIN3 and acquires the wave though the ADC input of PINA0. The setup with the appropriate short circuit between the two pins is presented in Figure 3.14a. The source code of the example is given in Figure 3.14b. The *analogWrite()* function inside the *setup()* function admits two arguments. The former determines the configuration of PWM pin and the latter, (which admits values from 0 to 255), determines the DC of the wave. In code example presented herein, the DC is set to 25%.

The main code inside the *loop()* function is the same as before and exploits PINA0 in order to acquire the wave on PIN3. Figure 3.14c illustrates a PWM wave of 25% and 75% DC, while Figure 3.14d illustrates a PWM wave of 75% and 50% DC, as acquired by the *Serial Plotter* tool.

The 25%, 50%, and 75% DC is defined by the decimal values 64, 128, and 192, respectively, which initialize the *analogWrite()* function. It is here noted that there is no need to set a pin as output, through the *pinmode()* function, before calling analogWrite() function. For a more accurate representation of the PWM wave we have acquired the PIN3 signal through an oscilloscope. The setup is given in Figure 3.15a, while the acquisition of the PWM wave of DC equal to 50%, 25%, and 75%, respectively, is represented by Figure 3.15b–d. We may observe that the mean value of the PWM wave in each case is measured 2.61 V, 1.34 V, and 3.89 V, while the period[5] of the signal for each particular case is equal to 2 ms.

Interrupt Pin Interface

In computing, the process of sampling the state of an input pin until an event occurs (e.g. an alternation from logical '1' to '0') is regularly referred to as *polling*, a process which requires particular devotion by the source code in order to detect a pulse. Microcontrollers often embed an additional function on some of the IO pins, which allows them to automatically response to an external event and suspend (without occupying) the normal execution of the processor unit. In an Arduino Uno board, those pins are the PIN2(INT0) and PIN3(INT1), as highlighted in Figure 3.16, and are also known as *external interrupt pins.*

5 It is noted that the period of the PWM wave can be changed by modifying the value of particular μC registers. However, the exploration of an abundant new information related to that particular case is away from the scope of the present example.

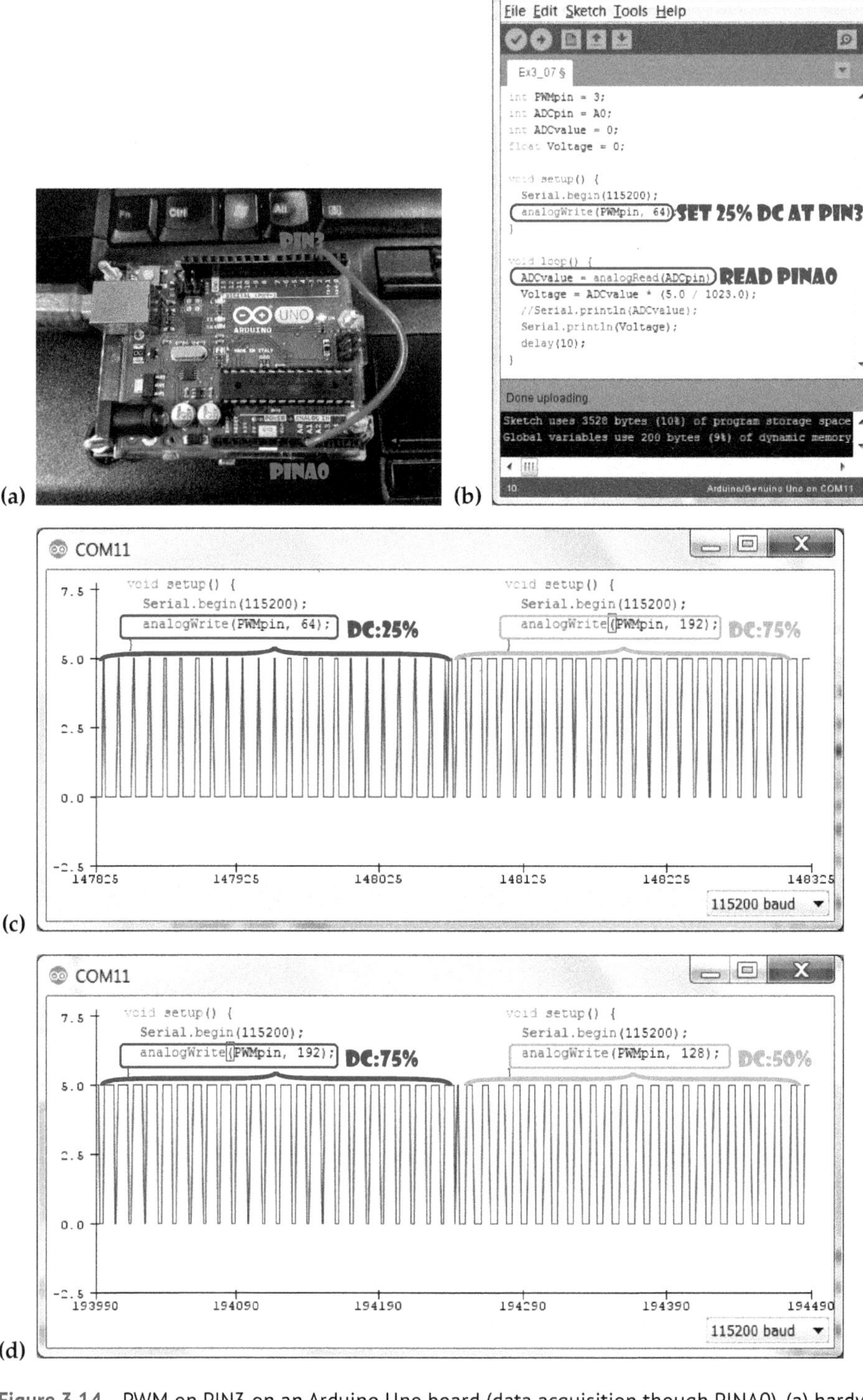

Figure 3.14 PWM on PIN3 on an Arduino Uno board (data acquisition though PINA0). (a) hardware setup, (b) example firmware, (c) plot of the PWM signal of DC 25% and 75%, (d) plot of the PWM signal of DC 75% and 50%. *Source:* Arduino Software.

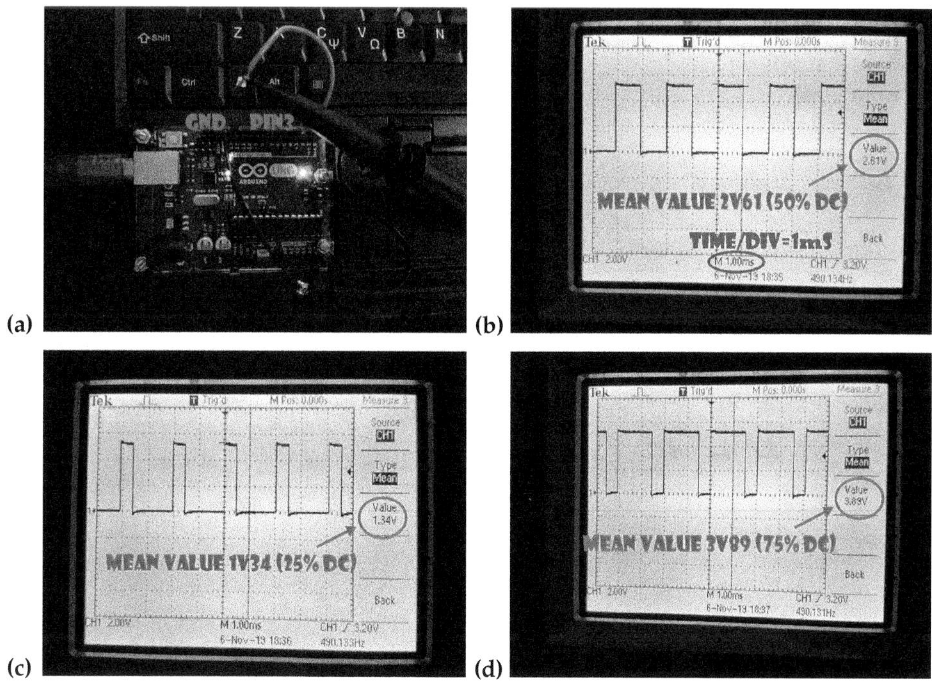

Figure 3.15 PWM on PIN3 (data acquisition though oscilloscope). (a) connection of the oscilloscope to PIN3 of the Arduino Uno, (b) mean value of PWM signal of 50% DC (2.61V), (c) mean value of PWM signal of 25% DC (1.34V), (d) mean value of PWM signal of 75% DC (3.89V). *Source:* Arduino Software.

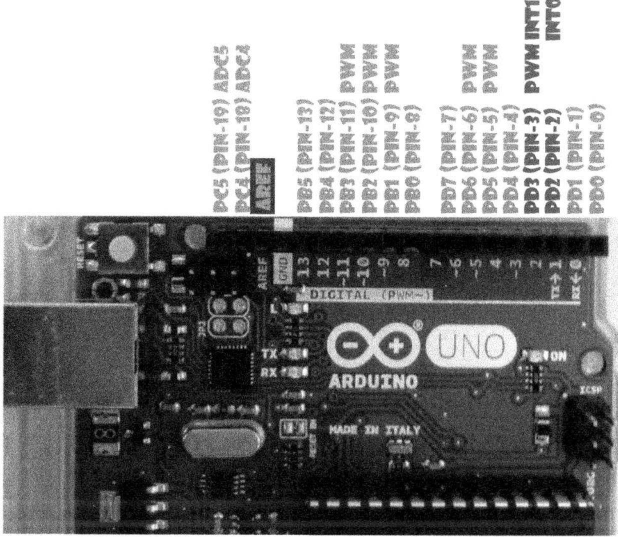

Figure 3.16 External interrupt pins on an Arduino Uno board.

The external interrupt pins respond very fast in the event of an external pulse and are triggered, for instance, when the interrupt pin goes from logical '1' to '0' (i.e. *falling edge*).

As soon as the trigger on the external interrupt pin takes place, the normal execution of the source code source is suspended and the μC performs a call to an interrupt handler, which is actually a small part of code, aka *interrupt service routine* (*ISR*). When the processor finishes the execution of the ISR it automatically resumes the control flow back to the position where the interrupt occurred. When an ISR is fetched, no other interrupts are possible until the execution of the current ISR. It should be noted that the Arduino *delay()* function requires interrupts to work, and hence, it cannot be called inside an ISR (after all, the ISR code should be as short and as fast as possible).

Arduino Ex.3.8

Figure 3.17a presents the example code that updates the DC of the PWM wave generated by PIN3. The updating process occurs when the user presses the push-button and triggers the external interrupt on PIN2 (while the DC cyclically obtains the values 25%, 50, and 75%).

The first instruction inside the *setup()* function activates the available internal pull-up resistor found on PIN2. This selection allows us to directly connect the push-button on the external interrupt pin without using an external pull-up resistor. The subsequent *attachInterrupt()* Arduino function admits three arguments. The first argument maps the external interrupt to the special function PIN2, the second argument declares the indicative label of the ISR that will be fetched upon triggering the external interrupt, and the third argument defines the type of the trigger that initiates an interrupt. The final argument admits one of following constants:

i) ***LOW*** triggers an interrupt when the pin state is found *LOW*;
ii) ***CHANGE*** triggers an interrupt whenever the pin state changes value;
iii) ***RISING*** triggers an interrupt when the pin changes from *LOW* to *HIGH*;
iv) ***FALLING*** triggers an interrupt when the pin changes from *HIGH* to *LOW*.

The subsequent two instructions inside the *setup()* function respectively activate the PWM wave on PIN3 (configured with DC = 25%), and initiate the serial port (with 115 200 baud rate). Next, the *loop()* function is repeated forever but incorporates no instructions. The user may insert any instruction that will be executed without needed to poll the input signal on external interrupt PIN2.

The control flow is inserted inside the ISR code, named *changeDC()*, as soon as the input signal on PIN2 changes from *HIGH* to *LOW*. At first, the decimal number 64 is added to the DC variable and if the latter result overflows (i.e. the 8-bit value of the DC variable is cleared) it adds again the 64 value inside the *if()* statement. This process forces the DC of PWM wave to cyclically obtain the values 25%, 50%, and 75%, and then, updates the PWM wave by the execution of the subsequent *analogWrite()* function. The last *Serial.println()* sends the current DC to the serial port.

Since it is not possible to use the debounce delay inside an ISR, the *Serial.println()* function allows to inspect the switch bounce phenomenon. When we press the push-button connected to PIN2 (Figure 3.18a) we anticipate one and single change to the DC value of the PWM wave. However, because the switch bounces (as acquired by the oscilloscope presented in Figure 3.18b), we may sometimes observe two or more changes.

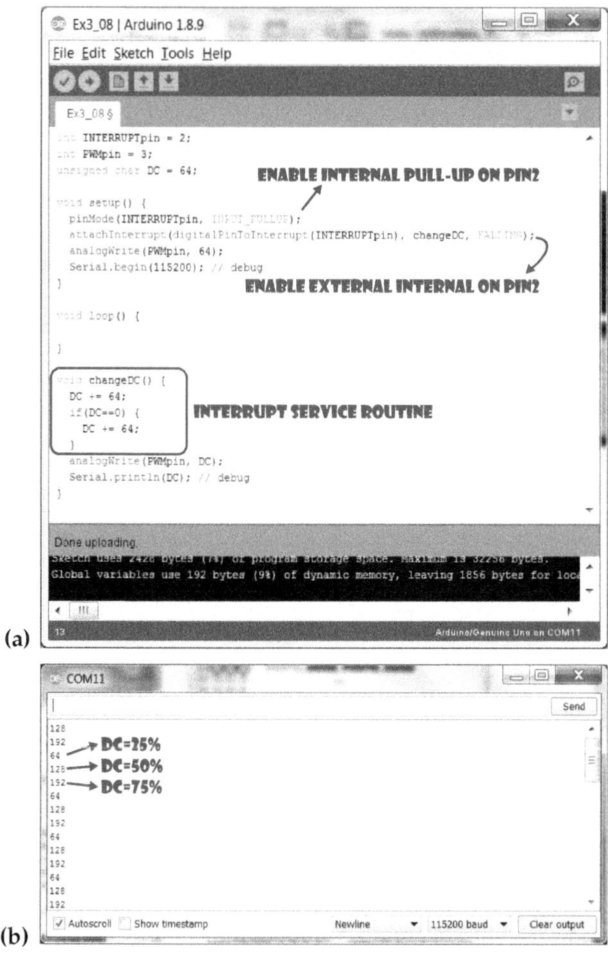

Figure 3.17 Update DC of the PWM wave on PIN3 via the external interrupt on PIN2. (a) example firmware, (b) printed results. *Source:* Arduino Software.

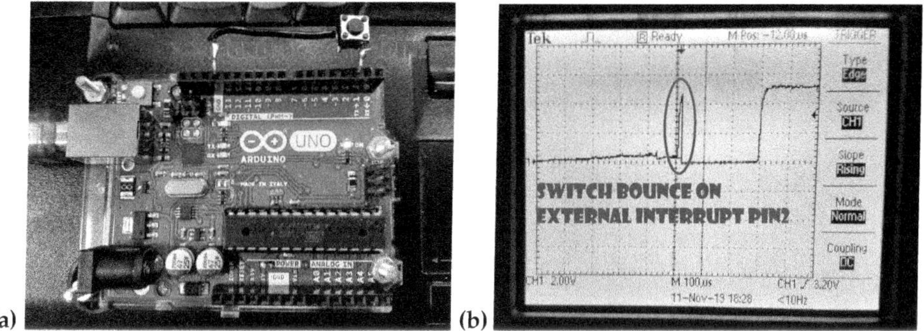

Figure 3.18 Switch bounce on the external interrupt PIN2. (a) hardware setup, (b) signal acquisition of PIN2 with the oscilloscope. *Source:* Arduino Software.

UART Serial Interface

Up until this point we have explored the digital and analog IO pin interface in essence of a parallel type of interface with the microcontroller device. The rather limited pins available by a μC chip often generate a demand for serial communication protocols, which establish the hardware interface between the microcontroller and the incorporated peripheral modules (of the user-defined application). One of the most popular serial interfaces in μC applications is the UART interface, which has extensively been exploited by the previous examples of the book (for the communication between the microcontroller and a personal computer through USB). Next, we provide some technical details in terms of the UART serial interface protocol.

The hardware interface through the UART is performed by two single-ended[6] signals. That is, the TX and RX wires where the former carries the transmitted data and the latter, the received data. In microcontroller applications the UART communication is regularly performed in half-duplex mode, that is, only one device is set to transmit/receive data over each transaction. The UART constitutes an asynchronous type of serial communication, and hence, the communicating devices work on their own internal synchronization clock (which must be configured to a matching transaction rate among the various devices), aka *baud rate.*

Figure 3.19a presents the connection between a μC to another μC device, or module, or personal computer, through the UART hardware interface. Figure 3.19b illustrates the

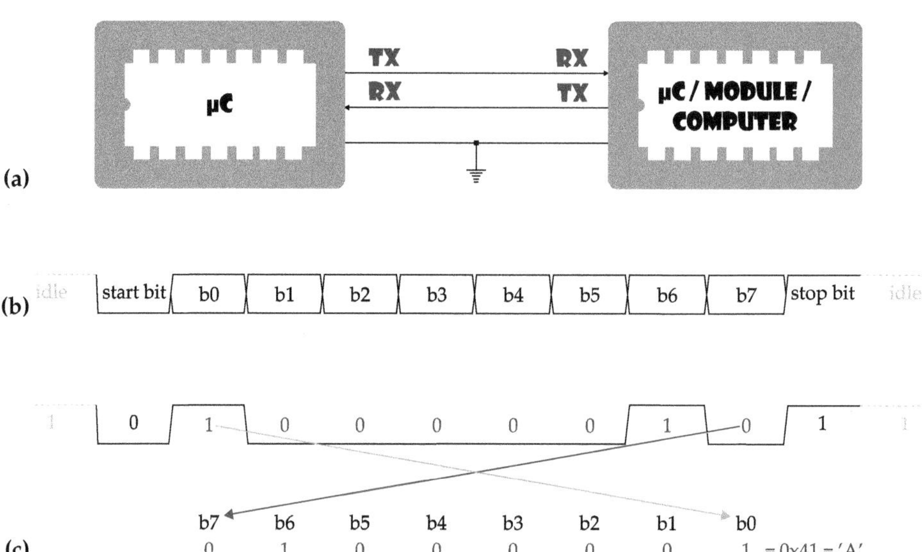

Figure 3.19 UART connection of a μC to a μC, module, or computer (8N1 frame). (a) connection diagram, (b) generalized data frame in UART communication, (c) transmission of the 'A' character in UART 8N1 format.

6 The most common way to transmit/receive signals over wires is the *single-ended* signaling, where one wire represents the signal and a second wire represents the reference voltage, usually the ground.

generalized data frame when transmitting/receiving bytes in the most common type of UART communication, that is, the 8N1. In detail, when no transaction is performed, the serial line is held to logic '1' in order to denote an *idle* state. The transmitter initiates a transaction by pulling TX line LOW for one-bit period, and that particular information is referred to as *start bit*. The *start bit* is used to synchronize the transmitter with the receiver, where the latter device starts to acquire the samples that follow the *start bit*. In 8N1 type of communication, the transmitter sends eight bits that incorporate the transmitted information, and then holds the line to state HIGH in order to denote a *stop bit*, which denotes the end of the transaction. Figure 3.19c presents the transmission of the 'A' character in 8N1 format. It should be noted that the serial transmission of the byte character starts from the least significant bit (LSB) (i.e. b0).

Arduino Ex.3.9

In the example presented herein we transmit the 'A' character through the dedicated UART hardware interface, available on an Arduino Uno board. In detail, Figure 3.20a presents the UART pin arrangement on the board. Figure 3.20b presents the source code that repeatedly

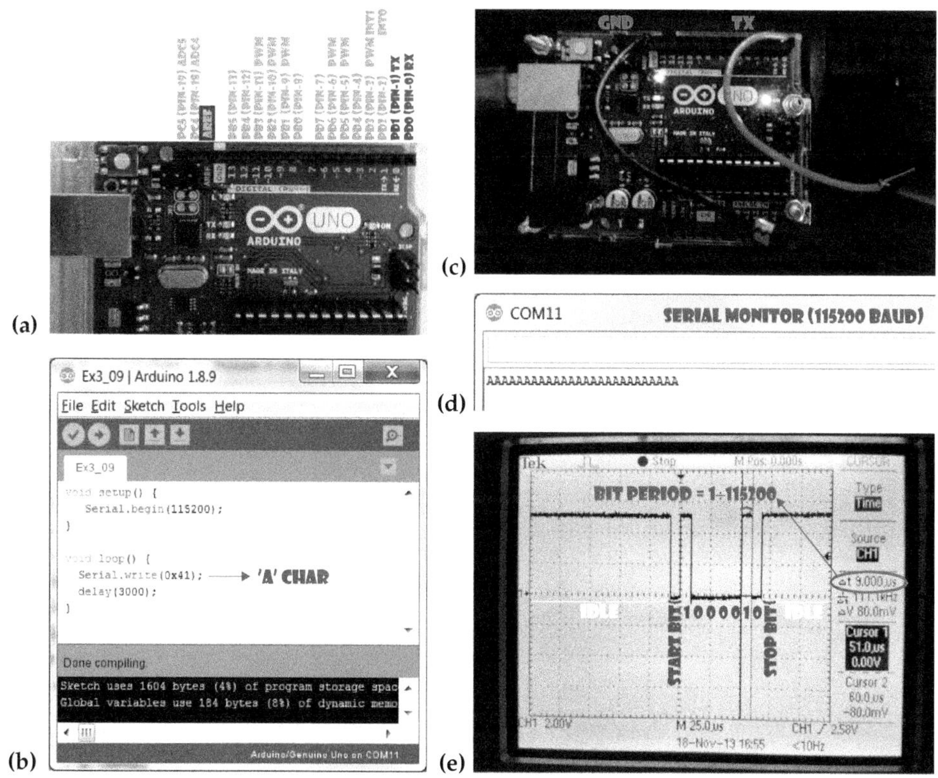

Figure 3.20 UART hardware interface on an Arduino Uno board (pinout and data framing). (a) UART pin arrangement on the hardware board, (b) example firmware, (c) signal acquisition of UART TX line with the oscilloscope, (d) printed results, (e) data frame of the transmitted 'A' character acquired by the oscilloscope. *Source:* Arduino Software.

transmits the 'A' character from Arduino Uno to a personal computer every three seconds, using a 115 200 baud rate. Figure 3.20c illustrates the connection of an oscilloscope on TX line of the UART hardware interface. Figure 3.20d depicts the transmitted 'A' character as acquired by the *Serial Monitor* tool of an Arduino IDE. Finally, Figure 3.20e presents the data frame of the transmitted 'A' character as acquired by the oscilloscope. It is worth noting that each bit period is equal to 1/baud, which in this particular case is calculated as follows: $1 \div 115\,200 = 8.7 \cdot 10^{-6}\,\text{s} \cong 9\,\mu\text{s}$.

Arduino Ex.3.10

In the following example we explore the UART hardware interface between the Arduino Uno and an external module, that is, the FT2232H chip delivered by FTDI Company. FT2232H constitutes a USB 2.0 to UART chip that can be used to transmit serial data to a personal computer. While this operation can be straightforwardly performed by the Arduino Uno board, the purpose of this example is to explore further the UART interface as well as the third-party tools that can be utilized in microcontroller (Arduino-based) applications.

The left part of Figure 3.21 depicts the *FTDI click* of MikroElektronika Corporation. The latter business provides an influx of boards (aka *click boards*) that can be used in microcontroller applications. The same company delivers the so-called *click shields* that render feasible the connection of a click board to a microcontroller board of different manufacturer. For instance, the *Arduino Uno click shield* depicted by the right part of Figure 3.21 renders feasible the connection of the MikroElektronika *click boards* to an Arduino Uno board. It is worth noting that the *FTDI click* can be controlled either by UART or I2C, as determined by the short circuit inside the blue frame of Figure 3.21.

The source code of Figure 3.22a is the exact same code that was explored in the previous chapter and addressed for printing to the Arduino Uno serial port the capital and small

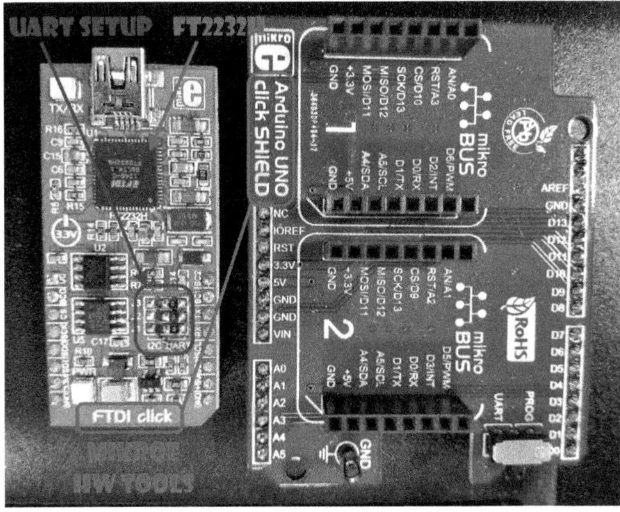

Figure 3.21 Third-party tools for the Arduino Uno board (FTDI click by MikroElektronika).

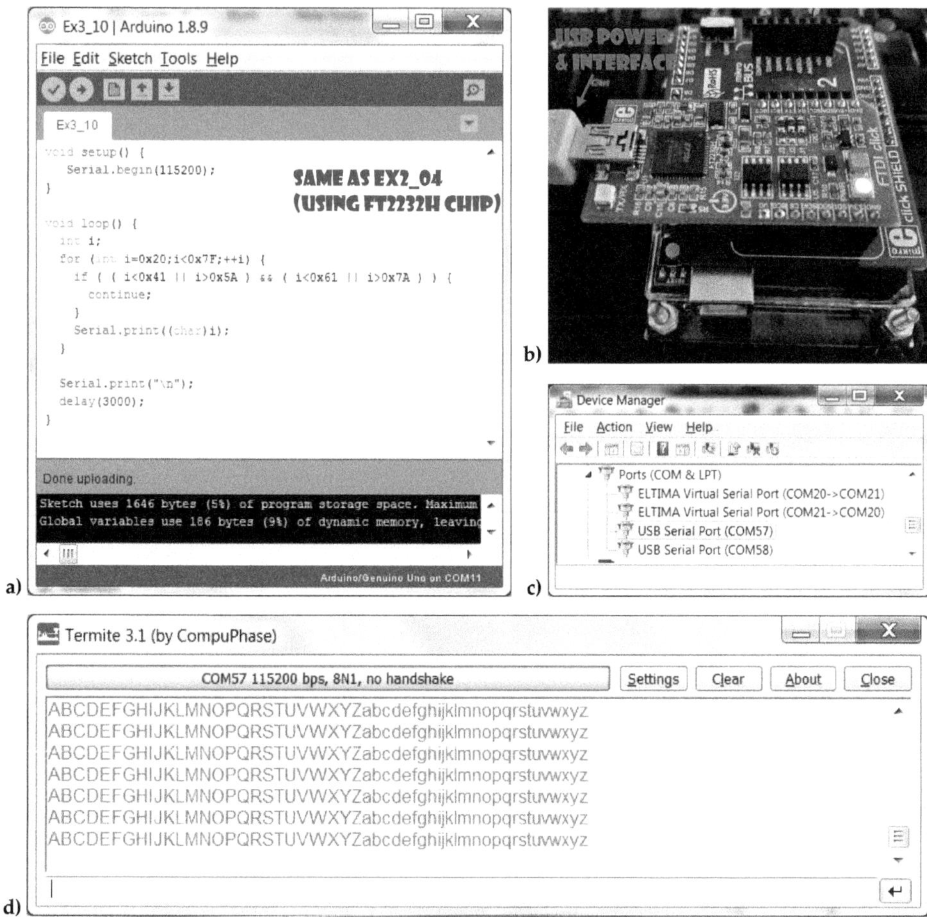


```
void setup() {
  Serial.begin(115200);
}

void loop() {
  int i;
  for (int i=0x20;i<0x7F;++i) {
    if ( ( i<0x41 || i>0x5A ) && ( i<0x61 || i>0x7A ) ) {
      continue;
    }
    Serial.print((char)i);
  }

  Serial.print("\n");
  delay(3000);
}
```

Figure 3.22 UART hardware interface on an Arduino Uno using third-party tools (by MikroElektronika). (a) example firmware, (b) hardware setup, (c) COM ports reserved by the FTDI chip, (d) printed results in Termite console. *Source:* Arduino Software.

letter of the English alphabet. However, the example presented herein sends to the USB port the same string, but through the serial port configured by the *FTDI click* (instead of the Arduino Uno board). As presented by the hardware setup given in Figure 3.22b, the Arduino Uno is powered by the *FTDI click*. Figure 3.23c illustrates the COM port reserved by the device driver installation of the FTDI chip. Finally, Figure 3.23d illustrates the printed results sent to the serial port through the *FTDI click*.

Arduino Ex.3.11

The aforementioned examples apply to the built-in UART interface of the μC device employed by an Arduino Uno board. However, it is possible to configure a software-implemented UART, through the Arduino *SoftwareSerial* library. This library is particular useful when

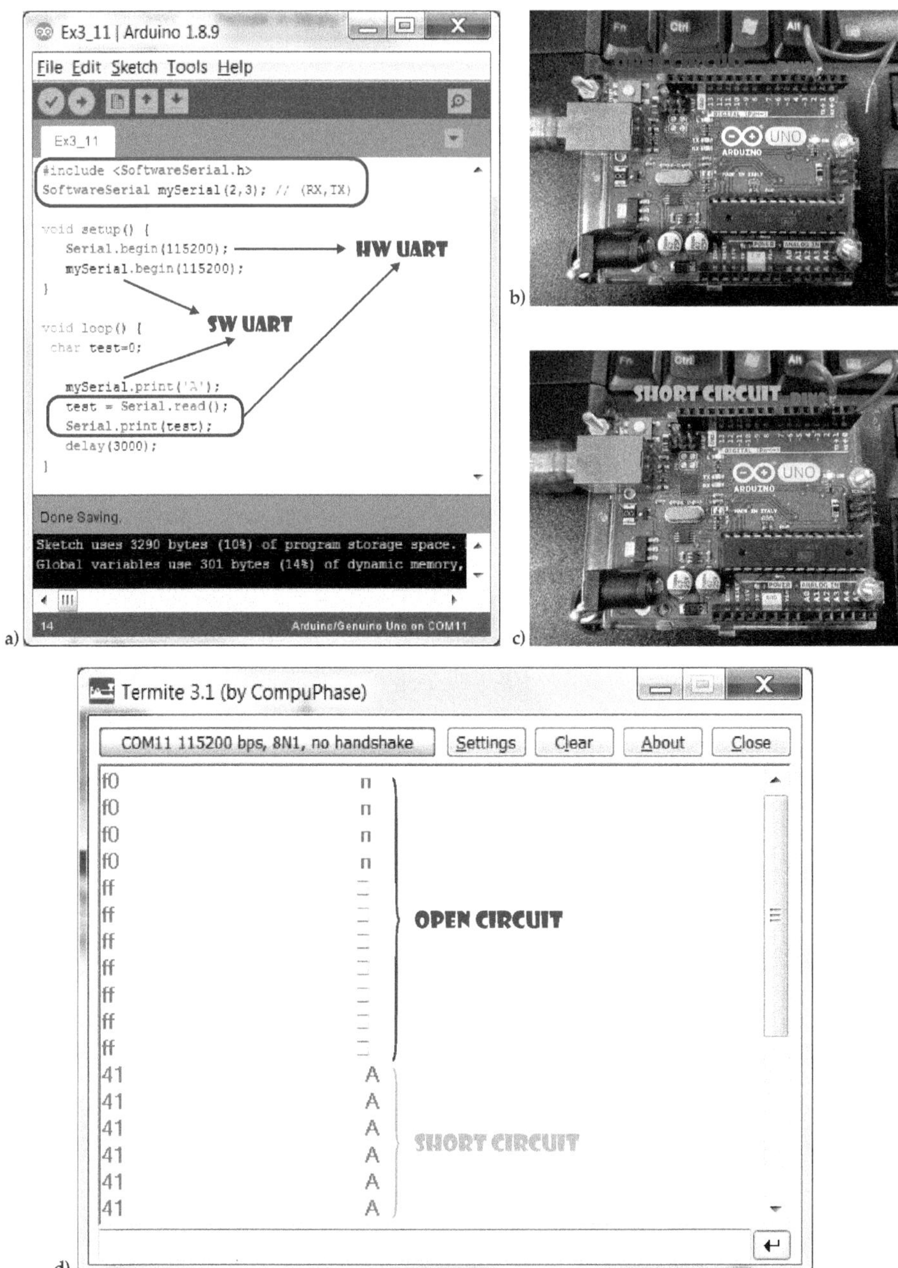

Figure 3.23 Software-implemented UART on an Arduino Uno board (*SoftwareSerial* library). (a) example firmware, (b) hardware setup: open circuit between the software-based and the hardware-based UART, (c) apply short circuit between software-based TX line and hardware-based RX line, (d) printed results in Termite console. *Source:* Arduino Software.

there is a need to add more UART interfaces on an Arduino Uno board, or implement a UART interface to a μC of no built-in UART hardware.

Figure 3.23a presents the source code that initializes and controls a software-implemented UART interface. The first two code lines include *SoftwareSerial* library and configure PIN2 and PIN3 as the RX and TX lines of the software-based UART, which is referenced by the indicative name *mySerial*.[7] Inside the *setup()* function both the hardware-based and software-based UART was set to 115 200 baud rate. Inside the *loop()* function with transmit over the TX line of the software-based UART the 'A' character. Next, we read the received character found on the hardware-based UART, and we send to the USB port of a personal computer the latter character acquired by the hardware-based UART. The overall process is repeated every three seconds.

To test the source code we address the hardware setup depicted by Figure 3.23b,c. In detail, we connect the TX line (i.e. PIN3) of the software-based UART to the RX line (i.e. PIN0) of the hardware-based UART. When the two lines are not connected together, we obtain no data on the terminal console (as presented by Figure 3.23d). A short circuit between the two lines causes the hardware-based UART to acquire the 'A' character transmitted by the software-based UART, and then, to forward the received character to the USB port of a personal computer.

SPI Serial Interface

Another popular serial interface in μC applications is the SPI, which is used for short-distance communications between the microcontroller and the peripheral modules. The hardware interface through the SPI is performed by four single-ended signals, that is, the *master output slave input (MOSI)*, *master input slave output (MISO)*, *serial clock (SCLK)*, and *slave select (SS)*. The SPI constitutes a synchronous type of serial communication (i.e. the SPI devices share the common SCLK synchronization clock) in full-duplex[8] mode, using master-slave architecture.

Figure 3.24a presents the hardware connection between a μC and an SPI module. The CS signal is used to activate/deactivate the peripheral module, and hence, to incorporate more than one modules in the SPI bus, we reserve one additional output pin in the μC device for each additional module (Figure 3.24b). It here noted that only one SPI module should be activated (through the corresponding CS line) on every single transaction.

Figure 3.24c,d present the single-byte read and write timing diagrams of the SPI protocol. The SPI supports four different modes of operation, determined by the *clock polarity* (*CPOL*) and the *clock phase* (*CPHA*). The clock polarity refers to the SCLK signal and when CPOL = 0, the clock signal idles at logical '0', while for CPOL = 1 the clock signal idles at logical '1'. Additionally to the CPOL, the CPHA determines the SPI mode of operation as presented in Table 3.1. In detail, in SPI mode 0 and 3 the data are shifted out on the falling edge and sampled on the rising edge of the clock pulse, where in SPI mode 0 the clock idles

7 It should be noted that the software-based UART is controlled by the same functions as the hardware-based UART, but using the referenced name instead of *Serial*, e.g. *mySerial.begin()* instead of *Serial.begin()*.
8 The full-duplex communication refers to the possibility of two commutating devices to transmit/receive data simultaneously.

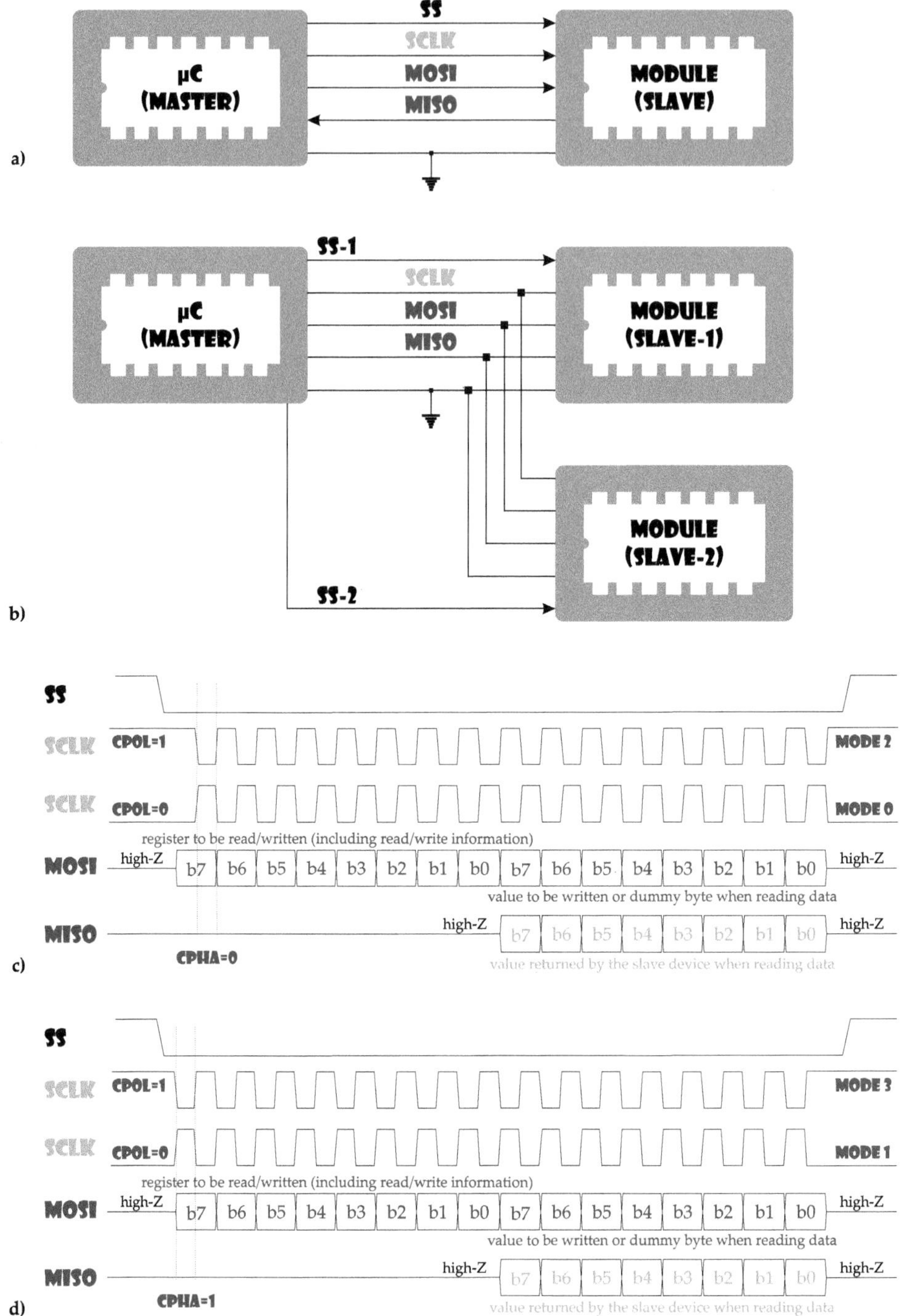

Figure 3.24 SPI hardware topology and read/write timing diagrams (four modes of operation). (a) connection diagram between the µC and a single SPI device, (b) connection diagram between the µC and two SPI devices, (c) single-byte read timing diagrams of the SPI protocol, (d) single-byte write timing diagrams of the SPI protocol.

Table 3.1 SPI modes of operation (Truth table).

CPOL	CPHA	SPI mode	Idle clock	Description
0	0	0	Logical '0'	Data are shifted out on the falling edge of SCLK and sampled on the rising edge of SCLK.
0	1	1	Logical '0'	Data are shifted out on the rising edge of SCLK and sampled on the falling edge of SCLK.
1	0	2	Logical '1'	Data are shifted out on the rising edge of SCLK and sampled on the falling edge of SCLK.
1	1	3	Logical '1'	Data are shifted out on the falling edge of SCLK and sampled on the rising edge of SCLK.

at '0' and in SPI mode 3 the clock idles at '1'. On the other hand, in SPI mode 1 and 2 the data are shifted out on the rising edge and sampled on the falling edge of the clock pulse, where in the SPI mode 1 the clock idles at '0' and in the SPI mode 2 the clock idles at '1'.

The (most common) timing diagrams for the SPI mode 0 and mode 2 are given in Figure 3.24c, while the timing diagrams for the SPI mode 1 and mode 3 are given in Figure 3.24d. The first byte transmitted by the MOSI signal (i.e. the byte highlighted in red color) incorporates the register address to be read or written along with the *read/write* (*R/W*) information, while the second byte (i.e. the byte highlighted in blue color) incorporates the value to be written to that register, or a dummy byte in case of reading data. In the latter case the data returned by the slave device are received by the MISO signal (i.e. the byte highlighted in green color).

It is also possible to perform multiple-byte R/W operations in the SPI protocol. However, the designer should always implement the code for the SPI transaction (either for single-byte or multiple-byte R/W operations) in consideration of the timing diagrams information given by the datasheet of the incorporated SPI module.

Arduino Ex.3.12

The following examples apply to the built-in SPI of the (μC device employed by) Arduino Uno board. While it is possible to configure a software-implemented SPI through the Arduino libraries, that particular information is not explored by the book (so as to avoid too much technical details, which are inappropriate for the novice learners). The SPI protocol is regularly addressed for the control of peripheral modules that apply to an interface, such as, wireless interface, Ethernet interface, and so forth, where speed and accuracy are of particular importance. Since the most common applications regularly incorporate a couple of such interfaces, it would be wise to address the μC's built-in SPI. Yet, we will explore later in the book the development of a custom-designed SPI library based on the aforementioned timing diagrams.

The following SPI examples apply to a sensor device, that is, the environmental sensor unit *BME280* by Bosch Sensortec. BME280 sensor device measures pressure, temperature, as well as humidity, and it is controlled either by SPI or I2C protocol. Hence, this sensor will be utilized to explore both the SPI and I2C protocols (the latter protocol will be explored in the subsequent section). The hardware setup that will be used by the code examples applies to the *Weather click*, as well as the *Arduino Uno click shield* delivered by the MikroElektronika Corporation.

A simple example to begin with is to perform an SPI read in order to acquire the chip identification number; hereafter referred to as *chip id*. Regularly, sensor devices incorporate a register that contains a *chip id* that can be used so as to identify the device. The *chip id* of BME280 sensor is the hexadecimal value 0x60 stored to the register address 0xD0. The following example performs an SPI read to the register 0xD0 and prints its content to the serial port. Figure 3.25 presents the source code, the hardware setup, as well as the data on the serial port when reading *chip id* of BME280 sensor.

In regard to the source code of Figure 3.25a, the first two lines incorporate the SPI Arduino library, and then, they define the SS pin location (that is, PIN10). The configuration of the built-in SPI is performed within the *setup()* function. The *SPI.begin()* function initializes the SPI bus by setting SCLK, MOSI, and SS to outputs, and by pulling *LOW* the SCK and MOSI signals, as well as by pulling *HIGH* the SS signal. Then, the *SPI.beginTransaction()* function initializes the SPI bus using the specified *SPISettings*. The *SPISettings* object admits three parameters, where the first determines the maximum speed of communication in Hz (this parameter can admit values up to *20 000 000*, where the latter number corresponds to 20MHz operating frequency), the second determines the order of the serially transmitted data (i.e. *MSBFIST* or *LSBFIRST*), and the third determines SPI mode of operation (i.e. *SPI_MODE0*, *SPI_MODE1*, *SPI_MODE2*, or *SPI_MODE3*). Herein, the function *SPI.beginTransaction()* initializes the SPI bus at 500KHz with the most significant bit (MSB) to be transmitted first (on each serial transaction) using the SPI mode[9] 0. Subsequent to the *SPI.beginTransaction()* function the code initializes the SS pin to output. While the latter code line is redundant because of the initialization performed earlier by the *SPI.begin()* function, we address that particular command in order to highlight the additional code lines that should be addressed in order to control more than one peripheral devices from the same SPI bus.

Inside the *loop()* function we implement the timing diagram of SPI mode 0 (presented earlier in Figure 3.24c). At first we assert the SS line low to activate the BME280 slave device, and then, the subsequent *SPI.transfer()* function transmits the hexadecimal value 0xD0, which has been previously subjected to a bitwise OR with the bitmask 0x80. The latter operation forces the MSB of the 0xD0 value to logical '1', which, according the datasheet[10] of BME280 device, holds the R/W information. In detail, the MSB = 1 information initiates an SPI read process, while the MSB = 0 initiates a write process. Since the current example initiates a read process of the 0xD0 register address, the subsequent *SPI.transfer()* function transfers a dummy byte of zero value, while simultaneously obtains the content of 0xD0 register (which is assigned to the *id* variable). After the SPI read procedure, we deassert SS pin in order to deactivate the SPI on the slave device.

The hardware setup of this example employs an Arduino Uno board, along with an *Arduino Uno click shield* as well as a *Weather click* (both delivered by MikroElektronika), as

9 It is here noted that the BME280 device is compatible with the SPI mode 0 as well as the SPI mode 3, and supports 10MHz maximum SPI clock frequency.
10 It should be noted that the SPI transaction may be different from device to device, and hence, the designer should always prepare the SPI data according to the datasheet of the employed device.

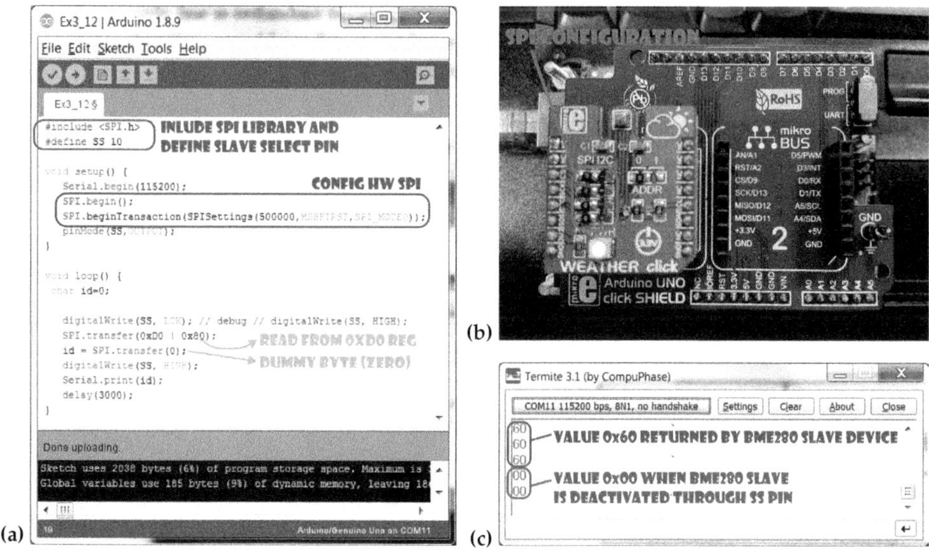

Figure 3.25 SPI single-byte read of the *chip id* of BME280 sensor device. (a) example firmware, (b) hardware setup: SPI configuration of the *Weather Click* module, (c) printed results in Termite console. *Source:* Arduino Software.

presented by Figure 3.25b. It should be noted that the latter board holds the BME280 sensor and determines the serial peripheral interface (SPI)/inter-integrated circuit (I2C) operation by three short-circuits (herein configured for the SPI mode of operation as highlighted by the red frame of Figure 3.25b). Figure 3.25c depicts the 0x60 hexadecimal value returned by the slave when the latter device is activated. If we deactivate the slave device by holing the SS signal permanently high (e.g. by replacing the first code line inside the main loop to the currently commented line), the slave returns zero value.

The timing diagrams of the aforementioned code can be captured by an oscilloscope of at least four channels. Figure 3.26a presents the pin location of the built-in SPI found in an Arduino Uno board. Figure 3.26b depicts how to connect a 4-channel oscilloscope in order to capture the SPI signals (i.e. SS, MISO, SPI clock signal [SCLK], MOSI) in the incorporated hardware setup. Finally, Figure 3.26c illustrates the timing diagrams of the SPI single-byte read, of the BME280 *chip id* (that is, reading from register 0xD0 the permanently stored value 0x60).

As presented by Figure 3.26c, the BME280 works on SPI mode 0, and hence, the serial data are shifted out on the falling edge of the clock pulse and sampled on the rising of the clock pulse. In addition, the SCLK idles at logical '0'. The SPI transaction starts by asserting SS signal (i.e. changing its state from *HIGH* to *LOW*) and terminates by deasserting SS signal (i.e. from *LOW* to *HIGH*). On the first eight clock pulses the MOSI signal transmits the register address with the MSB = 1 (i.e. 0xD0|0x80 = 0xD0) to denote a read process (MSB is transmitted first), as defined by the datasheet of BME sensor device. During those pulses the MOSI signals is kept at a *high-z* level. On the next eight clock pulses, the MOSI signal transmits a dummy byte (i.e. a zero value in this case) and at the same time, the MISO signal transmits simultaneously the content of 0xD0 register, that is, the 0x60 value (MSB first).

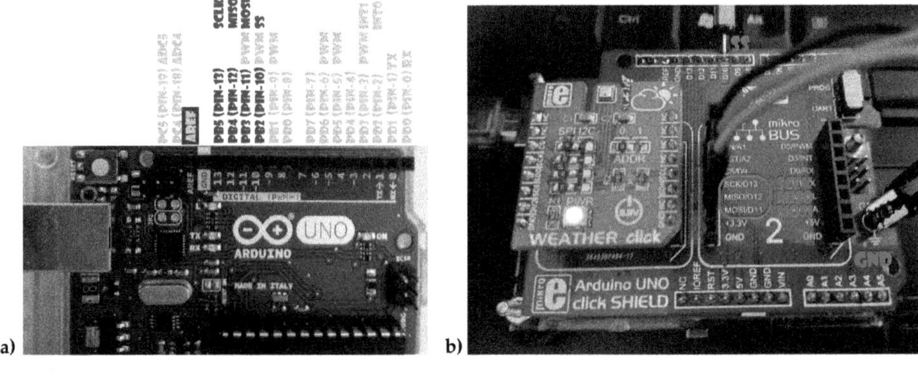

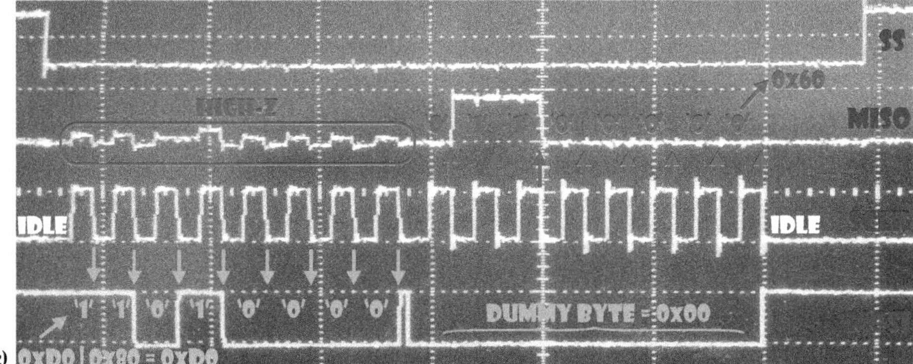

Figure 3.26 SPI single-byte read of the *chip id* of BME280 sensor device (timing diagrams). (a) SPI pin arrangement on the *Arduino Uno* board, (b) capture the SPI pins with an oscilloscope via the header of the *Arduino Uno click shield*, (c) acquisition of the SPI single-byte read timing diagrams with the oscilloscope.

An SPI single-byte write and read process is given in the code example of Figure 3.27a. The code performs a write and then a read process of the register address 0xF4 (i.e. *"ctrl_meas"*), which sets the pressure and temperature data acquisition options[11] of BME device. The SPI write process is encompassed within a custom-designed function, named *BME280_Init()*. The function first asserts SS and then transfers the register address 0xF4 with the MSB = 0 in order to denote a write process (as defined by the BME280 datasheet). It is here noted that the bitwise AND of the register 0xF4 to the bitmask 0x7F clears the MSB of 0xF4 value, while also leaving unchanged the rest of the bits (i.e. 0xF4&0x7F = 0x74). Then, the subsequent instruction of the *BME280_Init()* function transmits the value 0x27, which performs the desired configuration of the sensor device. Finally, the function deasserts the SS signal. Subsequent to the *BME280_Init()* function is the *setup()* function, which is the same as the one used by the previous example code. Next is the *loop()* function, which first performs a call to the *BME280_Init()* function, and then, it performs an SPI single-byte

11 To provide a control of the data rate, noise, response time, and current consumption, the BME sensor device supports a variety of oversampling modes. A detailed description of those modes (which can be found on the datasheet of the sensor) is away from the scope of this chapter.

a)

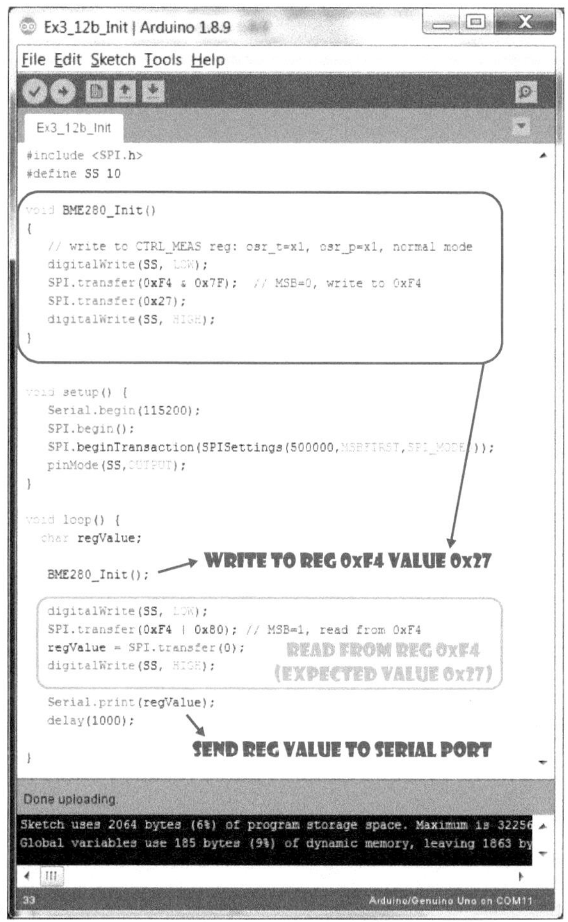

b)

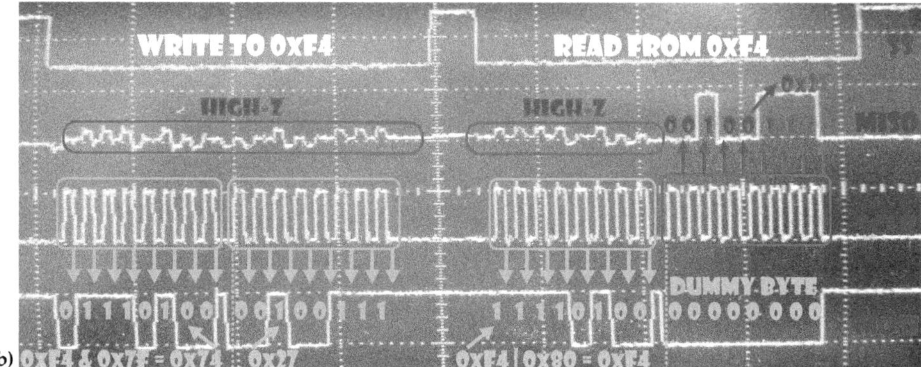

Figure 3.27 SPI single-byte write and read (BME280 register 0xF4, aka *"ctrl_meas"*). (a) example firmware, (b) acquisition of the SPI single-byte write/read timing diagrams with the oscilloscope. *Source:* Arduino Software.

read of the 0xF4 register (where the read process is equivalent to the one presented in the previous example code). Finally, the main loop sends the content of 0xF4 register to the serial port and then delays the process for one second (before repeating the loop).

Figure 3.27b depicts the timing diagrams of the example code. In detail, the SS is initially asserted, and then, the MOSI signal transmits on the first eight clock pulses, the 0xF4 register address along with the MSB = 0 (i.e. 0xF4&0x7F = 0x74), so as to denote a write process. On the next eight clock pulses the MOSI signal transmits the desired 0x27 byte to be assigned to *"ctrl_meas"* register. The SS signal is then deasserted in order to denote the end of the SPI write process. Afterward, the SS is asserted again to initiate the SPI read process, and then, the MOSI signal transmits on the first eight clock pulses the register address along with the MSB = 1 (i.e. 0xF4|0x80 = 0xF4) to denote a read process. On the next eight clock pulses, the MOSI signal transmits a dummy byte (i.e. a zero value) and at the same time, the MISO signal transmits simultaneously the content of 0xF4 register, that is, the 0x27 value that was previously assigned to *"ctrl_meas"* register. It is here noted that the user may verify the transmission using the *Termite* console; otherwise, the *Serial Monitor* tool prints a *single quotation mark (')*, which corresponds to the hexadecimal value 0x27.

The SPI protocol supports the R/W process of multiple bytes when reading/writing from/to consecutive register addresses. An SPI multiple-byte read process of the T *coefficients (coeffs)* of the BME280 device is presented in the example of Figure 3.28. It is here noted that when acquiring data from BME280 sensor, the device provides raw measurements (i.e. raw temperature, pressure, and humidity). To obtain compensated (true) measurements the user should perform particular mathematic operation (as described by the datasheet of BME280 sensor) using the unique – per device – coeffs for temperature, pressure, and humidity. Thereby, the user should read those coeffs once, and then perform the necessary arithmetic tasks either by the hardware, or by a software program running on a personal computer. The latter implementation allows us to perform a faster data acquisition of the raw data, and then, to assess the actual data offline by the personal computer.

Figure 3.28a presents the example code that reads and sends to the serial port the T coeffs of BME280 device, which are found on six successive memory addresses (that is, from 0x88 to 0x8D). The SPI multiple-byte read process is encompassed within a function named *BME280_ReadTcoeffs()*, as highlighted by the red frame. The process starts exactly as in the single-byte read and, hence, (after asserting the SS signal) the foremost *SPI.transfer()* function transmits the register address 0x88 (where the MSB is set to *HIGH* in order to denote the SPI read process). Afterward, instead of transmitting merely a single dummy byte through an *SPI.transfer()* which reads the content of 0x88 register, we encompass the *SPI.transfer()* function within a *for* loop. The loop is repeated six times and, hence, six successive dummy bytes of zero value are sent to the slave device. As a result, we obtain and store to the *dig_T[]* array the content of six successive registers, that is, the content of 0x88-0x8D. At the end of the transaction we deassert the SS signal.

The subsequent code of *BME280_ReadTcoeffs()*, which is highlighted by the green frame, prepares the T coeffs in the format recommended by BME280 the datasheet. Then, it prints to the serial port the values of those variables. For example, the first two bytes of *dig_T[]* array, which herein hold the values *dig_T[0] = 0x87* and *dig_T[1] = 0x6E*, are concatenated to the *unsigned short* variable *dig_T1*. As presented by Figure 3.28b, the content of *dig_T1*, herein, is equal to the 16-bit hex value 0x6E87.

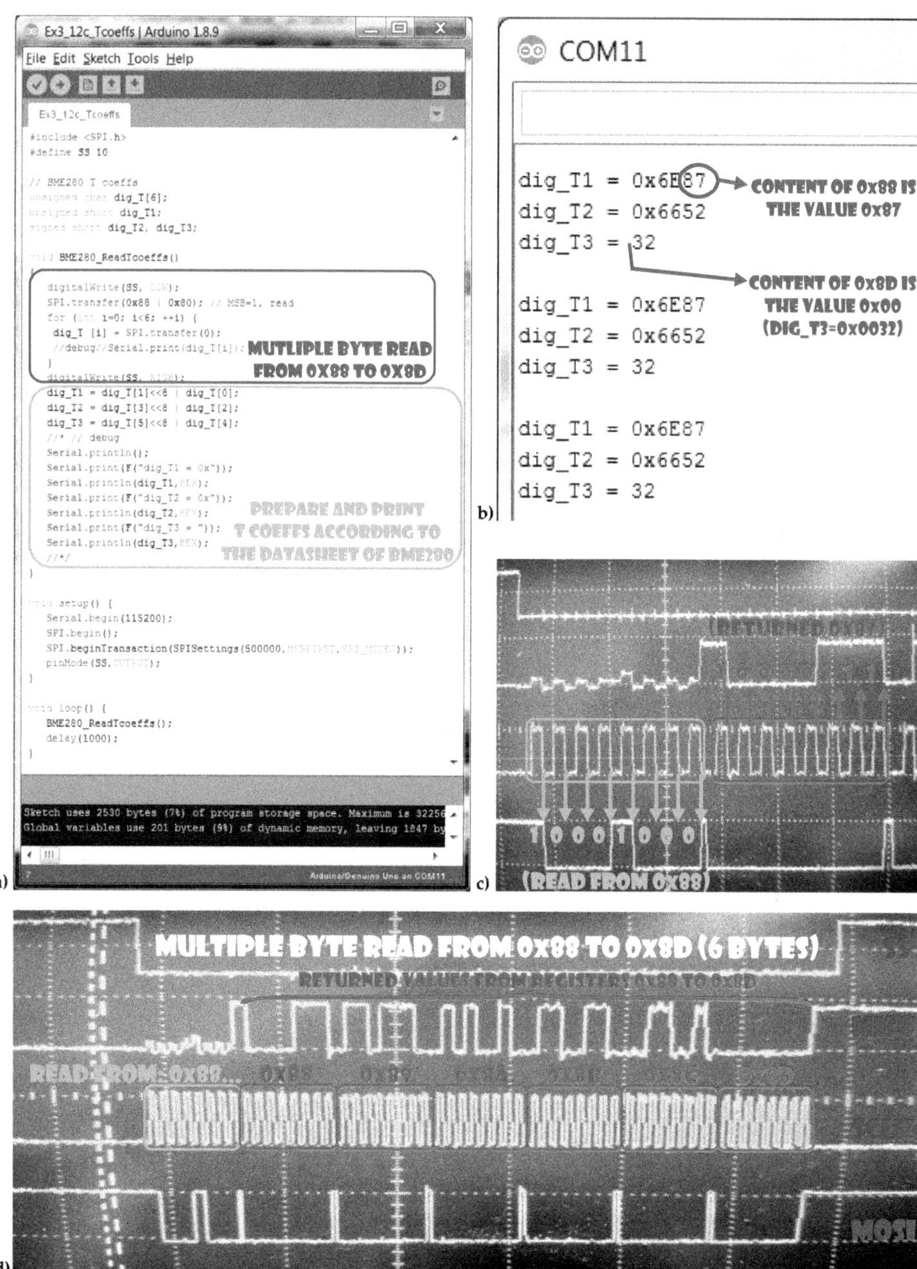

Figure 3.28 SPI multiple-byte read (BME280 temperature coefficients 0x88-0x8D). (a) example firmware, (b) printed results, (c) acquisition of the SPI multiple-byte read timing diagrams with the oscilloscope (1 of 2), (d) acquisition of the SPI multiple-byte read timing diagrams with the oscilloscope (2 of 2). *Source:* Arduino Software.

The *setup()* function is the same as in the previous example, while the *loop()* function performs a call to *BME280_ReadTcoeffs()* function and then delays the code for one second. Figure 3.28c presents the beginning of the SPI multiple-byte read for acquiring the content of 0x88 register (which is the same as the SPI single-byte read). The overall SPI multiple-byte read is presented in Figure 3.28d, which considerably decreases the requisite time for accessing six successive registers (compared to the process of incorporating six successive single-byte reads). The corresponding process for reading six registers via six successive single-byte reads, would require 2xN = 2x6 = 12 transferred bytes over the MOSI signal. The multiple-byte read task decreases the transferred bytes to N+1 = 6+1 = 7 bytes.

Arduino Ex.3.13

The following example prints to the serial port the compensated temperature, pressure, and humidity acquired by the BME280 sensor device, through the built-in SPI of Arduino Uno board. The code example, along with the corresponding flowchart and the printed results, are presented in Figure 3.29. The BME280 registers that are used by the previous and current code examples are presented in Table 3.2.

The code example of Figure 3.29a configures the built-in UART and SPI inside the *setup()* function. Inside the *loop()* function, the code initializes the sensor and obtains the compensation coefficients for temperature, pressure, and humidity. The subsequent *while()* loop first obtains the raw temperature data, and then, it compensates the latter value using the T coeffs. Next, the code prints to the serial port the float value of the compensated temperature, and afterward, it performs the corresponding process for the pressure as well as humidity. The overall process is repeated after a one second delay.

The BME280 functions are incorporated within the header file *__CH3_BME280spi.h,* which is stored inside the *libraries* folder (found in the root directory of the installed Arduino IDE). The corresponding code is given in Figure 3.30, Figure 3.31, and Figure 3.32.

Figure 3.30 presents the declaration of the global variable used by the header file as well the functions *BME280_Init()* and *BME280_ReadTcoeffs()*. The *BME280_Init()* function initializes the sensor through two SPI single-byte write tasks. The first SPI write controls the humidity, while the second controls the pressure and temperature measurements. If the humidity measurement is not required, the corresponding SPI write task can be omitted. However, if all measurements are required by the code, the configuration should be done in that particular order. That is, first the configuration of the humidity, and then the configuration of the pressure/temperature measurements. The *BME280_ReadTcoeffs()* function performs an SPI multiple-byte read of six bytes located at addresses 0x88–0x8D, which hold the coefficients for the conversion of the raw temperature data to a compensated value. Subsequent to the SPI multiple-byte read task, the code arranges the acquired data to the appropriate format recommended by the BME280 datasheet.

Figure 3.31 presents the functions *BME280_ReadPcoeffs()* and *BME280_ReadHcoeffs()* as well as the *BME280_Read_uncompensatedT()*, *BME280_Read_uncompensatedP()*, and *BME280_Read_uncompensatedH()*. The *BME280_ReadPcoeffs()* function performs an SPI multiple-byte read of 18 bytes located at addresses 0x8E–0x9F, which hold the

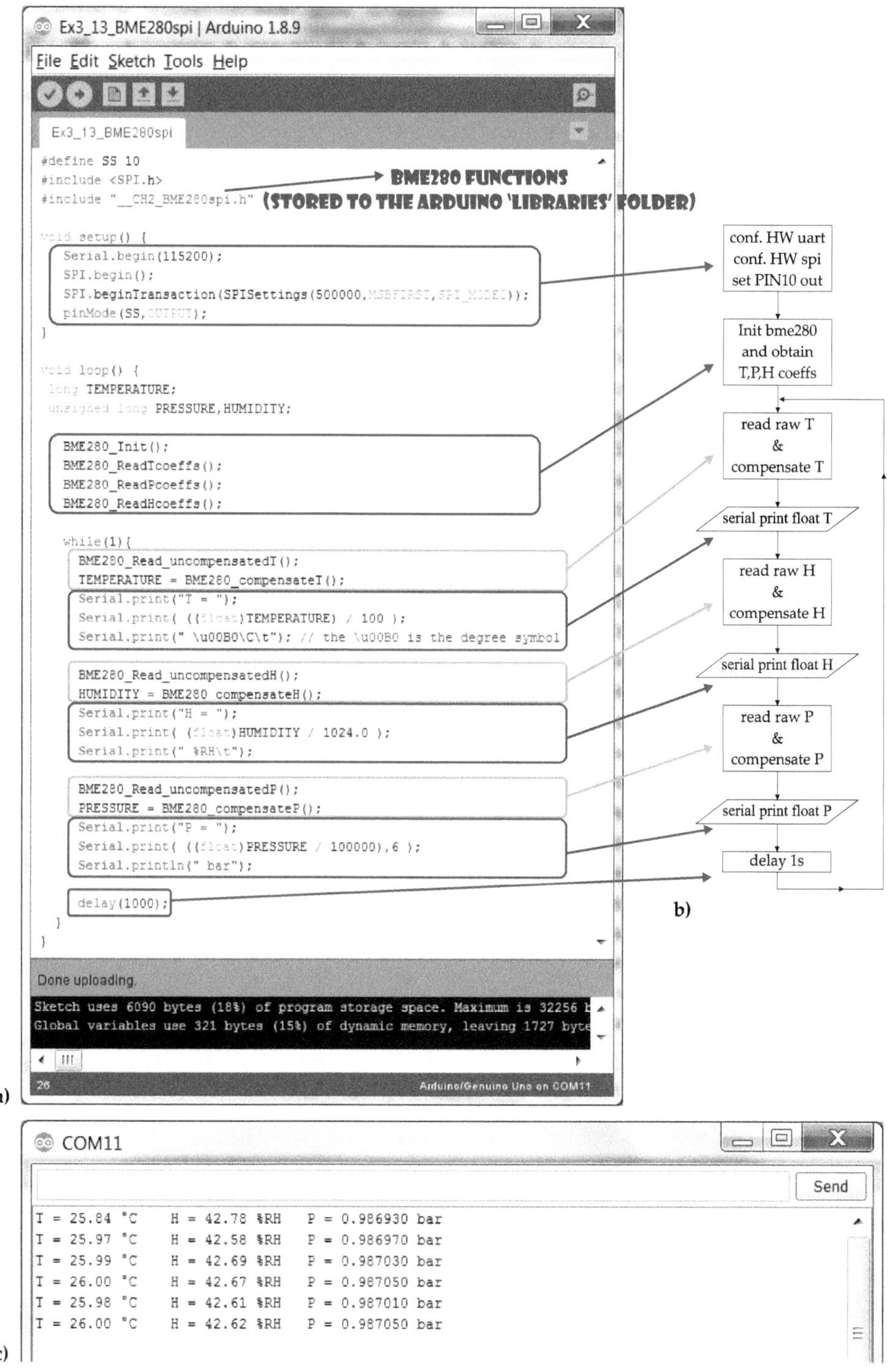

Figure 3.29 BME280: obtain temperature (T), humidity (H), and pressure (P) through the built-in SPI. (a) application firmware, (b) application flowchart, (c) printed results. *Source:* Arduino Software.

Table 3.2 BME280 registers used by the code examples.

Register address	Register content	Data type
0x88/0x89	dig_T1[7..0]/[15..8]	Unsigned short
0x8A/0x8B	dig_T2[7..0]/[15..8]	Signed short
0x8C/0x8D	dig_T3[7..0]/[15..8]	Signed short
0x8E/0x8F	dig_P1[7..0]/[15..8]	Unsigned short
0x90/0x91	dig_P2[7..0]/[15..8]	Signed short
0x92/0x93	dig_P3[7..0]/[15..8]	Signed short
0x94/0x95	dig_P4[7..0]/[15..8]	Signed short
0x96/0x97	dig_P5[7..0]/[15..8]	Signed short
0x98/0x99	dig_P6[7..0]/[15..8]	Signed short
0x9A/0x9B	dig_P7[7..0]/[15..8]	Signed short
0x9C/0x9D	dig_P8[7..0]/[15..8]	Signed short
0x9E/0x9F	dig_P9[7..0]/[15..8]	Signed short
0xA1	dig_H1[7..0]	Unsigned char
0xE1/0xE2	dig_H2[7..0]/[15..8]	Signed short
0xE3	dig_H3[7..0]	Unsigned char
0xE4/0xE5	dig_H4[11..4]/[3..0]	Signed short
0xE5/0xE6	dig_H5[3..0]/[15..4]	Signed short
0xE7	dig_H6[7..0]	Signed char
Register address	**Description**	
0xD0	Chip id [7..0] with permanent value 0x60	
0xF2	**ctrl_hum:** osrs_h[2..0]	
0xF4	**ctrl_meas:** osrs_t[7..5], osrs_p[4..2], mode[1 : 0]	
0xF7	Raw pressure msb [7..0]	
0xF8	Raw pressure lsb [7..0]	
0xF9	Raw pressure xlsb [7..4]	
0xFA	Raw temperature msb [7..0]	
0xFB	Raw temperature lsb [7..0]	
0xFC	Raw temperature xlsb [7..4]	
0xFD	Raw humidity msb [7..0]	
0xFE	Raw humidity lsb [7..0]	

coefficients for the conversion of the raw pressure data to a compensated value. Subsequent to the SPI multiple-byte read task, the code arranges the acquired data to the appropriate format recommended by the (BME280) datasheet. Because the coefficients for humidity are not all found in successive order (in the sensor's memory), the *BME280_ReadHcoeffs()* function first performs an SPI single-byte read of memory

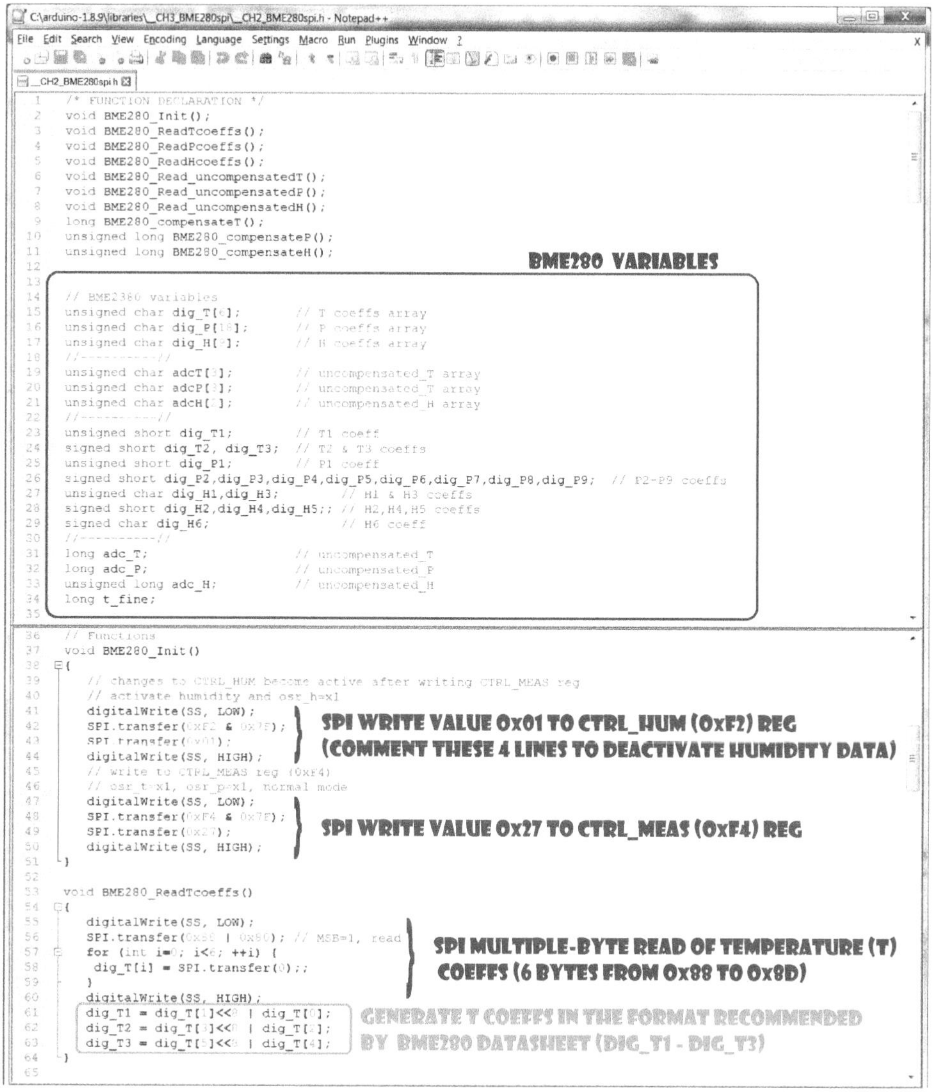

C:\arduino-1.8.9\libraries_CH3_BME280spi_CH2_BME280spi.h - Notepad++

```
/* FUNCTION DECLARATION */
void BME280_Init();
void BME280_ReadTcoeffs();
void BME280_ReadPcoeffs();
void BME280_ReadHcoeffs();
void BME280_Read_uncompensatedT();
void BME280_Read_uncompensatedP();
void BME280_Read_uncompensatedH();
long BME280_compensateT();
unsigned long BME280_compensateP();
unsigned long BME280_compensateH();


// BME2380 variables
unsigned char dig_T[6];        // T coeffs array
unsigned char dig_P[18];       // P coeffs array
unsigned char dig_H[9];        // H coeffs array
//-----------//
unsigned char adcT[3];         // uncompensated_T array
unsigned char adcP[3];         // uncompensated_T array
unsigned char adcH[2];         // uncompensated_H array
//-----------//
unsigned short dig_T1;         // T1 coeff
signed short dig_T2, dig_T3;   // T2 & T3 coeffs
unsigned short dig_P1;         // P1 coeff
signed short dig_P2,dig_P3,dig_P4,dig_P5,dig_P6,dig_P7,dig_P8,dig_P9;  // P2-P9 coeffs
unsigned char dig_H1,dig_H3;          // H1 & H3 coeffs
signed short dig_H2,dig_H4,dig_H5;;   // H2,H4,H5 coeffs
signed char dig_H6;                   // H6 coeff
//-----------//
long adc_T;                    // uncompensated_T
long adc_P;                    // uncompensated_P
unsigned long adc_H;           // uncompensated_H
long t_fine;

// Functions
void BME280_Init()
{
    // changes to CTRL_HUM become active after writing CTRL_MEAS reg
    // activate humidity and osr_h=x1
    digitalWrite(SS, LOW);
    SPI.transfer(0xF2 & 0x7F);
    SPI.transfer(0x01);
    digitalWrite(SS, HIGH);
    // write to CTRL_MEAS reg (0xF4)
    // osr_t=x1, osr_p=x1, normal mode
    digitalWrite(SS, LOW);
    SPI.transfer(0xF4 & 0x7F);
    SPI.transfer(0x27);
    digitalWrite(SS, HIGH);
}

void BME280_ReadTcoeffs()
{
    digitalWrite(SS, LOW);
    SPI.transfer(0x88 | 0x80); // MSB=1, read
    for (int i=0; i<6; ++i) {
     dig_T[i] = SPI.transfer(0);;
    }
    digitalWrite(SS, HIGH);
    dig_T1 = dig_T[1]<<8 | dig_T[0];
    dig_T2 = dig_T[3]<<8 | dig_T[2];
    dig_T3 = dig_T[5]<<8 | dig_T[4];
}
```

BME280 VARIABLES

SPI WRITE VALUE 0x01 TO CTRL_HUM (0xF2) REG (COMMENT THESE 4 LINES TO DEACTIVATE HUMIDITY DATA)

SPI WRITE VALUE 0x27 TO CTRL_MEAS (0xF4) REG

SPI MULTIPLE-BYTE READ OF TEMPERATURE (T) COEFFS (6 BYTES FROM 0x88 TO 0x8D)

GENERATE T COEFFS IN THE FORMAT RECOMMENDED BY BME280 DATASHEET (DIG_T1 - DIG_T3)

Figure 3.30 BME280 SPI functions (1 of 3). *Source:* Arduino Software.

address 0xA1, and then, it performs an SPI multiple-byte read of seven bytes located at addresses 0xE1–0xE7. The overall eight memory addresses hold the coefficients for the conversion of the raw humidity data to a compensated value, and hence, subsequent to the SPI multiple-byte read task the code arranges the acquired data to the appropriate format recommended by the datasheet.

The subsequent three functions that read the uncompensated temperature (T), pressure (P), and humidity (H) data, perform SPI multiple-byte read tasks of the memory locations 0xFA–0xFC, 0xF7–0xF9, and 0xFD–0xFE, respectively. Subsequent to each SPI read task,

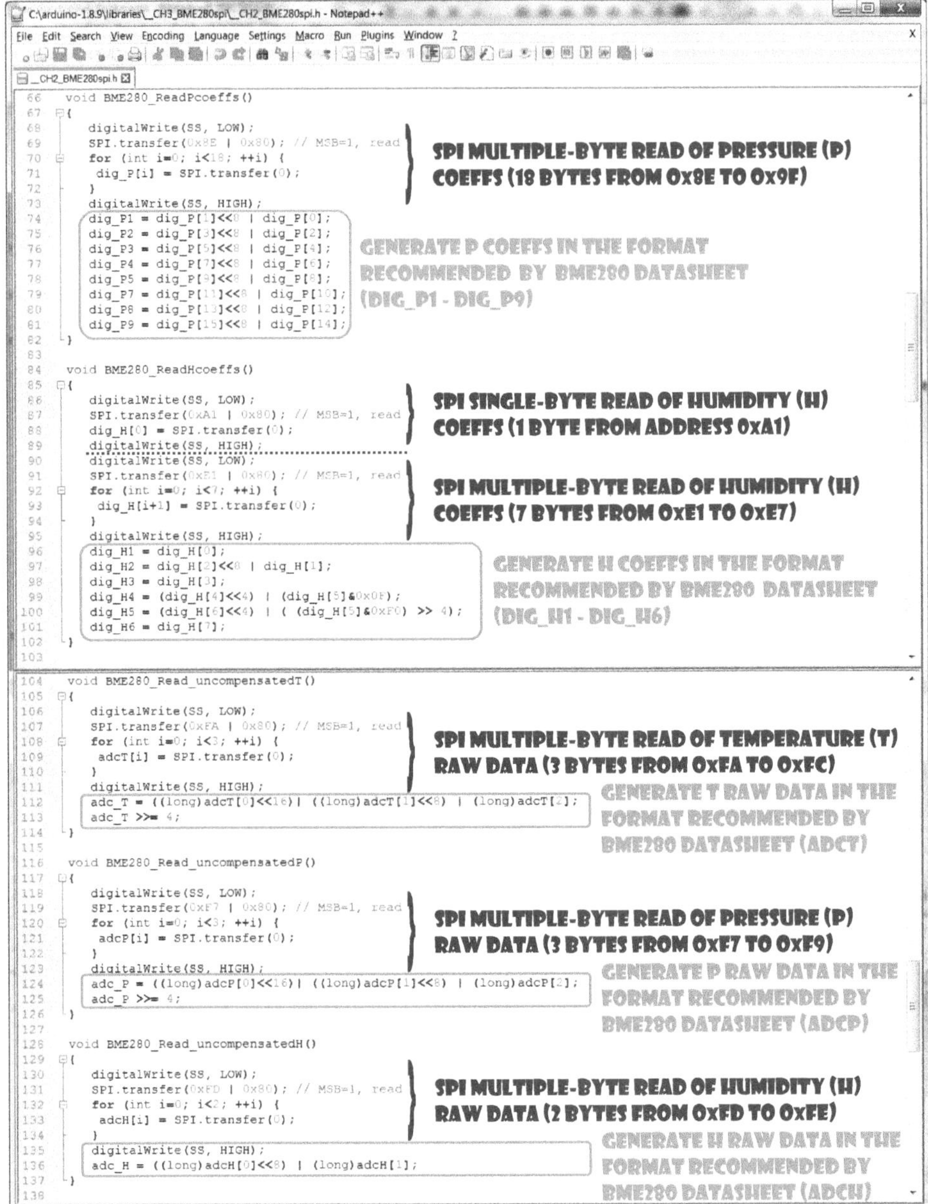

Figure 3.31 BME280 SPI functions (2 of 3). *Source:* Arduino Software.

the raw data are arranged to the appropriate format recommended by the datasheet and assigned to the variables adc_T, adc_P, and adc_H, respectively. Finally, Figure 3.32 presents the functions *BME280_compensateT()*, *BME280_compensatedP()*, and *BME280_compensatedH()*, which hold the appropriate arithmetic operations for the calculation of the compensated value of temperature, pressure, and humidity, respectively. For more

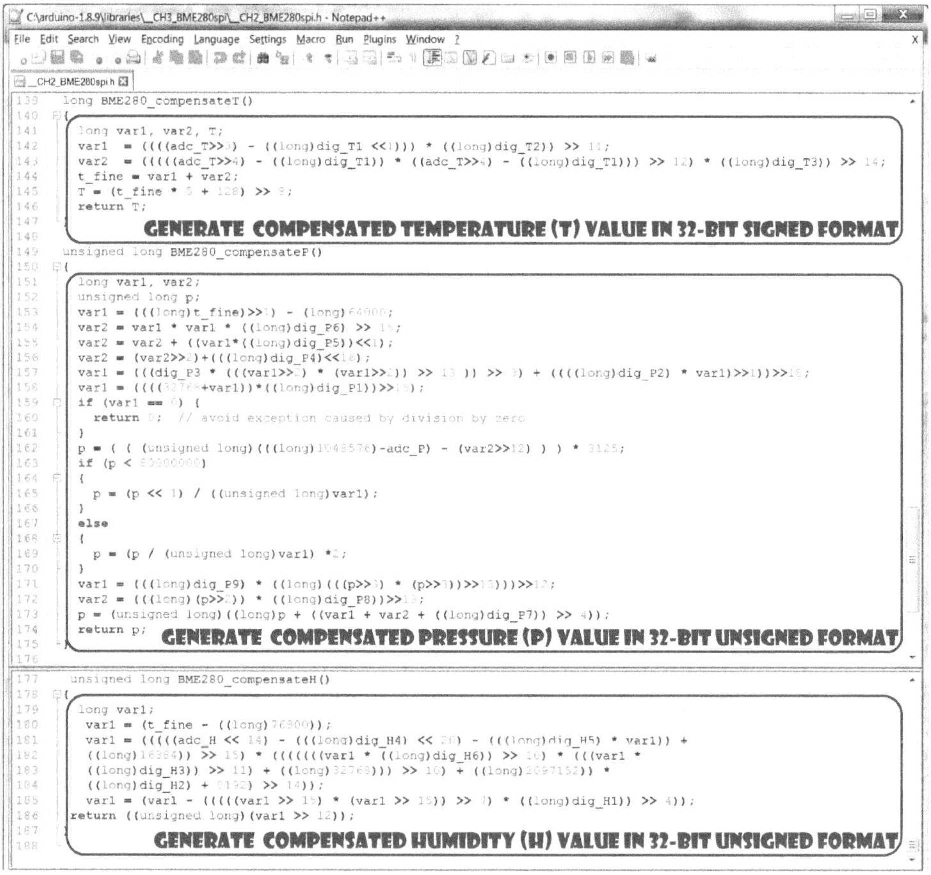

```
long BME280_compensateT()
{
  long var1, var2, T;
  var1  = ((((adc_T>>3) - ((long)dig_T1 <<1))) * ((long)dig_T2)) >> 11;
  var2  = (((((adc_T>>4) - ((long)dig_T1)) * ((adc_T>>4) - ((long)dig_T1))) >> 12) * ((long)dig_T3)) >> 14;
  t_fine = var1 + var2;
  T = (t_fine * 5 + 128) >> 8;
  return T;
}
// GENERATE COMPENSATED TEMPERATURE (T) VALUE IN 32-BIT SIGNED FORMAT

unsigned long BME280_compensateP()
{
  long var1, var2;
  unsigned long p;
  var1 = (((long)t_fine)>>1) - (long)64000;
  var2 = var1 * var1 * ((long)dig_P6) >> 15;
  var2 = var2 + ((var1*((long)dig_P5))<<1);
  var2 = (var2>>2)+(((long)dig_P4)<<16);
  var1 = (((dig_P3 * (((var1>>2) * (var1>>2)) >> 13 )) >> 3) + ((((long)dig_P2) * var1)>>1))>>18;
  var1 = ((((32768+var1))*((long)dig_P1))>>15);
  if (var1 == 0) {
    return 0;  // avoid exception caused by division by zero
  }
  p = ( ( (unsigned long)(((long)1048576)-adc_P) - (var2>>12) ) ) * 3125;
  if (p < 0x80000000)
  {
    p = (p << 1) / ((unsigned long)var1);
  }
  else
  {
    p = (p / (unsigned long)var1) *2;
  }
  var1 = (((long)dig_P9) * ((long)(((p>>3) * (p>>3))>>13)))>>12;
  var2 = (((long)(p>>2)) * ((long)dig_P8))>>13;
  p = (unsigned long)((long)p + ((var1 + var2 + ((long)dig_P7)) >> 4));
  return p;
}
// GENERATE COMPENSATED PRESSURE (P) VALUE IN 32-BIT UNSIGNED FORMAT

unsigned long BME280_compensateH()
{
  long var1;
   var1 = (t_fine - ((long)76800));
   var1 = (((((adc_H << 14) - (((long)dig_H4) << 20) - (((long)dig_H5) * var1)) +
   ((long)16384)) >> 15) * (((((((var1 * ((long)dig_H6)) >> 10) * (((var1 *
   ((long)dig_H3)) >> 11) + ((long)32768))) >> 10) + ((long)2097152)) *
   ((long)dig_H2) + 8192) >> 14));
   var1 = (var1 - (((((var1 >> 15) * (var1 >> 15)) >> 7) * ((long)dig_H1)) >> 4));
 return ((unsigned long)(var1 >> 12));
}
// GENERATE COMPENSATED HUMIDITY (H) VALUE IN 32-BIT UNSIGNED FORMAT
```

Figure 3.32 BME280 SPI functions (3 of 3). *Source:* Arduino Software.

information about the arithmetic tasks, the reader may refer to the sensor's datasheet as well as the example code, available from: https://github.com/BoschSensortec/BME280_driver/blob/master/bme280.c.

Arduino Ex.3.14

The aforementioned examples apply to the built-in SPI protocol of the μC device employed by the Arduino Uno board. While there are several shareable libraries over the internet, which can be used for the implementation of a software-implemented SPI for Arduino, we herein explore the development of a custom-designed SPI library based on the SPI timing diagrams that were explored before. Such a library is particular useful when there is a need to control SPI peripheral devices from a μC of no built-in SPI hardware, or perhaps to use SPI pins other than the dedicated pins available by the μC device.

Arduino libraries are regularly written in C++ and there are basically three different files for each library. That is, (i) the *header (.h)* file that holds the definitions of the library,

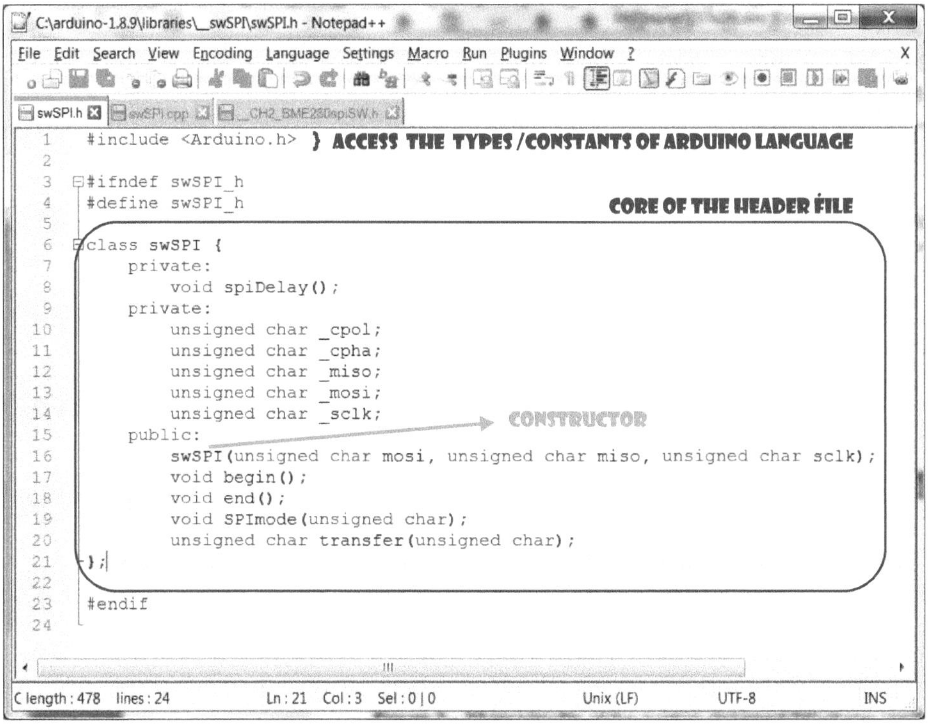

```
#include <Arduino.h>   } ACCESS THE TYPES/CONSTANTS OF ARDUINO LANGUAGE

#ifndef swSPI_h
#define swSPI_h                                   CORE OF THE HEADER FILE

class swSPI {
    private:
        void spiDelay();
    private:
        unsigned char _cpol;
        unsigned char _cpha;
        unsigned char _miso;
        unsigned char _mosi;
        unsigned char _sclk;                  CONSTRUCTOR
    public:
        swSPI(unsigned char mosi, unsigned char miso, unsigned char sclk);
        void begin();
        void end();
        void SPImode(unsigned char);
        unsigned char transfer(unsigned char);
};

#endif
```

Figure 3.33 Header file of the custom-designed SPI library (*swSPI.h*). *Source:* Arduino Software.

(ii) the *source code (.cpp)* file that holds the actual Arduino code, and (iii) a *text (.txt)* file of the precise name *keywords.txt* that notifies the Arduino IDE which of the user-defined *keywords* to highlight during the code-writing process. For the custom-designed SPI library we use the file names *swSPI.h* and *swSPI.cpp.*

Figure 3.33 presents the code of the *swSPI.h* header file. The core of the file is a *class* (as defined by the corresponding keyword), which incorporates the definition of the library's functions and variables. The *public* and *private* keywords determine which functions/variables can be accessed either by the user (i.e. *public*), or only by the class itself (i.e. *private*). For instance, the *spiDelay()* function cannot be invoked by the application code. Rather, it is only accessible by the *swSPI.cpp* source code of the library.

The *swSPI* class incorporates the private function *spiDelay()* and the private variables *_cpol*, *_cpha*, *_miso*, *_mosi*, *_sclk* of *unsigned char* type. The *underscore (_)* to the start of each name is a regular approach to indicate private variables but also distinguish variables from function arguments. Subsequent to the variables is the C++ *constructor* of the *swSPI* class. Each class has a special function called *constructor*, which has the same name as the class, it is always *public*, it may admit arguments, but has no return value. The *constructor* creates an instance of the class (allowing the user to pass the desired arguments to the *constructor* function), and it is automatically called whenever a new object of the class is created (allowing the class to initialize its member variables). Subsequent to the C++ constructor are the (accessible by the user) public functions of the *swSPI* class. That is the *begin()*, *end()*, *SPImode()*, and *transfer()* functions.

The foremost code line of *swSPI.h* header file includes the *Arduino.h,* which provides access to the standard Arduino functions. It should be noted that the class definition is wrapped around a weird looking construct, within the following preprocessor directives:

```
#ifndef swSPI_h
#define swSPI_h
        "CLASS definition"
#endif
```

This syntax prevents problems in case the user accidentally includes the library twice.

The source code of the *swSPI.cpp* file is given in Figures 3.34 and 3.35. Immediately after the inclusion of the *swSPI.h* file at the top of Figure 3.34, we place the *constructor* of the

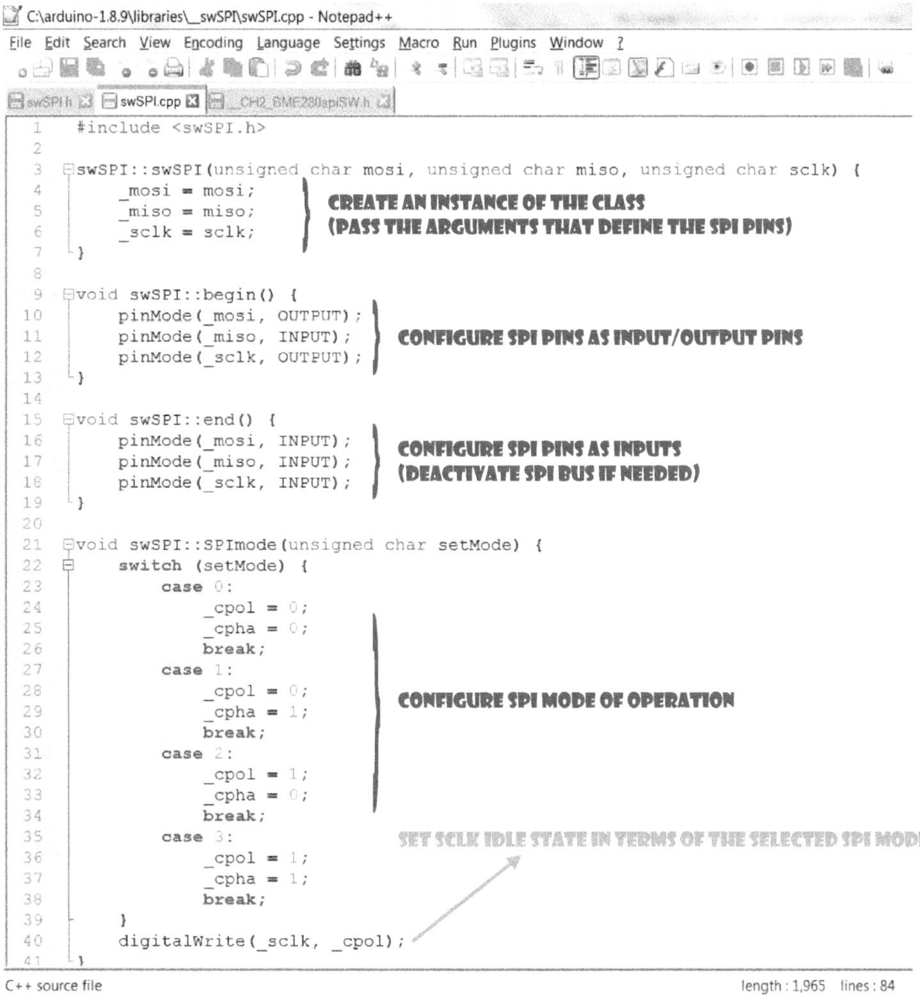

C:\arduino-1.8.9\libraries_swSPI\swSPI.cpp - Notepad++

File Edit Search View Encoding Language Settings Macro Run Plugins Window ?

swSPI.h | swSPI.cpp | _CH2_BME280spiSW.h

```
#include <swSPI.h>

swSPI::swSPI(unsigned char mosi, unsigned char miso, unsigned char sclk) {
      _mosi = mosi;
      _miso = miso;
      _sclk = sclk;
}

void swSPI::begin() {
      pinMode(_mosi, OUTPUT);
      pinMode(_miso, INPUT);
      pinMode(_sclk, OUTPUT);
}

void swSPI::end() {
      pinMode(_mosi, INPUT);
      pinMode(_miso, INPUT);
      pinMode(_sclk, INPUT);
}

void swSPI::SPImode(unsigned char setMode) {
      switch (setMode) {
            case 0:
                  _cpol = 0;
                  _cpha = 0;
                  break;
            case 1:
                  _cpol = 0;
                  _cpha = 1;
                  break;
            case 2:
                  _cpol = 1;
                  _cpha = 0;
                  break;
            case 3:
                  _cpol = 1;
                  _cpha = 1;
                  break;
      }
      digitalWrite(_sclk, _cpol);
}
```

CREATE AN INSTANCE OF THE CLASS (PASS THE ARGUMENTS THAT DEFINE THE SPI PINS)

CONFIGURE SPI PINS AS INPUT/OUTPUT PINS

CONFIGURE SPI PINS AS INPUTS (DEACTIVATE SPI BUS IF NEEDED)

CONFIGURE SPI MODE OF OPERATION

SET SCLK IDLE STATE IN TERMS OF THE SELECTED SPI MODE

C++ source file   length : 1,965   lines : 84

Figure 3.34 Source code file of the custom-designed SPI library (*swSPI.cpp* – 1 of 2). *Source:* Arduino Software.

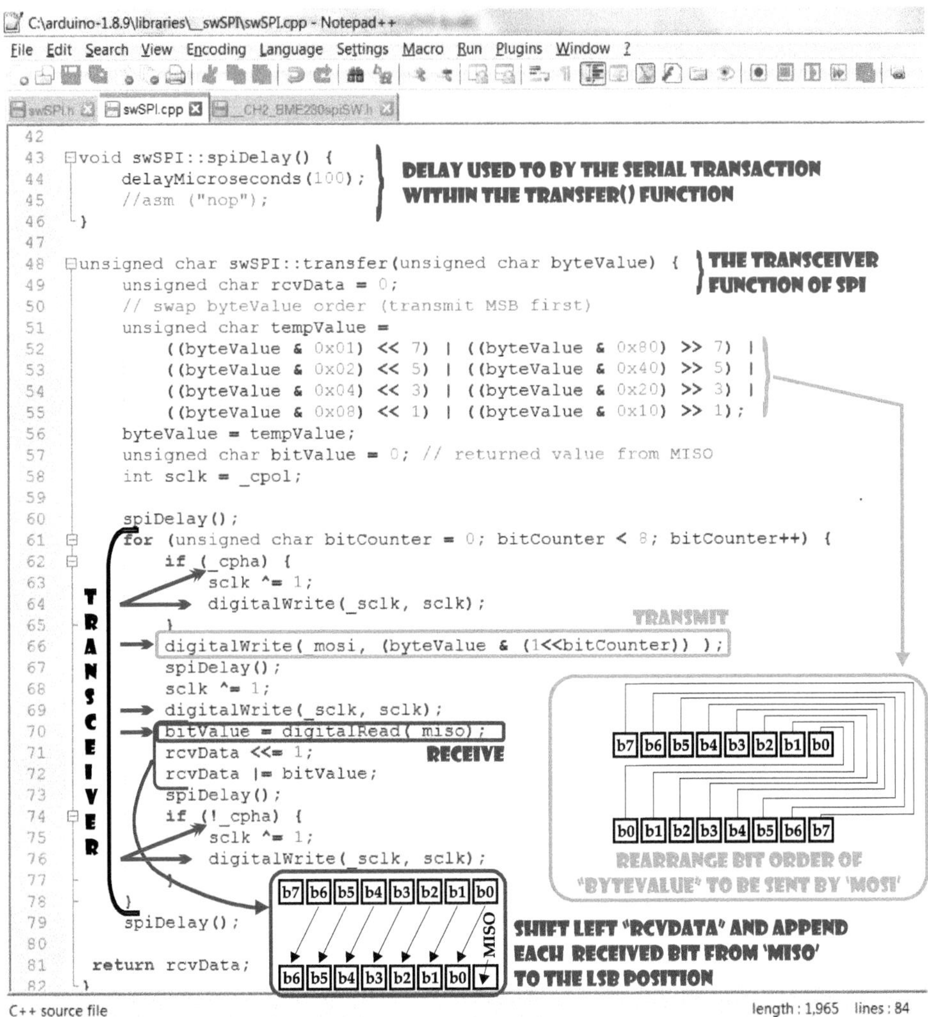

Figure 3.35 Source code file of the custom-designed SPI library (*swSPI.cpp* – 2 of 2). *Source:* Arduino Software.

class. When the user creates an instance of the class (within the application code) through that constructor, he/she defines the desired pins that control the software-implemented SPI bus (which are controlled by the corresponding private variables of the *swSPI* class). Next in the code, the *begin()* function configures MOSI and SCLK as output pins, and MISO as input pin. The *end()* function is optionally used to configure the SPI pins as inputs in case the user desires to deactivate the SPI bus (somewhere within the application code). The subsequent *SPImode()* function admits arguments 0,1,2 or 3 and sets the corresponding SPI mode of operation. The *digitalWrite()* command at the end of the *SPImode()* function configures the *idle* state of the SCLK signal in accordance with the SPI mode of operation selected by the application code. It is here noted that the *_cpol* and *_cpha* private variables are exploited later by the *transfer()* function that implements the SPI transceiver (in accordance with the SPI R/W timing diagram that were explored by Figure 3.24).

The source code continues to the Figure 3.35, where the *spiDelay()* function generates a delay of 100 μs. The latter delay is used by the serial transceiver, which controls the SPI pins. It is here noted that if there is a need to configure a software-implemented SPI bus of higher speed, the user may comment the Arduino *delayMicroseconds()* function and uncomment the subsequent code line. The *asm* directive of the latter code line forces the compiler to incorporate the assembly mnemonic *nop* into the C++ code. The *nop* mnemonic is the smallest possible delay that can be executed by a microcontroller device, which is normally executed within a single clock cycle and performs *no operation*.

The concluding the *transfer()* function describes the transceiver tasks of the software-implemented SPI bus. The function admits a single argument that is the byte to be transferred to the slave device. To facilitate the serial byte transfer, which is configured to start from the MSB, the bit order of the argument is rearranged via the *tempValue* variable (as depicted by the wide green frame of Figure 3.35). Next, the *transfer()* function performs a call to the private *spiDelay()* function, and then goes the *for()* loop, which is repeated eight successive times.

At first, the *for()* loop explores the value of *_cpha* private variable and if *_cpha = 1*, then the code toggles the state of the SCLK pin through a bitwise XOR and a subsequent execution of the *digitalWrite()* function. Otherwise, the state of the SCLK pin remains unchanged and the code transmits the first bit (i.e. the MSB) of the function argument (i.e. of the *byteValue*). The latter decision is accomplished through a *shift left (<<)* operation of a logical '1' (where the number of the shifts is determined by the current value of the *loop counter*), and then a *bitwise AND (&)* of the latter value to the *byteValue* argument. The bitwise AND operation forces all bits of the *byteValue* variable to be cleared, except the bit denoted by the current value of the *loop counter* (that is, the *bitCounter*). Hence, the code writes to the MOSI pin a logical '1' or '0' according to the current bit of the *byteValue* argument, which has not be forced to logical '0' by the bitwise AND operation. For instance, if *bitCounter = 5* the code writes to the MOSI pin the value of the *byteValue[5]* and if *byteValue[5] = 1*, the MOSI is set to logical '1'; otherwise, the MOSI is set to logical '0'. The MOSI state is changed through the *digitalWrite()* function within the narrow green frame (denoted TRANSMIT).

After forcing the MOSI signal to a state identical to the current bit of the *byteValue* argument, which is serially transmitted to the slave device, the *for()* loop performs a call to the private *spiDelay()*. Then, it toggles the state of the SCLK pin through a bitwise XOR and a subsequent execution of the *digitalWrite()* function. The next command (within the narrow blue frame denoted RECEIVE) reads the state of the MISO signal and assigns the result to the *bitValue* variable. The next two code lines first perform a *shift left (<<)* task to the content of *rcvData* variable, and then append to the LSB of the latter variable the state of the current MISO signal (held by the *bitValue* variable). Hence, the content of the *rcvData* variable after eight successive repetitions of the *for* loop will hold the byte received by the slave device. After obtaining the current state of MISO signal, the *for* loop calls the private *spiDelay()* function. Then, it explores the value of *_cpha* private variable and if *_cpha = 0*, the code toggles the state of the SCLK pin through a bitwise XOR and a subsequent execution of the *digitalWrite()* function.

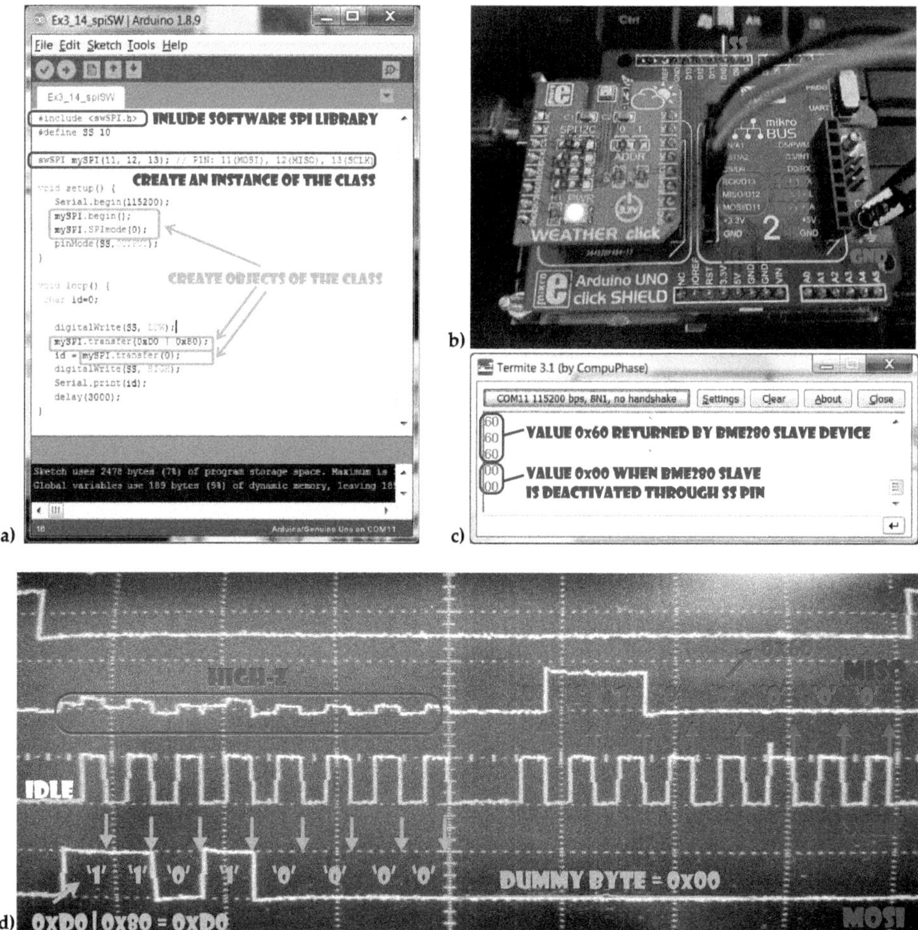

Figure 3.36 *swSPI* single-byte read of the *chip id* of BME280 sensor device. (a) example firmware: software-implemented SPI, (b) capture the SPI pins with an oscilloscope (the software-implemented SPI pins are intentionally assigned to the location where the built-in SPI pins are originally traced), (c) printed results in Termite console, (d) acquisition of the software-based SPI single-byte read timing diagrams with the oscilloscope. *Source:* Arduino Software.

Immediately after the *for* loop, the *transfer()* function calls the *spiDelay()* in order to shape the proper SPI timing diagrams. Finally, the *transfer()* function returns *rcvData, a* byte-size variable, which holds the data that were serially acquired by the slave device.

Figure 3.36 presents a single-byte read of the *chip id* of BME280 sensor (that is the value 0x60), which is stored to the register 0xD0. The serial transaction is accomplished with the custom-designed SPI library, while the current example is identical to the code example Ex.3.12 that was implemented with the built-in hardware SPI. The differences of the custom-designed SPI library are highlighted in the source code of Figure 3.36a. In detail, the foremost code line within the red frame includes the custom-designed

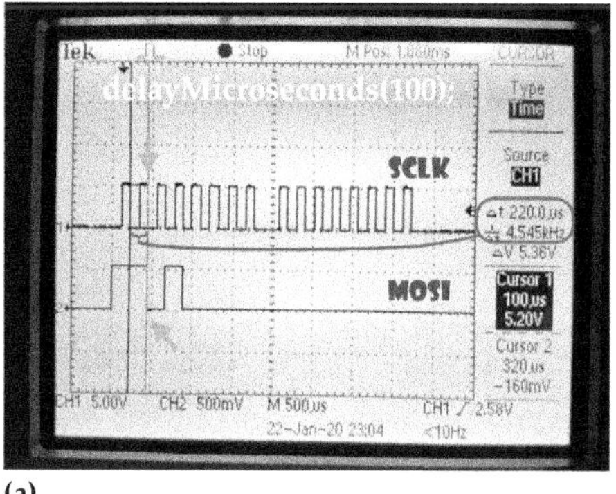

(a)

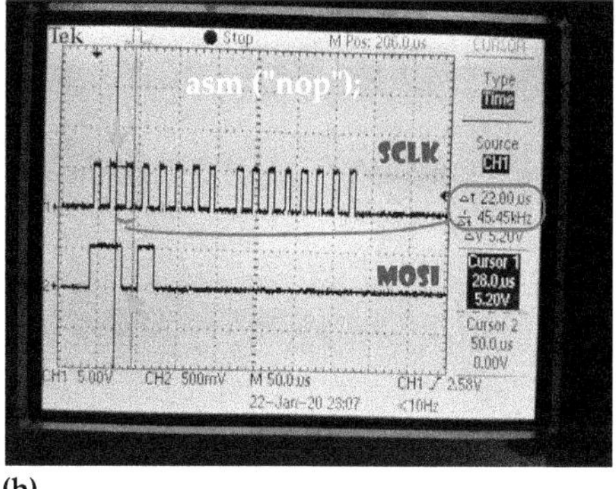

(b)

Figure 3.37 SCLK period of the swSPI library. (a) SPI clock period using 100us delay, (b) SPI clock period using *nop* assembly mnemonic (minimum delay).

swSPI.h library. Then, the code line within the blue frame creates an instance of the class, named *mySPI*, and determines the arguments of the *constructor* function, that is, the MOSI, MISO, and SCLK pins of the software-implemented SPI library. To support hardware compatibility of the current code with the previous examples (of the built-in hardware SPI), the pin location is the same as before. However, the MOSI, MISO, and SCLK pins are currently controlled by the *swSPI.h* library instead of the hardware SPI. Finally, the green frames of the code denote the establishment of the *objects* of *mySPI* class.

Figure 3.37 illustrates the SLCK period of the software SPI library when using a 100 μs delay (Figure 3.37a), or the minimum possible delay achieved with the *nop* assembly mnemonic (Figure 3.37b). In the former case the SPI clock frequency is approximately 4.5KHz, while in the latter case the frequency is approximately 45KHz. It is worth noting that the short delay among the transitions of the SPI pins (as accomplished by the *nop* mnemonic) renders noticeable the transition between the SCLK and MOSI signals of Figure 3.37b (as denoted by the green arrows). On the other hand, the 100 μs delay is much longer than the small delay between the successive transitions of the SCLK and MOSI signals, and hence, the state of the SCLK and MOSI signal appears as being instantaneous (as denoted by the green arrows of Figure 3.37a).

Arduino Ex.3.15

The example below is identical to the Ex.3.13, which obtains temperature, humidity, and pressure measurements from BME280 sensor, except that the current example exploits the custom-designed SPI library (instead of the hardware SPI). The red frames of Figure 3.38a illustrate the differences from the original example (of the built-in hardware SPI). To be able to include the exact same library of BME280 sensor (i.e. *__CH3_BME280spi.h*), it is important to substitute the *SPI* keyword with *mySPI* (which is the name that was used to create the instance of *swSPI* class). The printed results of the current example are given in Figure 3.38b.

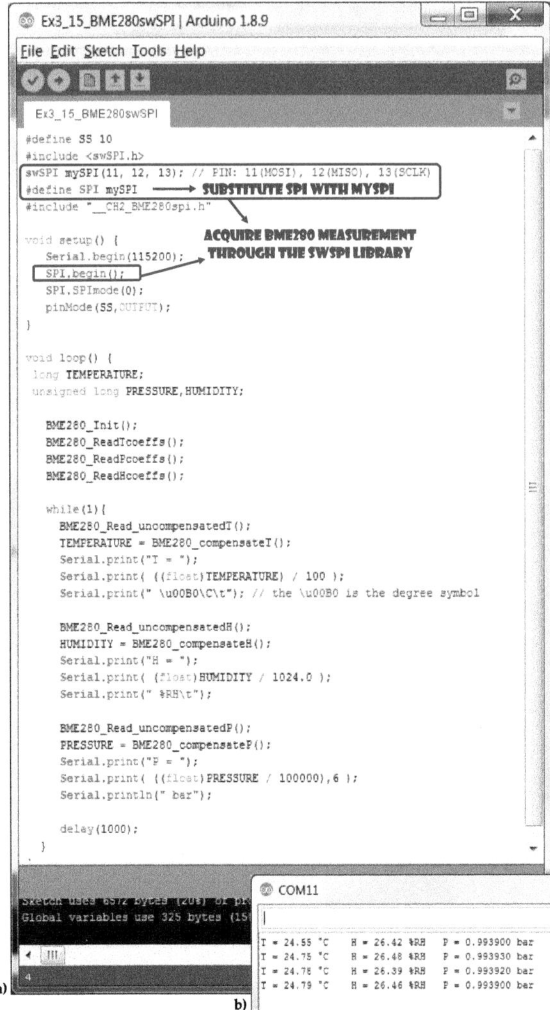

Figure 3.38 BME280: obtain temperature, humidity, and pressure through the *swSPI* library. (a) application firmware, (b) printed results. *Source:* Arduino Software.

I2C Serial Interface

The I2C serial interface constitutes the most common communication between a μC and the peripheral sensor modules. The hardware interface is performed merely by two single-ended signals, regularly referred to as I2C, or, *I2C clock signal (SCL)* and *serial data (SDA)*. The SDA line is bidirectional and, hence, it minimizes the hardware signals required for the communication between the μC and the incorporated I2C modules. The I2C relies on a synchronous type of serial communication (i.e. the I2C devices share a common synchronization clock) in half-duplex mode (i.e. only one device is set to transmit/receive data over each transaction), using master-slave architecture.

Figure 3.39a presents the hardware setup between the μC and an I2C module, while Figure 3.39b presents the hardware setup between the master μC and two (or more) slave modules. Each I2C slave device features an identical address that is used to identify and respond to the master's call. Thereby, each slave connected to the bus should feature a unique address. The slave device regularly features a 7-bit address that allows us to attach up to $2^7 = 128$ different slave devices on the same I2C bus. Because the I2C drivers consist of open-drain type of outputs, they are able to pull the signal lines *LOW* but cannot drive them *HIGH*. Thereby, the pull-up resistors on the SCL and SDA signals assert them to a *HIGH* state, when the incorporated devices let the lines float. This kind of physical layer eliminates conflicts and potential damages of the I2C drivers when the associated devices attempt to drive the bus simultaneously.

Figure 3.39c presents the most common timing diagram of a successfully accomplished write transaction to a single slave register. The serial data in blue color illustrate the information sent by the master device, while the data in green color depict the slave's response. In detail, the master initiates the so-called *START* condition by pulling the SDA line *LOW*, while the SCL line is held *HIGH*.[12] Then, the master transmits three serial bytes starting from the MSB, while the new serial data on the SDA line appear on falling edge of the SCL signal. At the end of each transmitted byte, the master releases the SDA line and waits the slave's response in order to confirm the successful reception of the transmitted information by the slave device.

In regard to the write transaction of the slave register, the first byte incorporates the 7-bit slave address along with the LSB = 0 in order to denote a *write* condition. The master then releases the SDA line and its control is undertaken by the slave device. If the slave has received the data sent by the master, it asserts the SDA line *LOW* in order to *acknowledge* (*ACK*) a successfully reception. Otherwise, the slave sends *no-acknowledge* (*NACK*) by releasing the SDA line, and in that case, the master should terminate the transmission. The second byte incorporates the register address of the slave that will be modified, followed by an ACK bit sent by the slave. The third byte incorporates the new data that will update the register content, also followed by an ACK bit.

Every I2C transaction terminates with the so-called STOP condition sent by the master device. In detail, the master releases the SCL line and then performs a LOW to HIGH transition to the SDA signal (i.e. it releases the SDA line). It is worth mentioning that most standard frequency of the I2C clock is 100KHz.

Figure 3.39d presents the most common timing diagram of a successfully accomplished read transaction from a single slave register. The two first bytes are the same as in the write transaction. In detail, the master initiates the transaction by sending a *START* condition and then it transmits the 7-bit slave address appended with the LSB = 0 (i.e. preparation of a *write* condition). After receiving an ACK bit from the slave device, the master sends the register address. After receiving again an ACK bit from the slave device, the master initiates a *STOP* condition, and subsequently a *START* condition. It is here noted that the master may initiate a *Repeated START* condition instead of the consecutives *STOP* and *START* conditions. The *Repeated START* condition occurs in case the SDA line is asserted to *HIGH*

12 It is here noted that both lines are kept floated (that is, *HIGH* state) as long as there is no transaction on the I2C bus.

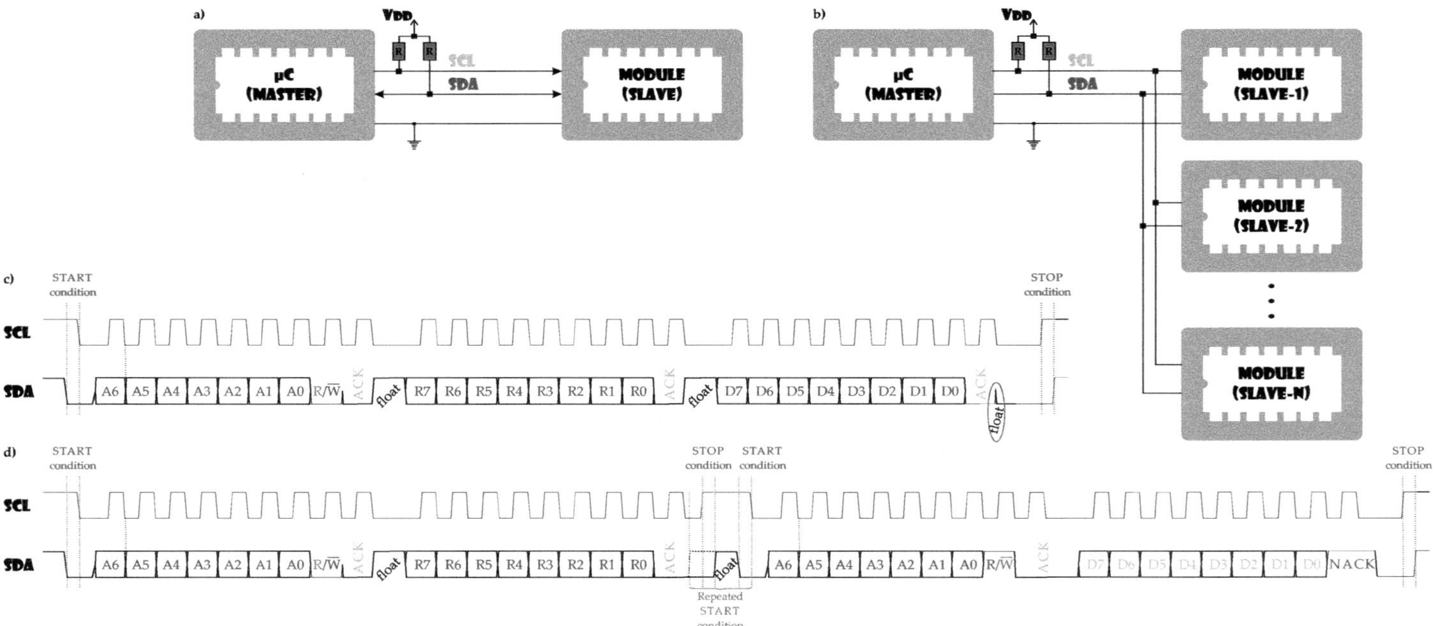

Figure 3.39 I2C hardware topology and read/write timing diagrams. (a) connection diagram between the μC and a single I2C device, (b) connection diagram between the μC and multiple I2C devices, (c) timing diagrams of a successful I2C write transaction to a single slave register, (d) timing diagrams of a successful I2C read transaction to a single slave register.

state while the SCL signal is kept *LOW* (as presented by the dash line of Figure 3.39d, which illustrates the appropriate timing for setting SDA signal to *HIGH* state). However, the examples presented hereafter apply to the consecutives STOP and START conditions. The subsequent two bytes (after the second START condition) are as follows: (i) The master transmits the 7-bit slave address appended with an LSB = 1 (i.e. preparation of a *READ* condition); (ii) After receiving an ACK bit from the slave device, the master acquires the register's 8-bit content sent by the slave device; (iii) Finally, the master terminates the transaction by sending a NACK bit to the slave immediately before the STOP condition.

The latter curious transmission of the NACK bit is performed because of the possibility of reading multiple and consecutive registers from a slave device, through a single transaction only. To do so, the master should successively obtain an ACK bit after each reception of the content of the auto-incremented register location, and after reading the concluding register, the master should terminate the multiple-byte read process by sending NACK bit to the slave device. To perform multiple-byte write/read transactions, the designer should consider the datasheet of the employed I2C slave device.

Arduino Ex.3.16

The following examples apply to the built-in I2C of the μC device employed by Arduino Uno board. Later in the chapter we will additionally explore the development of a custom-designed I2C library based on the aforementioned timing diagrams. As before, the examples apply to the BME280 sensor, but using an I2C control of the device instead of SPI control. Figure 3.40a presents the pinout of the build-in I2C on Arduino Uno board, while Figure 3.40b presents the corresponding pinout on the *Arduino Uno click shield* which holds the BME280 sensor. The latter figure illustrates the three short-circuits on the *Weather click* that configure the I2C control of the sensor device.

a)

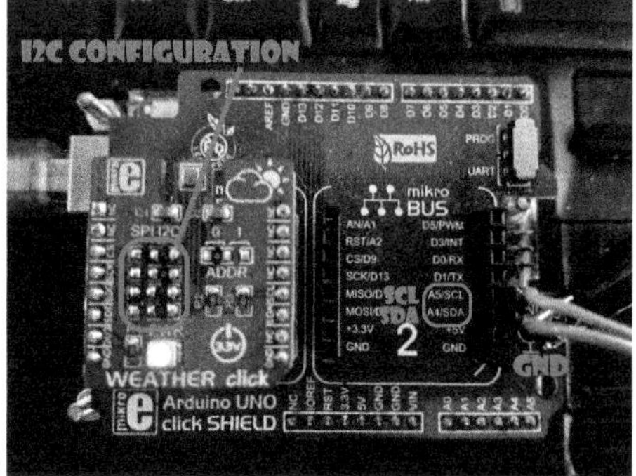

b)

Figure 3.40 I2C hardware interface (Arduino Uno and BME280/click shield pinout). (a) I2C pin arrangement on the Arduino Uno board, (b) hardware setup: I2C configuration of the *Weather Click* module.

The following code example obtains the *chip id* of BME280 sensor through the built-in I2C interface (Figure 3.41b) using a procedure, equivalent to previous one that was based on the built-in SPI interface (Figure 3.41a). In regard to Figure 3.41b, the first two code lines incorporate the Arduino I2C library and define the slave address of the *Weather click* (holding the BME280 sensor), that is, the hexadecimal 0x76 I2C address. The *Wire.begin()* function inside the *setup()* initiates the Arduino I2C library (i.e. the Wire library). The *chip id* of BME280 sensor is acquired inside the *loop()* through five distinctive functions of the Wire library, where the hexadecimal value of the *chip id* (that is, the 0x60) is repeatedly transmitted to the serial port every three seconds (Figure 3.41c).

In consideration of the timing diagrams given in Figure 3.41d, the *Wire.beginTransmission()* initiates an *I2C START* condition and then transmits the 7-bit slave address (that is, the 0x76) appended with a logical '0' in the LSB position of the transmitted byte (i.e. denoting a *write* condition). After validating the ACK bit (i.e. logical '0') sent by the slave, the *Wire.write()* transmits the second byte, which incorporates the register address (i.e. 0xD0) of the slave device that should be read. After validating the second ACK bit sent by the slave, the *Wire.endTransmission()* initiates an *I2C STOP* condition. Next, the *Wire.requestFrom()* resends the 7-bit slave address, which is now appended with a logical '1' in the LSB position of the transmitted byte (i.e. denoting a *read* condition). It should be noted that the second argument of the *Wire.requestFrom()* function determines the number of bytes to request from the slave device. After validating the third ACK bit (i.e. a logical '0') sent by the slave, the *Wire.read()* obtains the content of the 0xD0 register address (that is the hex value 0x60), and sends a NACK bit (i.e. logical '1') to the slave so as to prevent a multiple read process. Finally, it initiates the concluding *I2C STOP* condition. It is worth mentioning that the default operating frequency of the built-in I2C is 100KHz.

An I2C single-byte write and read process is given in the example of Figure 3.42. The code performs a write and then a read process of the register address 0xF4 (i.e. the *"ctrl_meas"* register) for setting the pressure and temperature data acquisition options. The code example is equivalent to the previous one that addressed SPI control over the BME280 sensor (Figure 3.42a), but currently the control is passed to the I2C interface (Figure 3.42b). The code repeatedly writes the value 0x27 to the register 0xF4 and then read the content of the latter register address. The latter value is sent to the serial port and the printed results are given in Figure 3.42c.

The I2C single-byte write process is performed by the custom-designed function *BME280_Init()*. In consideration of the code example Figure 3.42b as well as the timing diagrams of Figure 3.43a, the *BME280_Init()* function performs a call to the functions of the Wire library as follows: (i) First, the *Wire.beginTransmission()* initiates an *I2C START* condition and then transmits the 7-bit slave address (that is, the 0x76) appended with a logical '0' in the LSB position of the transmitted byte (denoting a *write* condition); (ii) After validating the ACK bit (i.e. logical '0') sent by the slave, the *Wire.write()* transmits the second byte, which incorporates the register address (i.e. 0xF4) of the slave device that should be written; (iii) After validating the second ACK bit sent by the slave, the second *Wire.write()* transmits the byte 0x27 to be assigned to the *"ctrl_meas"* register of the slave device; (iv) After validating the third ACK bit, the concluding call of the *Wire.endTransmission()* function initiates an *I2C STOP* condition, which finishes the I2C single-byte write process.

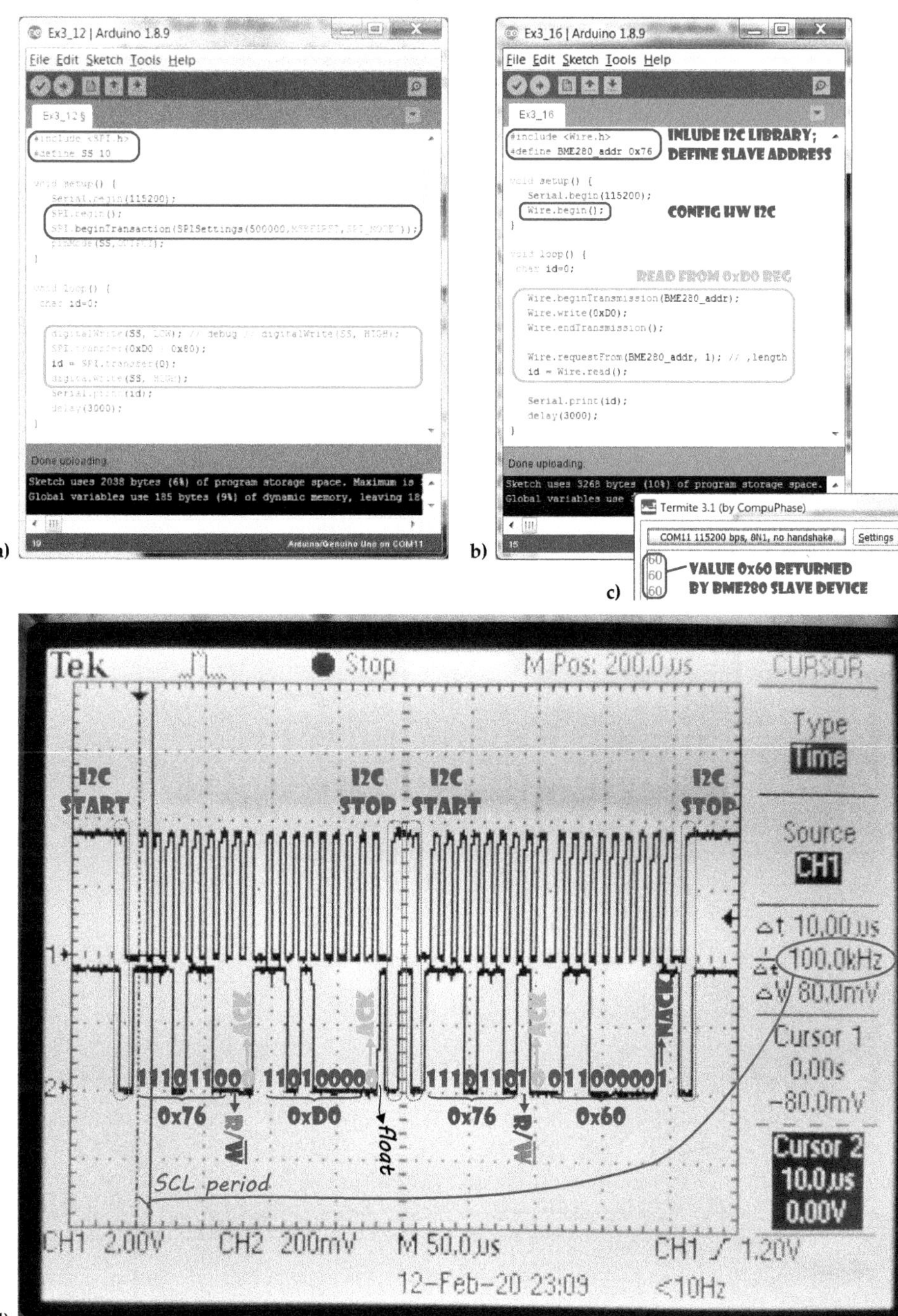

Figure 3.41 I2C single-byte read of the *chip id* of BME280 sensor (code and timing diagrams). (a) the previous single-byte SPI read firmware to highlight code differences with the I2C, (b) single-byte I2C read example firmware, (c) printed results, (d) acquisition of the I2C single-byte read timing diagrams with the oscilloscope. *Source:* Arduino Software.

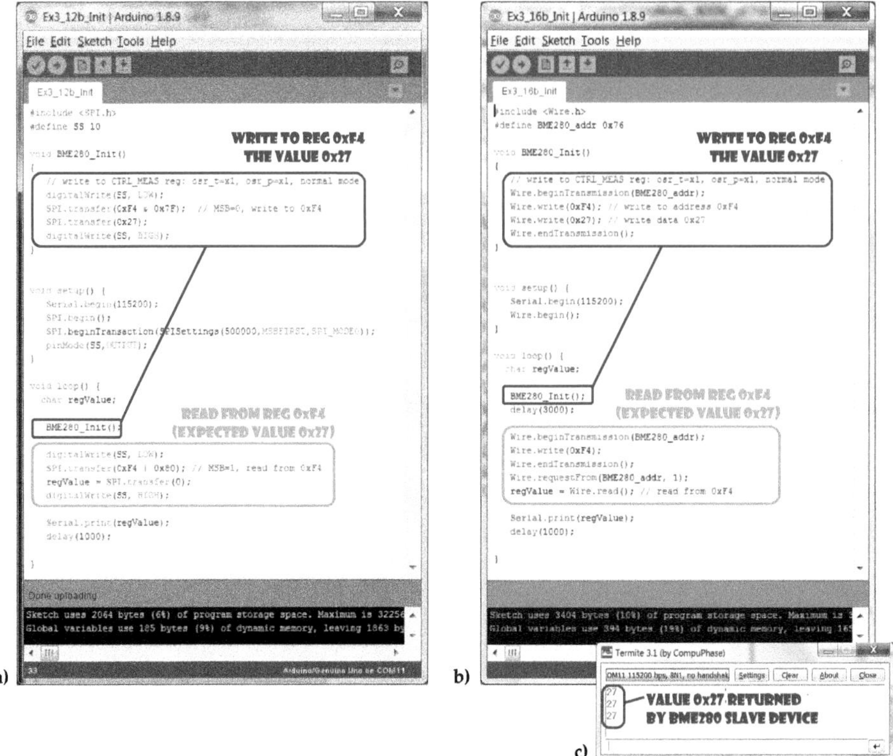

Figure 3.42 I2C single-byte write and read code (BME280 register 0xF4, aka *"ctrl_meas"*). (a) the previous single-byte SPI write/read firmware to highlight code differences with the I2C, (b) single-byte I2C write/read example firmware, (c) printed results. *Source:* Arduino Software.

The I2C single-byte read is identical to the one explored before as presented by Figure 3.43b (i.e. reading the value 0x27 from *"ctrl_meas"* reg).

An I2C multiple-byte read process of the T *coefficients (coeffs)* of the BME280 device is presented in the example of Figure 3.44. As mentioned earlier, to obtain compensated (true) measurements the user should perform particular mathematic operation using the unique – per device – coeffs for temperature, pressure, and humidity. The code example is equivalent to the previous one that addressed SPI control over the BME280 sensor (Figure 3.44a), but currently the control is passed to the I2C interface (Figure 3.44b). The code repeatedly reads and sends to the serial port the T coeffs of BME280 device, which are found on six successive memory addresses (that is, from 0x88 to 0x8D). The I2C multiple-byte read process is encompassed within the function *BME280_ReadTcoeffs()*, as highlighted by the red frame. The *Wire.requestFrom()* function defines six bytes to request from the slave device, and then, the *Wire.read()* successively obtains the six bytes within a *while()* loop, which evaluates the *Wire.available()* function.

The process starts exactly as in the single-byte read example, but as soon as the master obtains the content of the first memory address (i.e. the content of 0x88 address, which is equal to the value 0x87), it sends to the slave an ACK bit, denoted as *acknowledge master (ACK(M))*. The latter task forces the slave to transmit to the master the content of the next

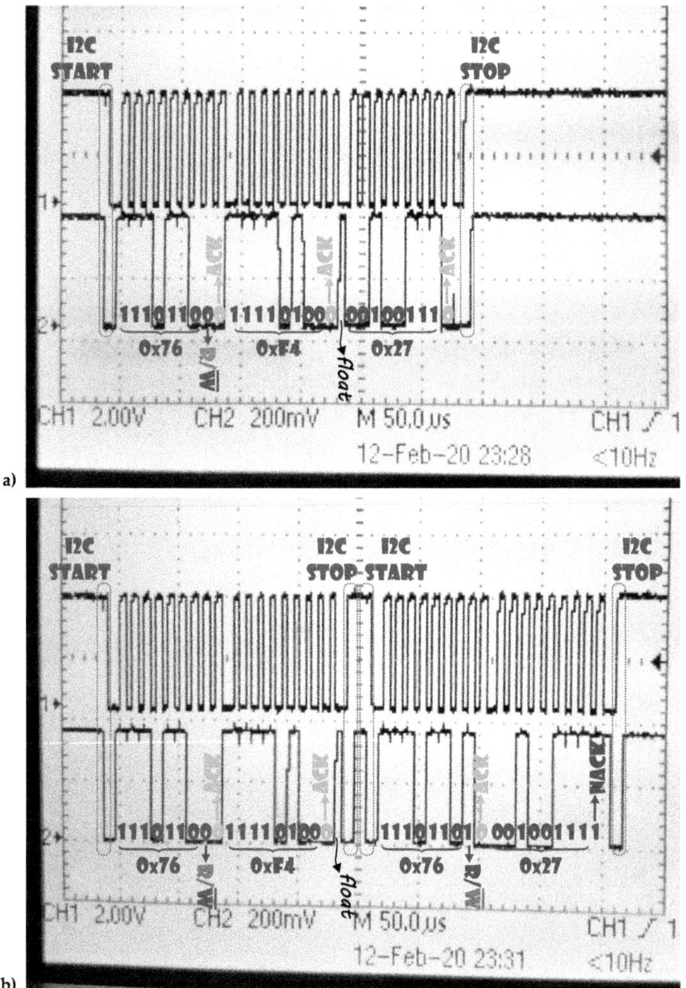

Figure 3.43 I2C single-byte write and read timing diagrams (BME280 *"ctrl_meas"* register). (a) acquisition of the I2C single-byte write timing diagrams with the oscilloscope, (b) acquisition of the I2C single-byte read timing diagrams with the oscilloscope.

memory location (i.e. the content 0x89 address, which is equal to the value 0x6E). The process is repeated until the slave obtains a *no-acknowledge master* (*NACK(M)*), which is followed by the concluding *I2C STOP* condition. Figure 3.44c presents the printed results of the BME280 *T coeffs*, while the timing diagrams of the example are illustrated in Figure 3.44d. It is here noted that the data in red color depict the information sent by the master, while the data in green color depict the information sent by the slave. To distinguish the ACK bit sent by the slave, the latter is referred to as *acknowledge slave*[13] (*ACK(S)*).

---

13 Accordingly, we could address the term *no-acknowledge slave (NACK(S))* in order to denote no response by the slave device.

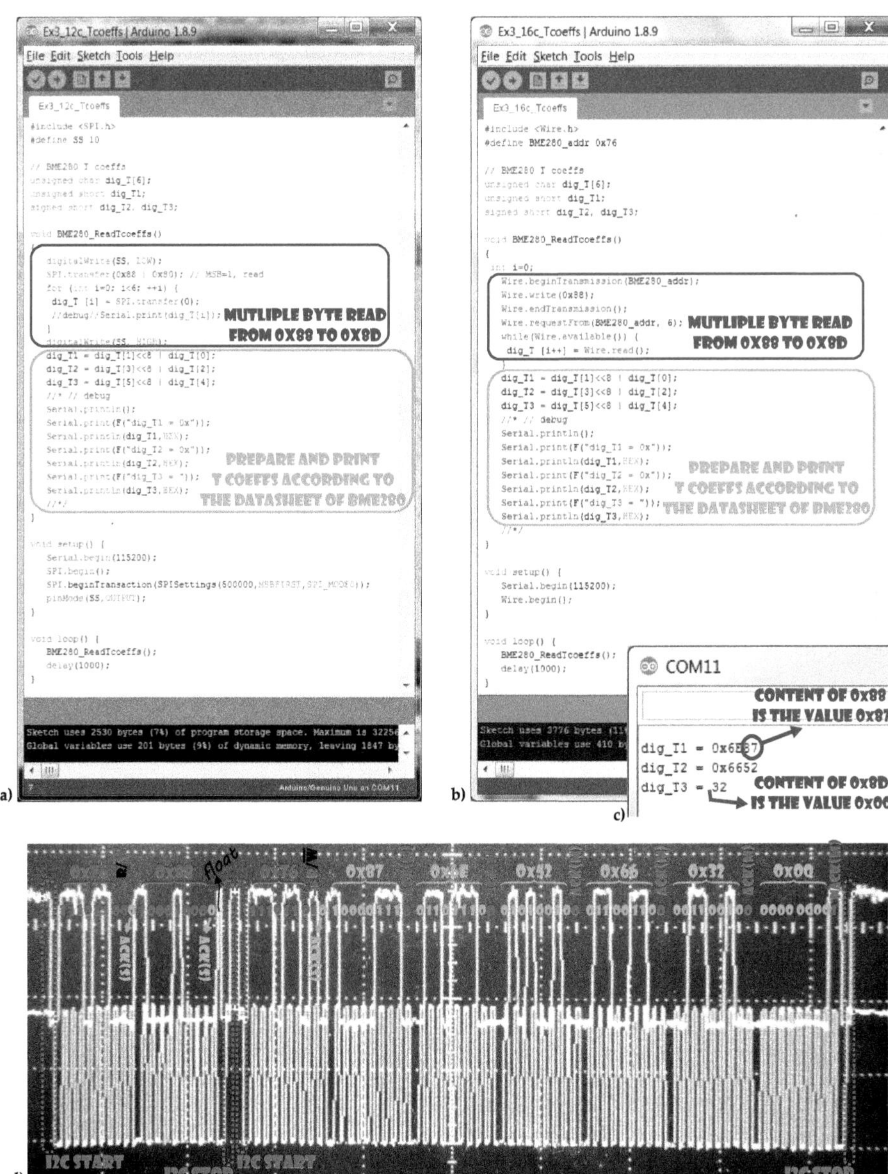

Figure 3.44 I2C multiple-byte read (BME280 temperature coefficients 0x88–0x8D). (a) the previous multiple-byte SPI read firmware to highlight code differences with the I2C, (b) multiple-byte I2C read example firmware, (c) printed results, (d) acquisition of the I2C multiple-byte read timing diagrams with the oscilloscope. *Source:* Arduino Software.

Arduino Ex.3.17

The following example prints to the serial port the compensated temperature, pressure, and humidity acquired by BME280 sensor device, through the built-in I2C of Arduino Uno board. The code example is equivalent to the previous one, which addressed SPI control

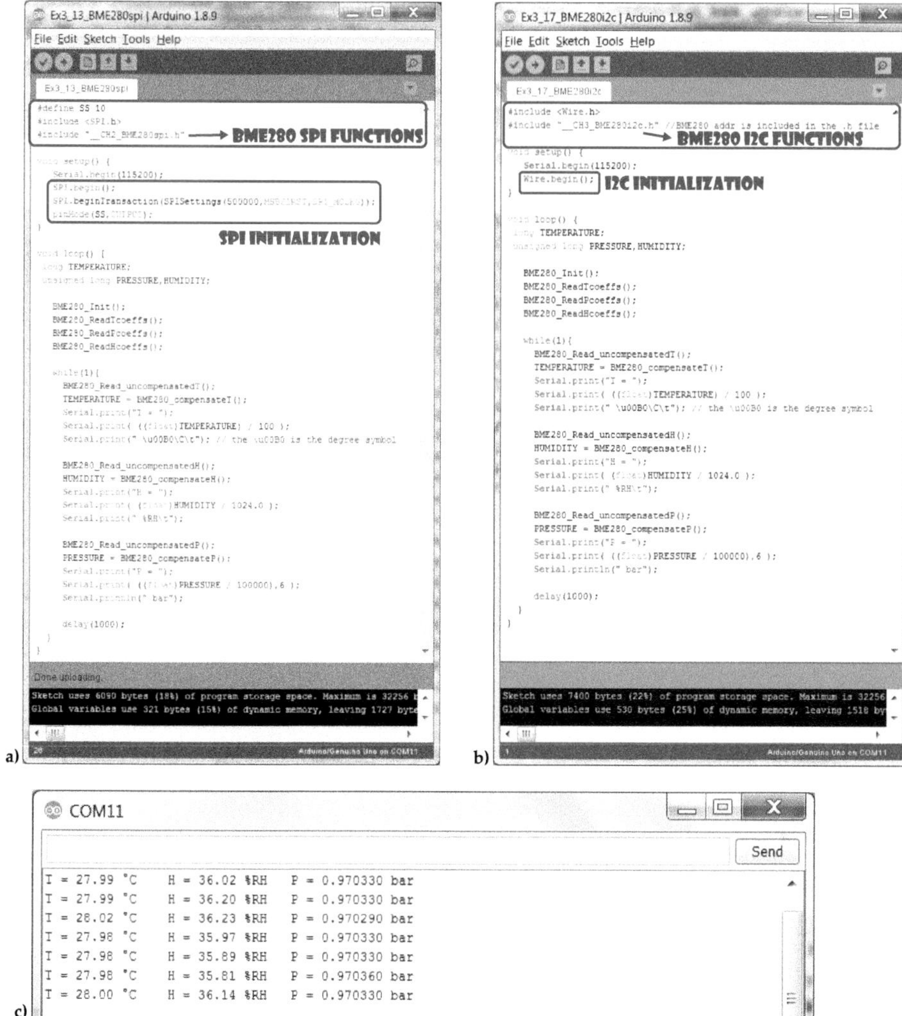

Figure 3.45 BME280: obtain Temperature (T), Humidity (H), and Pressure(P) through the built-in I2C. (a) the previous application firmware of SPI configuration to highlight code differences with the I2C, (b) application firmware of I2C configuration, (c) printed results. *Source:* Arduino Software.

over the BME280 sensor (Figure 3.45a), but currently the control is passed to the I2C interface (Figure 3.45b). The differences between the files are highlighted by red color and are related to (i) the inclusion of the SPI or I2C library, (ii) the inclusion of the user-defined functions for the BME280 sensor, and (iii) the initialization of the corresponding interface protocol. The printed results are given in Figure 3.45c.

The BME280 functions controlled by the I2C interface are incorporated within the file *__CH3_BME280i2c.h,* which is stored to the *libraries* folder found in the root directory of the installed Arduino IDE. The corresponding code is given in Figures 3.46, 3.47, and 3.48. The red rectangle of dash line represents the corresponding SPI functions that were used earlier in order to highlight the differences between the two interface protocols (i.e. I2C and SPI).

C:\arduino-1.8.9\libraries_CH3_BME280_CH3_BME280i2c.h - Notepad++

File Edit Search View Encoding Language Settings Macro Run Plugins Window ?

_CH3_BME280i2c.h

```
#define BME280_addr 0x76

/* FUNCTION DECLARATION */
void BME280_Init();
void BME280_ReadTcoeffs();
void BME280_ReadPcoeffs();
void BME280_ReadHcoeffs();
void BME280_Read_uncompensatedT();
void BME280_Read_uncompensatedP();
void BME280_Read_uncompensatedH();
long BME280_compensateT();
unsigned long BME280_compensateP();
unsigned long BME280_compensateH();


// BME2380 variables
unsigned char dig_T[6];          // T coeffs array
unsigned char dig_P[18];         // P coeffs array
unsigned char dig_H[9];          // H coeffs array
//----------//
unsigned char adcT[3];           // uncompensated_T array
unsigned char adcP[3];           // uncompensated_T array
unsigned char adcH[2];           // uncompensated_H array
//----------//
unsigned short dig_T1;           // T1 coeff
signed short dig_T2, dig_T3;     // T2 & T3 coeffs
unsigned short dig_P1;           // P1 coeff
signed short dig_P2,dig_P3,dig_P4,dig_P5,dig_P6,dig_P7,dig_P8,dig_P9;  // P2-P9 coeffs
unsigned char dig_H1,dig_H3;           // H1 & H3 coeffs
signed short dig_H2,dig_H4,dig_H5;;    // H2,H4,H5 coeffs
signed char dig_H6;                    // H6 coeff
//----------//
long adc_T;                      // uncompensated_T
long adc_P;                      // uncompensated_P
unsigned long adc_H;             // uncompensated_H
long t_fine;

// Functions
void BME280_Init()
{
    // changes to CTRL_HUM become active after writing CTRL_MEAS reg
    // activate humidity and osr_h=x1
    Wire.beginTransmission(BME280_addr);
    Wire.write(0xF2);
    Wire.write(0x01);
    Wire.endTransmission();
    // write to CTRL_MEAS reg (0xF4)
    // osr_t=x1, osr_p=x1, normal mode
    Wire.beginTransmission(BME280_addr);
    Wire.write(0xF4); // write to address 0xF4
    Wire.write(0x27); // write data 0x27
    Wire.endTransmission();
}

void BME280_ReadTcoeffs()
{
   int i=0;
    Wire.beginTransmission(BME280_addr);
    Wire.write(0x88);
    Wire.endTransmission();
    Wire.requestFrom(BME280_addr, 6);
    while(Wire.available()) {
     dig_T [i++] = Wire.read();
    }
    dig_T1 = dig_T[1]<<8 | dig_T[0];
    dig_T2 = dig_T[3]<<8 | dig_T[2];
    dig_T3 = dig_T[5]<<8 | dig_T[4];
}
```

```
void BME280_Init()
{
   // changes to CTRL_HUM become active after w
   // activate humidity and osr_h=x1
   digitalWrite(SS, LOW);
   SPI.transfer(0xF2 & 0x7F);
   SPI.transfer(0x01);
   digitalWrite(SS, HIGH);
   // write to CTRL_MEAS reg (0xF4)
   // osr_t=x1, osr_p=x1, normal mode
   digitalWrite(SS, LOW);
   SPI.transfer(0xF4 & 0x7F);
   SPI.transfer(0x27);
   digitalWrite(SS, HIGH);
}

void BME280_ReadTcoeffs()
{
   digitalWrite(SS, LOW);
   SPI.transfer(0x88 | 0x80); // MSB=1, read
   for (int i=0; i<6; ++i) {
    dig_T[i] = SPI.transfer(0);;
   }
   digitalWrite(SS, HIGH);
   dig_T1 = dig_T[1]<<8 | dig_T[0];
   dig_T2 = dig_T[3]<<8 | dig_T[2];
   dig_T3 = dig_T[5]<<8 | dig_T[4];
}
```

C++ source file   length : 6,080   lines : 205   Ln : 1   Col : 1   Sel : 0 | 0   Windows (CR LF)   UTF-8   INS

Figure 3.46 BME280 I2C functions (1 of 3). *Source:* Arduino Software.

Figure 3.47 BME280 I2C functions (2 of 3). *Source:* Arduino Software.

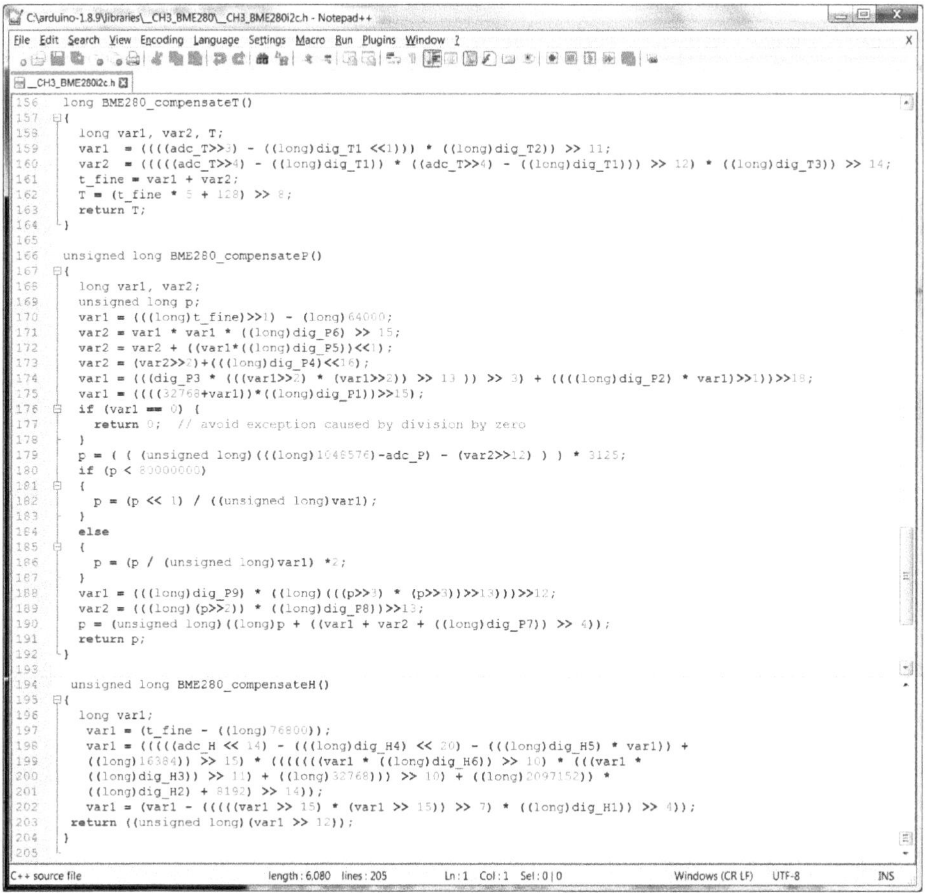

C:\arduino-1.8.9\libraries_CH3_BME280_CH3_BME280i2c.h - Notepad++

File Edit Search View Encoding Language Settings Macro Run Plugins Window ?

_CH3_BME280i2c.h

```
long BME280_compensateT()
{
  long var1, var2, T;
  var1  = ((((adc_T>>3) - ((long)dig_T1 <<1))) * ((long)dig_T2)) >> 11;
  var2  = (((((adc_T>>4) - ((long)dig_T1)) * ((adc_T>>4) - ((long)dig_T1))) >> 12) * ((long)dig_T3)) >> 14;
  t_fine = var1 + var2;
  T = (t_fine * 5 + 128) >> 8;
  return T;
}

unsigned long BME280_compensateP()
{
  long var1, var2;
  unsigned long p;
  var1 = (((long)t_fine)>>1) - (long)64000;
  var2 = var1 * var1 * ((long)dig_P6) >> 15;
  var2 = var2 + ((var1*((long)dig_P5))<<1);
  var2 = (var2>>2)+(((long)dig_P4)<<16);
  var1 = (((dig_P3 * (((var1>>2) * (var1>>2)) >> 13 )) >> 3) + ((((long)dig_P2) * var1)>>1))>>18;
  var1 = ((((32768+var1))*((long)dig_P1))>>15);
  if (var1 == 0) {
    return 0;  // avoid exception caused by division by zero
  }
  p = ( ( (unsigned long)(((long)1048576)-adc_P) - (var2>>12) ) ) * 3125;
  if (p < 80000000)
  {
    p = (p << 1) / ((unsigned long)var1);
  }
  else
  {
    p = (p / (unsigned long)var1) *2;
  }
  var1 = (((long)dig_P9) * ((long)(((p>>3) * (p>>3))>>13)))>>12;
  var2 = (((long)(p>>2)) * ((long)dig_P8))>>13;
  p = (unsigned long)((long)p + ((var1 + var2 + ((long)dig_P7)) >> 4));
  return p;
}

 unsigned long BME280_compensateH()
{
  long var1;
   var1 = (t_fine - ((long)76800));
   var1 = (((((adc_H << 14) - (((long)dig_H4) << 20) - (((long)dig_H5) * var1)) +
   ((long)16384)) >> 15) * (((((((var1 * ((long)dig_H6)) >> 10) * (((var1 *
   ((long)dig_H3)) >> 11) + ((long)32768))) >> 10) + ((long)2097152)) *
   ((long)dig_H2) + 8192) >> 14));
   var1 = (var1 - (((((var1 >> 15) * (var1 >> 15)) >> 7) * ((long)dig_H1)) >> 4));
 return ((unsigned long)(var1 >> 12));
}
```

C++ source file   length : 6,080 lines : 205   Ln : 1 Col : 1 Sel : 0 | 0   Windows (CR LF) UTF-8 INS

Figure 3.48 BME280 I2C functions (3 of 3). *Source:* Arduino Software.

Figure 3.46 highlights the differences between the functions of the two interface protocols (i.e. the I2C and SPI) for the control of BME280 sensor. As before, the *BME_Init()* initializes the sensor through two I2C single-byte write tasks for the control of the humidity, as well as the pressure and temperature measurements, respectively. The *BME_ReadTcoeffs()* performs an I2C multiple-byte read of six bytes located at addresses 0x88–0x8D, which are later used to compensate the raw temperature data.

Figure 3.47 highlights the differences between the next functions of the two interface protocols (i.e. the I2C and SPI) for the control of BME280 sensor. As before, the *BME280_ReadPcoeffs()* function performs an I2C multiple-byte read of 18 bytes located at addresses 0x8E–0x9F, which are later used to compensate the raw pressure data. Because the coefficients for humidity are not all found in successive order (in the sensor's memory), the *BME280_ReadHcoeffs()* function first performs an I2C single-byte read of memory address 0xA1, and then performs an I2C multiple-byte read of seven bytes located at addresses 0xE1–0xE7. The data of the overall eight memory addresses are later used to compensate the raw humidity data. The subsequent three functions that read the uncompensated temperature (T), pressure (P), and humidity (H) data perform I2C multiple-byte read tasks of the memory locations 0xFA–0xFC, 0xF7–0xF9, and 0xFD–0xFE, respectively.

Finally, Figure 3.48 presents the functions *BME280_compensateT()*, *BME280_compensatedP()*, and *BME280_compensatedH()*, which hold the appropriate arithmetic operations for the calculation of the compensated value of temperature, pressure and humidity, respectively. Since the aforementioned functions incorporate no I2C transactions, they hold the exact same code as before (i.e. as in the corresponding SPI example).

Arduino Ex.3.18

The aforementioned examples apply to the built-in I2C protocol of the μC device employed by the Arduino Uno board. While there are several shareable libraries over the internet that can be used for the implementation of a software-implemented I2C for the Arduino, we herein explore the development of a custom-designed I2C library based on the I2C timing diagrams that were explored before. Such a library is particular useful when there is a need to control I2C peripheral devices from a μC of no built-in I2C hardware, or perhaps to use I2C pins other than the built-in pins available by the μC device.

As before (in the example of the SPI custom-designed library), we create three different files to hold the custom-designed I2C library. That is, (i) the *header* file named *swWire.h,* which holds the *C++ class*, (ii) the *source code* file named *swWire.cpp,* which incorporates the library functions, and (iii) the *text* file named *keywords.txt,* which is used to notify Arduino IDE which user-defined keywords to highlight.

Figure 3.49 presents the code of the *swWire.h* header file. The *C++ class* of the file incorporates the function and variable definitions. The *swWire* class incorporates the following private functions (which cannot be invoked by the application code):

i) ***I2C_pinLOW()*** drives the corresponding I2C pin (SDA or SCL) to *LOW* state;
ii) ***I2C_pinHIGH()*** sets the I2C pin to *HIGH* state (i.e. keeps the pin floating);
iii) ***I2C_Delay()*** invokes a delay for the generation of the I2C timing diagrams;
iv) ***I2C_Start()*** initiates an *I2C START* condition;
v) ***I2C_Stop()*** initiates an *I2C STOP* condition;
vi) ***I2C_Write()*** creates the proper timing diagrams to write data to the I2C slave;
vii) ***I2C_Read()*** creates the proper timing diagrams to read data from the I2C slave.

Next, the *_scl* and *_sda* private variables are addressed to control the I2C pins. The latter are configured by the application code as soon as the instance of the C++ class is generated through the (subsequent) *swWire* C++ constructor of the class.

The subsequent public functions, i.e. ***begin()***, ***beginTransmission()***, ***write()***, ***requestFrom()***, ***read()***, and ***endTransmission()***, are the I2C functions used by the application code for the control of the slave device. The functions are similar to the corresponding functions used by the built-in I2C example. However, the ***requestFrom()*** function admits one argument only, that is, the slave address (while the second argument of the built-in I2C code, which determines the number of bytes to request, is not currently used). In addition, the ***read()*** function of the current example admits an argument, which is related to the ACK/NACK information sent by the master, aka *ACK(M)/ACK(M)*. The latter features of the ***requestFrom()*** and ***read()*** functions create a need for some amendments in the

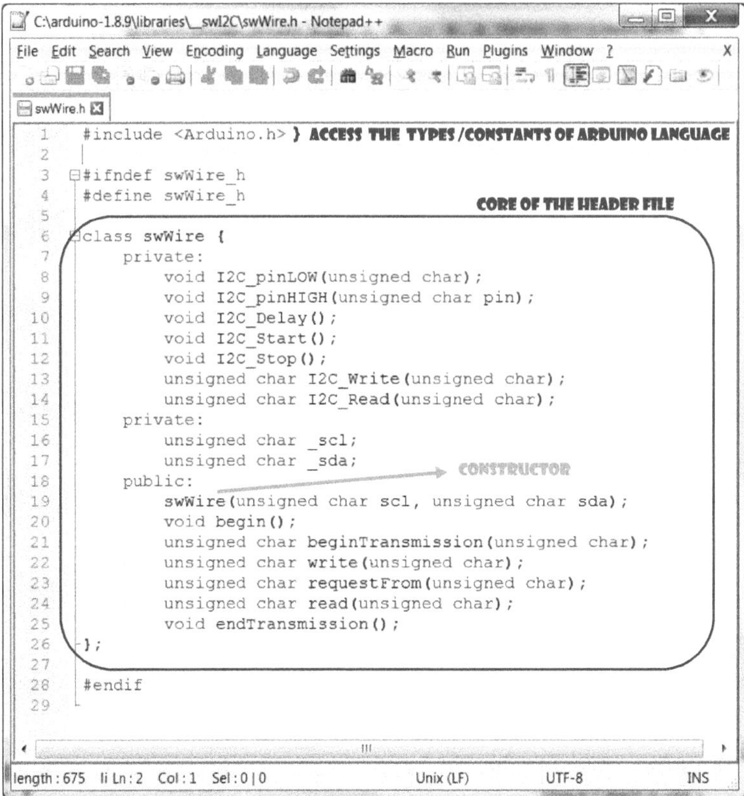

```
#include <Arduino.h>

#ifndef swWire_h
#define swWire_h

class swWire {
    private:
        void I2C_pinLOW(unsigned char);
        void I2C_pinHIGH(unsigned char pin);
        void I2C_Delay();
        void I2C_Start();
        void I2C_Stop();
        unsigned char I2C_Write(unsigned char);
        unsigned char I2C_Read(unsigned char);
    private:
        unsigned char _scl;
        unsigned char _sda;
    public:
        swWire(unsigned char scl, unsigned char sda);
        void begin();
        unsigned char beginTransmission(unsigned char);
        unsigned char write(unsigned char);
        unsigned char requestFrom(unsigned char);
        unsigned char read(unsigned char);
        void endTransmission();
};

#endif
```


Figure 3.49 Header file of the custom-designed I2C library (*swWire.h*). *Source:* Arduino Software.

application code, in comparison to the corresponding example of the built-in I2C. One of them relies on the fact that the *Wire.available()* function is not required anymore (but more details will be explored later in the example).

For simplicity reasons, the custom-designed I2C library explored herein does not make use the so-called *I2C clock stretching* mechanism. That particular mechanism allows the slave device to reduce the bus speed when it is not able to respond to the clock speed determined by the master. On every I2C data transfer, the slave is allowed to pull the SCL line low so as to stop the I2C communication for a while. The BME280 sensor explored by this example (as many other I2C devices) does not support the *I2C clock stretching* mechanism. Because we are not able to explore the mechanism in practice, we have herein omitted the examination of that issue.

The source code of the *swWire.cpp* file is given in Figures 3.50, 3.51, and 3.53. In regard to Figure 3.50, we first place the *constructor* of the class (immediately after the inclusion of the *swWire.h*) so as to define the desired pins of the software-implemented I2C bus (controlled by the corresponding private variables of the *swWire* class). Next in the code, the *begin()* function configures I2C pins as inputs and, hence, the I2C lines are kept floating. The subsequent *I2C_pinLOW()* and *I2C_pinHIGH()* functions are used for convenience during the code development process, in order to minimize the effort related to the procedure of clearing or setting the corresponding I2C pin (i.e. SCL or SDA). On the other hand,

C:\arduino-1.8.9\libraries_swI2C\swWire.cpp - Notepad++

```
#include <swWire.h>

swWire::swWire(unsigned char scl, unsigned char sda){
    _scl = scl;
    _sda = sda;
}

void swWire::begin() {
    pinMode(_scl, INPUT);
    pinMode(_sda, INPUT);
}

void swWire::I2C_pinLOW(unsigned char pin) {
    digitalWrite(pin, LOW);
    pinMode(pin, OUTPUT);
}

void swWire::I2C_pinHIGH(unsigned char pin) {
    pinMode(pin, INPUT);
}

void swWire::I2C_Delay() {
    delayMicroseconds(100);
    //asm ("nop");                               // ~50KHz MAX SCL period on Uno
}

void swWire::I2C_Start(void) {
   I2C_Delay();
   I2C_pinLOW(_sda);
   I2C_Delay();
   I2C_pinLOW(_scl);
   I2C_Delay();
   I2C_Delay();
}
```

CREATE AN INSTANCE OF THE CLASS (PASS THE ARGUMENTS TO DEFINE I2C PINS)

CONFIGURE I2C LINES AND KEEP THEM FLOATING

FORCE I2C LINE TO STATE LOW

FORCE I2C LINE TO STATE HIGH

CONFIGURE DELAY TO AFFECT I2C TIMING DIAGRAMS

INITIATE AN I2C START CONDITION

C++ s length : 2,513 lines : 125 Ln : 23 Col : 5 Sel : 0 | 0 Windows (CR LF) UTF-8 INS

Figure 3.50 Source code file of the custom-designed I2C library (*swWire.cpp* – 1 of 4). *Source:* Arduino Software.

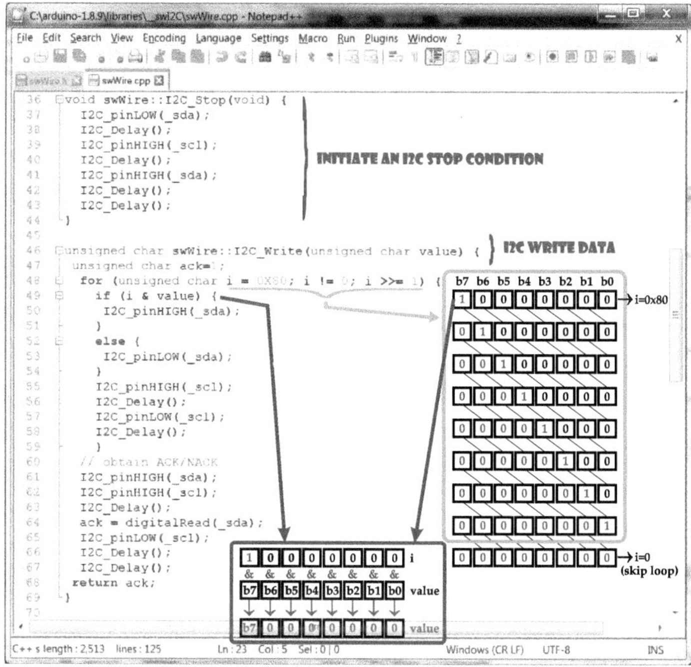

```
void swWire::I2C_Stop(void) {
  I2C_pinLOW(_sda);
  I2C_Delay();
  I2C_pinHIGH(_scl);
  I2C_Delay();
  I2C_pinHIGH(_sda);
  I2C_Delay();
  I2C_Delay();
}

unsigned char swWire::I2C_Write(unsigned char value) {
 unsigned char ack=1;
  for (unsigned char i = 0X80; i != 0; i >>= 1) {
    if (i & value) {
     I2C_pinHIGH(_sda);
    }
    else {
     I2C_pinLOW(_sda);
    }
    I2C_pinHIGH(_scl);
    I2C_Delay();
    I2C_pinLOW(_scl);
    I2C_Delay();
    }
  // obtain ACK/NACK
  I2C_pinHIGH(_sda);
  I2C_pinHIGH(_scl);
  I2C_Delay();
  ack = digitalRead(_sda);
  I2C_pinLOW(_scl);
  I2C_Delay();
  I2C_Delay();
 return ack;
}
```


Figure 3.51 Source code file of the custom-designed I2C library (*swWire.cpp* – 2 of 4). *Source:* Arduino Software.

the *I2C_Delay()* function is used to reduce the I2C clock speed. It is worth noting that using the minimum possible delay (determined by the *nop* assembly instruction) in the Arduino Uno board, we obtain an I2C clock frequency of approximately 50KHz (that is, half of the most standard frequency of the I2C clock). The subsequent *I2C_Start()* initiates the proper *START* condition as described earlier in the chapter.

The subsequent code of the library continues in Figure 3.51, where the *I2C_Stop()* function initiates the proper *STOP* condition of the I2C communication protocol. Next, the *I2C_Write()* function writes a single byte to the slave device in consideration of the I2C timing diagrams that were explored earlier in this chapter. The inside code of the function starts with a *for* loop, where the loop counter *i* is initially assigned to the hex value 0x80. The latter value sets the most significant bit of the counter and clears the rest of the bits. Every time the loop is repeated, the content of the counter is shifted one position to the right until its content is set to zero. The successive values that are assigned to the *i* counter on every repetition of the loop, are depicted inside the green frame at the right of the figure (while the final value $i = 0$, which is placed outside the frame, causes the termination of the *for* loop).

The *if()* statement inside the loop evaluates the result of the bitwise operation between the current content of the loop counter *i* to the function argument *value*, which holds byte to be sent to the slave device. Hence, the foremost time the loop is executed, the bitwise operation *i & value* clears all but the MSB (i.e. b7) of the function argument, as depicted by the blue frame at the bottom of the figure. If the remaining bit value equals to logical '1' (i.e. b7 = 1) the body of the *if* statement sets SDA line, otherwise, it clears SDA line. Next in the *for* loop and immediately before its repetition, the code triggers the I2C clock (i.e. its sets the SCL line, delays for a while, clears the SCL line, and delays again). The second time the loop is repeated, the code transmits the b6 of the function argument, and the process is repeated until all bits of the argument are sent to the slave device.

Subsequent to the *for* loop are placed the instructions, which obtain the ACK/NACK information from the slave device. In detail, the master leaves the SDA and SCL lines floating, delays for a while, and then reads the information present on the SDA line (which incorporates the ACK/NACK value sent by the slave). The latter information is loaded to the *ack* variable, and next, the master clears the SCL line so as to generate the ninth clock pulse of the I2C write operation. After calling the *I2C_Delay()* function twice, the master terminates the *I2C write* operation while returning the content of the *ack* variable. Thereby, the designer could be straightforwardly notified by the firmware code whether the slave responds to the master's call or not.

The subsequent code of the library continues in Figure 3.52, which depicts the *I2C_Read()* function. The function returns one byte (which is the information returned by the slave device), and admits one argument to hold the ACK(M) or NACK(M) bit sent by the master. The latter feature allows us to use the function either for single-byte, or multiply-byte read transaction. At first, the function code releases the SDA line to allow the slave device to undertake the control of data line. Next, the *for* loop is repeated eight successive times in order to obtain the eight bits serially transmitted by the slave device. The byte-size information is organized and held by the *acquiredDATA* variable. To minimize the programming instructions, the code assumes that the upcoming bit to be received by the slave is of zero value. Thereby, the *acquiredDATA* variable is shifted one position to the left (*acquiredDATA<< = 1*). The shifting process causes a logical '0' to be loaded to the LSB

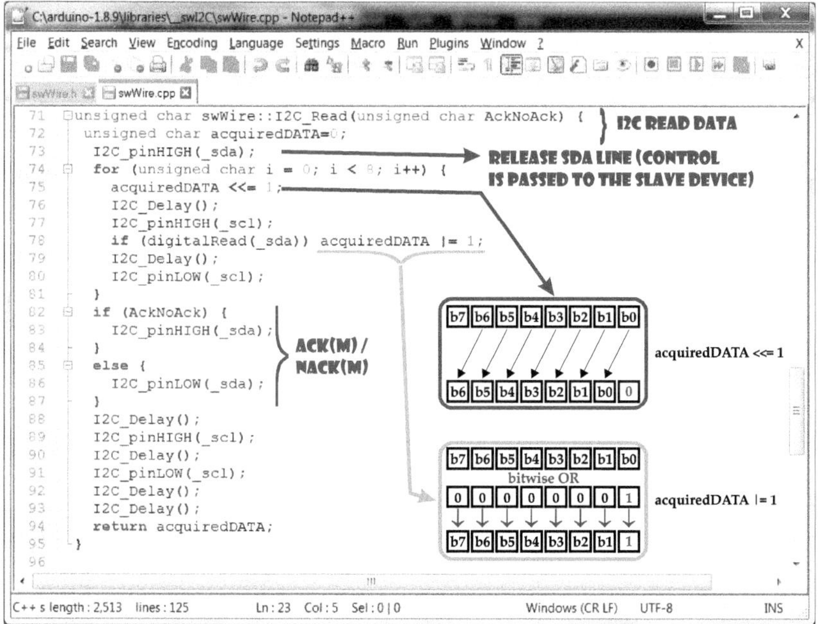

Figure 3.52 Source code file of the custom-designed I2C library (*swWire.cpp* – 3 of 4). *Source:* Arduino Software.

position of the variable, as presented by the blue frame of the figure. The code prepares the SCL line, and then, it evaluates the SDA line (currently controlled by the slave) through the Arduino *digitalRead()* function. If the logical state of the line is found *HIGH*, a bitwise OR is performed to the content of the *acquiredDATA* variable with the decimal value 1 (*acquiredDATA*| = *1*). The latter operation replaces the logical '0' that was previously assigned to the LSB position of the variable, with a logical '1' to (as presented by the green frame of the figure). Then, the code delays for a while, clears the SCL line, and repeats the *for()* loop in order to obtain the next upcoming bit from the slave device.

After eight successive times, the *for* loop is terminated and the master prepares the ACK/NACK information, through an *if(). . .else* expression, which evaluates the argument[14] of the *I2C_Read()* function. Subsequent to the *if(). . .else* expression, the code prepares the ninth clock pulse of the I2C read operation, along with the necessary delay calls that generate the transaction.

The code of *swWire.cpp* library concludes in Figure 3.53, which illustrates the public functions used by the main application code. The figure highlights the operations that are performed beyond a simple call to the private I2C functions. In detail, the code highlights the additional tasks performed the *beginTransmission()* and *requestFrom()* functions. The former function transmits the slave address before writing data to the slave device, while the latter transmits the slave address before reading data from the slave device. In both

14 It is here noted that the *AckNoAck* argument of the *I2C_Read()* function should admit either a logical '0' or '1', which corresponds to ACK(M) or NACK(M).

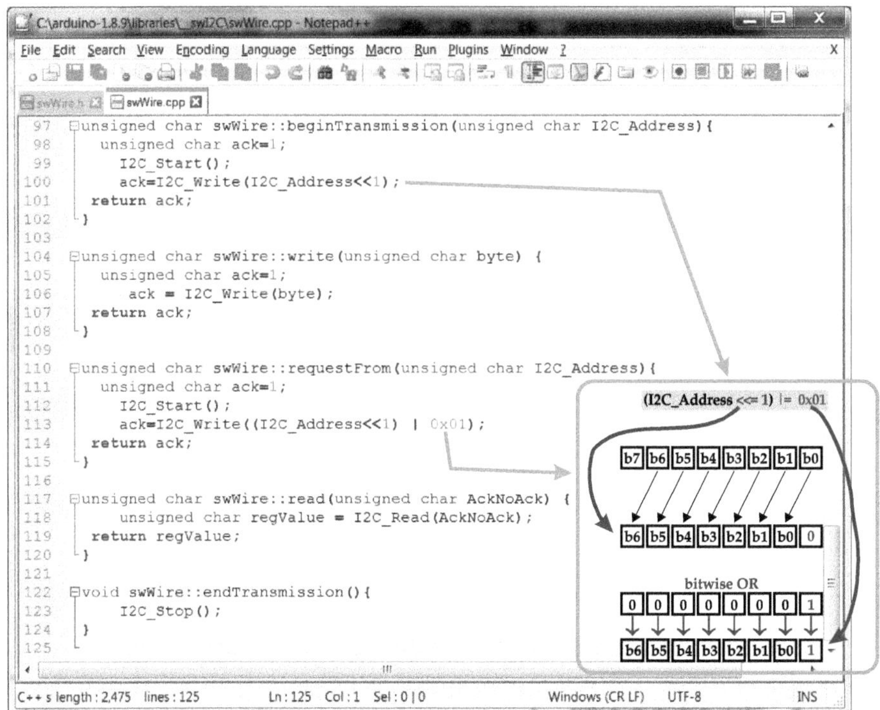

Figure 3.53 Source code file of the custom-designed I2C library (*swWire.cpp* – 4 of 4). *Source:* Arduino Software.

cases, the R/W information is held in the LSB of the transmitted byte (as described earlier in the chapter), while the corresponding tasks performed by those functions are depicted by the green frame at the right bottom area of the figure.

Figure 3.54 presents a single-byte read of the *chip id* of BME280 sensor, that is, the value 0x60 stored to the 0xD0 register. The serial transaction is accomplished with the custom-designed I2C library. The differences of the custom-designed I2C library (Figure 3.54a), in comparison to the corresponding code of the built-in I2C example (Figure 3.54b), are highlighted in the figure. In consideration of Figure 3.54a, the code line within the red frame includes the custom-designed *swWire.h* library. Then, the code lines within the blue frames, first create an instance of the class (named *myWire)* and determine the position of the I2C pins. Then, they substitute the name *Wire* with the name *myWire* so as to provide compatibility between the two examples (in terms of the code-writing process).

The program instructions within the green frame illustrate the differences of the main code between the custom-designed I2C library (Figure 3.54a) and the built-in library (Figure 3.54b). As mentioned earlier, the *Wire.requestFrom()* admits one argument only, and hence, the designer should specify the number of iterations (in consideration of the requested data) through the main code. Because of that feature of the revised *Wire.requestFrom()* function, the also revised *Wire.read()* function incorporates the ACK(M)/NACK(M) information when attempting *single-*, or *multiple-byte* read of data from the slave device. According to the aforementioned revisions, the custom-designed I2C library requires using

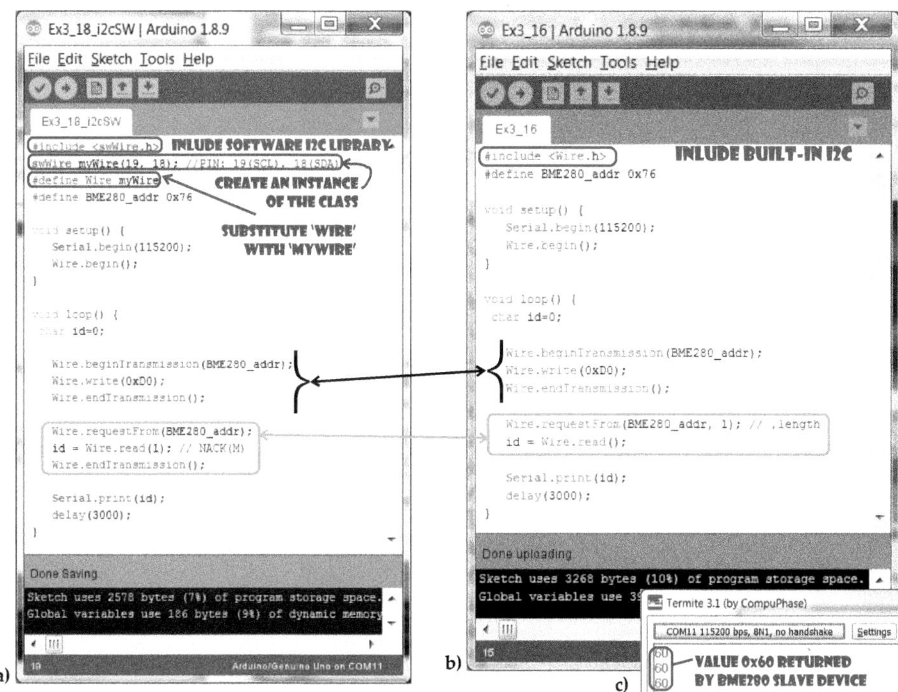

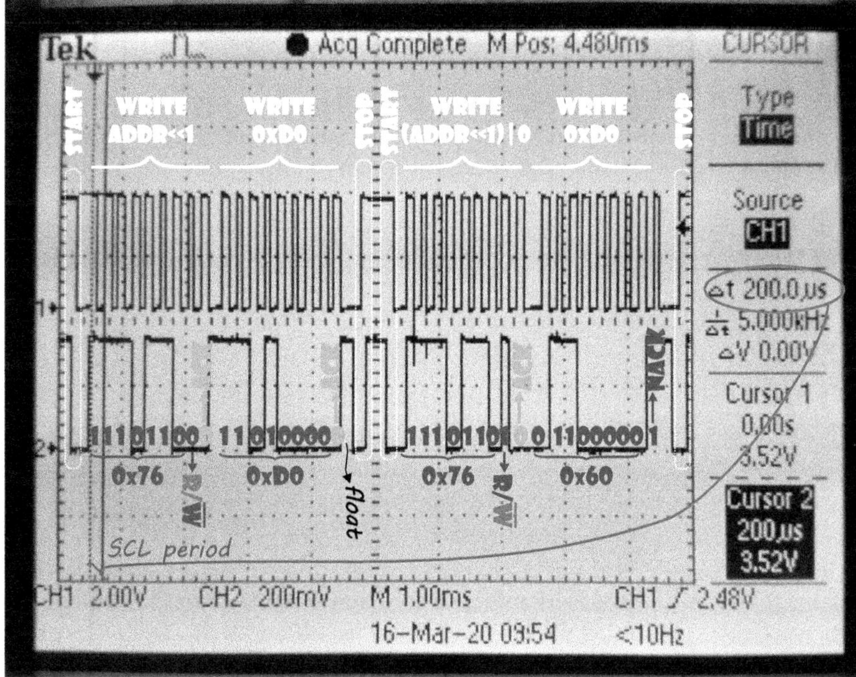

Figure 3.54 *swWire* single-byte read of the *chip id* of BME280 sensor device. (a) example firmware: software-implemented I2C, (b) the previous example firmware of built-in I2 configuration to highlight code differences with the software-implemented I2C, (c) printed results, (d) acquisition of the software-based I2C single-byte read timing diagrams with the oscilloscope. *Source:* Arduino Software.

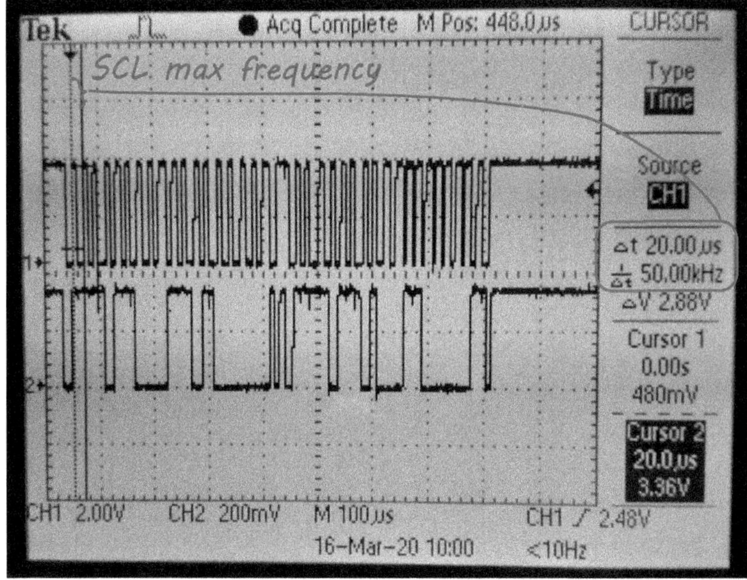

Figure 3.55 SCL max frequency of the *swWire* library on an Arduino Uno board.

the *Wire.endTransmission()* after concluding the *single-*, or *multiple-byte* read of data from the slave device, so as to accomplish the *I2C STOP* condition.

Figure 3.54c presents the printed results on the serial port when the code is running on an Arduino Uno board, while Figure 3.54d illustrates the I2C timing diagrams acquired by the oscilloscope. It is worth mentioning that the slow clock period of the SCL line (i.e. 200 μs) is determined by the *I2C_Delay()* function of the *swWire* library. The process of writing/reading data to/from the slave device, as illustrated by the SDA line, is identical to the one explored by the example of the built-in I2C. Figure 3.55 presents the maximum SCL frequency custom-designed I2C library, when using the *nop* instruction within the *I2C_Delay()*.

Arduino Ex.3.19

The code example of Figure 3.56 obtains temperature, humidity, and pressure measurements from BME280 sensor through the custom-designed I2C library that was explored before. The red frames of Figure 3.56a,b illustrate the differences between the two application codes (i.e. the code using the custom-designed I2C library and the code exploiting the built-in I2C, respectively). It is should be noted that, because of the revised functions *Wire.requestFrom()* and *Wire.read()* used by the current example, it is also required a revision of the driver that was previously used to control BME280 sensor. The name of the revised driver is *__CH3_BME280i2cSWi.h* (also stored inside the *libraries* folder of the Arduino IDE root directory). The printed results of the current example are given in Figure 3.56c.

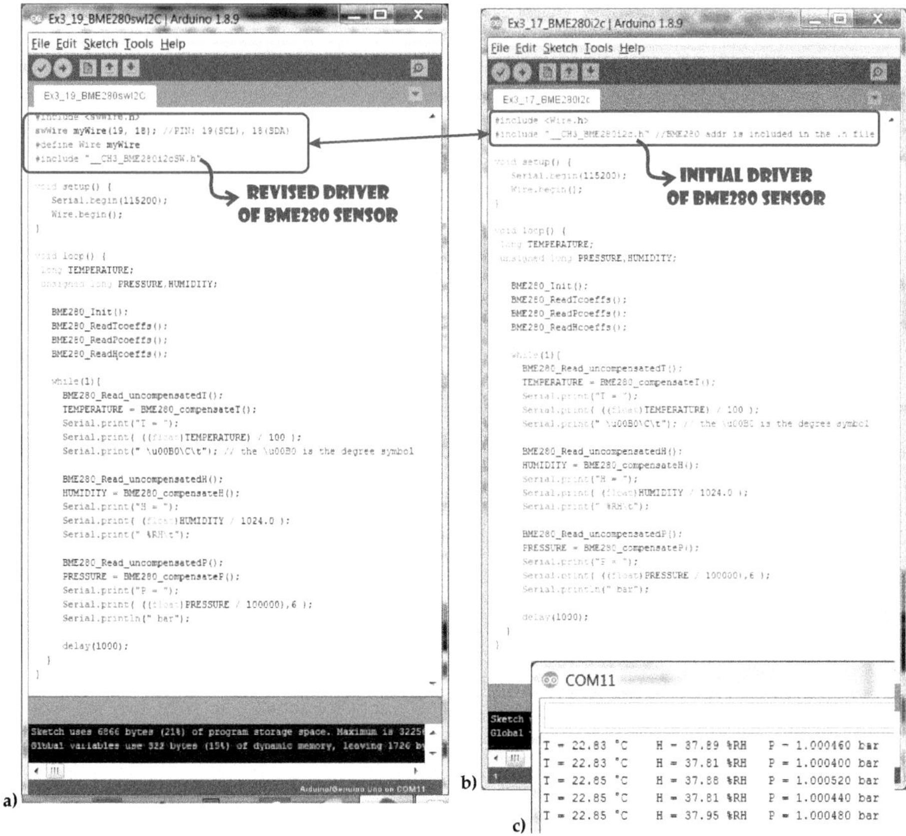

Figure 3.56 BME280: obtain temperature, humidity, and pressure through the *swI2C* library. (a) application firmware of the software-implemented I2C, (b) the previous application firmware of built-in I2 configuration to highlight code differences with the software-implemented I2C, (c) printed results. *Source:* Arduino Software.

Figures 3.57, 3.58, and 3.59 present the revised driver for BME280 sensor when addressing the custom-designed I2C library, while also highlighting the differences between the two libraries (i.e. the software-implemented I2C and the built-in I2C). The differences are found in the code lines where the master attempts to read data from the slave device. In regard to Figure 3.57, the revised *BME280_ReadTcoeffs()* addresses a *for* loop, which runs six successive times. In order to perform a multiple-byte read, the *if(). . .else* statement (within the loop) invokes the *Wire.read(1)* function the first five successive repetitions of the loop (that is, for $i = 0$ to $i = 4$). During the final repetition, the *if(). . .else* statement invokes the *Wire.read(0)* function, and hence, the master concludes the transactions by sending an ACK(M). Thereafter, the code initiates an *I2C STOP* condition through the call of the *Wire.endTransmission()*.

In regard to Figure 3.58, similar revisions are accomplished to the revised functions *BME280_ReadPcoeffs()*, *BME280_ReadHcoeffs()*, *BME280_Read_uncompensatedT()*,

C:\arduino-1.8.9\libraries_CH3_BME280_CH3_BME280i2cSW.h - Notepad++

File Edit Search View Encoding Language Settings Macro Run Plugins Window ?

_CH3_BME280i2cSW.h

```
#define BME280_addr 0x76

/* FUNCTION DECLARATION */
void BME280_Init();
void BME280_ReadTcoeffs();
void BME280_ReadPcoeffs();
void BME280_ReadHcoeffs();
void BME280_Read_uncompensatedT();
void BME280_Read_uncompensatedP();
void BME280_Read_uncompensatedH();
long BME280_compensateT();
unsigned long BME280_compensateP();
unsigned long BME280_compensateH();


// BME2380 variables
unsigned char dig_T[6];        // T coeffs array
unsigned char dig_P[18];       // P coeffs array
unsigned char dig_H[9];        // H coeffs array
//----------//
unsigned char adcT[3];         // uncompensated_T array
unsigned char adcP[3];         // uncompensated_T array
unsigned char adcH[2];         // uncompensated_H array
//----------//
unsigned short dig_T1;         // T1 coeff
signed short dig_T2, dig_T3;   // T2 & T3 coeffs
unsigned short dig_P1;         // P1 coeff
signed short dig_P2,dig_P3,dig_P4,dig_P5,dig_P6,dig_P7,dig_P8,dig_P9;  // P2-P9 coeffs
unsigned char dig_H1,dig_H3;         // H1 & H3 coeffs
signed short dig_H2,dig_H4,dig_H5;;  // H2,H4,H5 coeffs
signed char dig_H6;                  // H6 coeff
//----------//
long adc_T;                    // uncompensated_T
long adc_P;                    // uncompensated_P
unsigned long adc_H;           // uncompensated_H
long t_fine;

// Functions
void BME280_Init()
{
    // changes to CTRL_HUM become active after writing CTRL_MEAS reg
    // activate humidity and osr_h=x1
    Wire.beginTransmission(BME280 addr);
    Wire.write(0xF2);
    Wire.write(0x01);
    Wire.endTransmission();
    // write to CTRL_MEAS reg (0xF4)
    // osr_t=x1, osr_p=x1, normal mode
    Wire.beginTransmission(BME280_addr);
    Wire.write(0xF4); // write to address 0xF4
    Wire.write(0x27); // write data 0x27
    Wire.endTransmission();
}

void BME280_ReadTcoeffs()
{
    Wire.beginTransmission(BME280_addr);
    Wire.write(0x88);
    Wire.endTransmission();
    Wire.requestFrom(BME280_addr);
    for (int i=0; i<6; i++) {
     if(i!=6-1) dig_T[i] = Wire.read(0); // M(ACK)
     else  dig_T[i] = Wire.read(1); // M(NACK)
    }
    Wire.endTransmission();
    dig_T1 = dig_T[1]<<8 | dig_T[0];
    dig_T2 = dig_T[3]<<8 | dig_T[2];
    dig_T3 = dig_T[5]<<8 | dig_T[4];
}
```

C++ source file   length : 6,581   lines : 210   Ln : 2   Col : 1   Sel : 0 | 0

```
void BME280_ReadTcoeffs()
{
 int i=0;
  Wire.beginTransmission(BME280_addr);
  Wire.write(0x88);
  Wire.endTransmission();
  Wire.requestFrom(BME280_addr, 6);
  while(Wire.available()) {
   dig_T [i++] = Wire.read();
  }
  dig_T1 = dig_T[1]<<8 | dig_T[0];
  dig_T2 = dig_T[3]<<8 | dig_T[2];
  dig_T3 = dig_T[5]<<8 | dig_T[4];
}
```

Figure 3.57 BME280 revised driver (1 of 3). *Source:* Arduino Software.

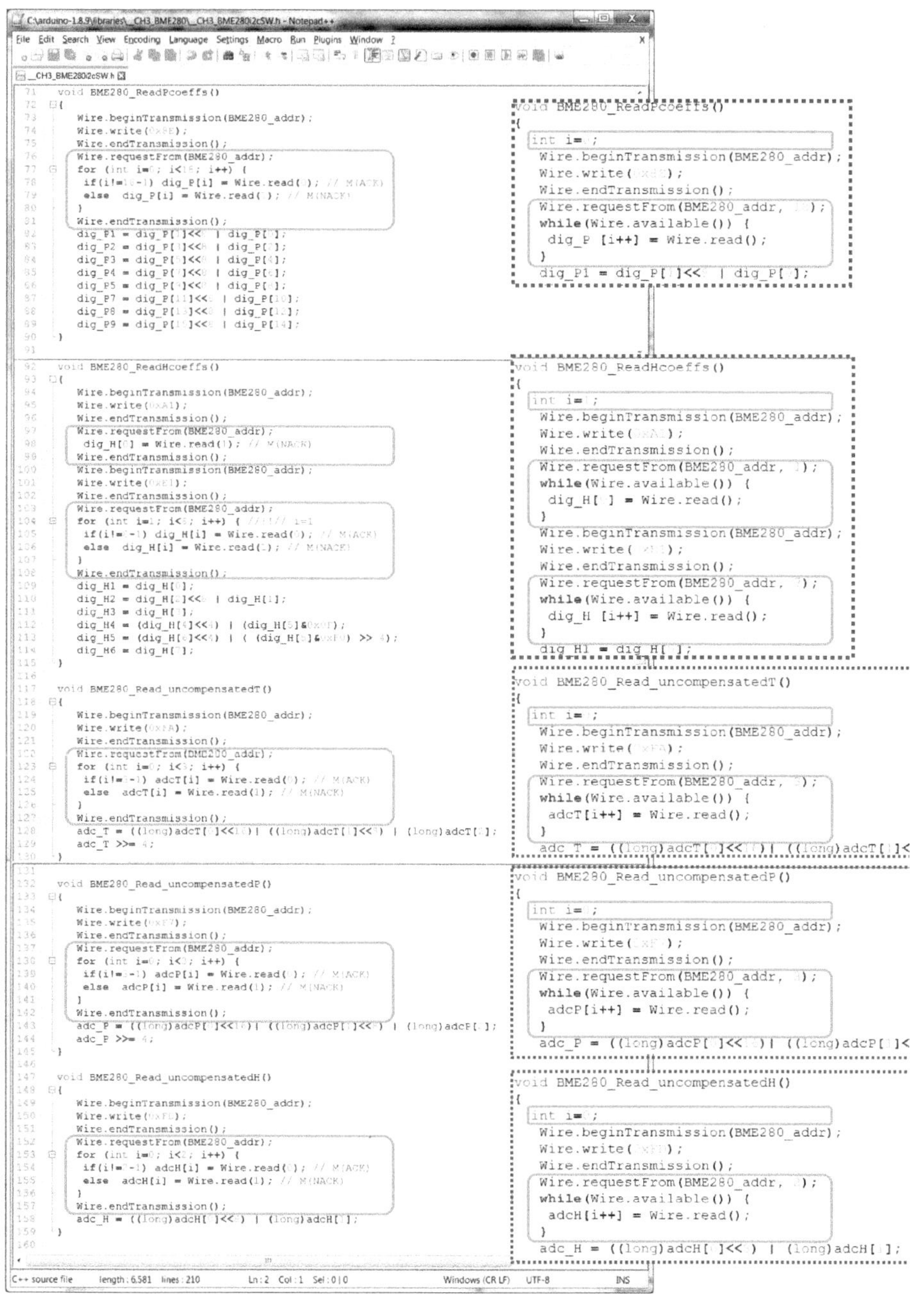

Figure 3.58 BME280 revised driver (2 of 3). *Source:* Arduino Software.

```
long BME280_compensateT()
{
  long var1, var2, T;
  var1  = ((((adc_T>>3) - ((long)dig_T1 <<1))) * ((long)dig_T2)) >> 11;
  var2  = (((((adc_T>>4) - ((long)dig_T1)) * ((adc_T>>4) - ((long)dig_T1))) >> 12) * ((long)dig_T3)) >> 14;
  t_fine = var1 + var2;
  T = (t_fine * 5 + 128) >> 8;
  return T;
}

unsigned long BME280_compensateP()
{
  long var1, var2;
  unsigned long p;
  var1 = (((long)t_fine)>>1) - (long)64000;
  var2 = var1 * var1 * ((long)dig_P6) >> 15;
  var2 = var2 + ((var1*((long)dig_P5))<<1);
  var2 = (var2>>2)+(((long)dig_P4)<<16);
  var1 = (((dig_P3 * (((var1>>2) * (var1>>2)) >> 13 )) >> 3) + ((((long)dig_P2) * var1)>>1))>>18;
  var1 = ((((32768+var1))*((long)dig_P1))>>15);
  if (var1 == 0) {
    return 0;  // avoid exception caused by division by zero
  }
  p = ( ( (unsigned long)(((long)1048576)-adc_P) - (var2>>12) ) ) * 3125;
  if (p < 80000000)
  {
    p = (p << 1) / ((unsigned long)var1);
  }
  else
  {
    p = (p / (unsigned long)var1) *2;
  }
  var1 = (((long)dig_P9) * ((long)(((p>>3) * (p>>3))>>13)))>>12;
  var2 = (((long)(p>>2)) * ((long)dig_P8))>>13;
  p = (unsigned long)((long)p + ((var1 + var2 + ((long)dig_P7)) >> 4));
  return p;
}

 unsigned long BME280_compensateH()
{
  long var1;
   var1 = (t_fine - ((long)76800));
   var1 = (((((adc_H << 14) - (((long)dig_H4) << 20) - (((long)dig_H5) * var1)) +
   ((long)16384)) >> 15) * (((((((var1 * ((long)dig_H6)) >> 10) * (((var1 *
   ((long)dig_H3)) >> 11) + ((long)32768))) >> 10) + ((long)2097152)) *
   ((long)dig_H2) + 8192) >> 14));
   var1 = (var1 - (((((var1 >> 15) * (var1 >> 15)) >> 7) * ((long)dig_H1)) >> 4));
 return ((unsigned long)(var1 >> 12));
}
```

Figure 3.59 BME280 revised driver (3 of 3). *Source:* Arduino Software.

BME280_Read_uncompensatedP(), and *BME280_Read_uncompensatedH()*. However, no revisions are observed in the code Figure 3.59 as the incorporated functions do not perform any I2C transactions.

Conclusion

This chapter has provided a thorough examination of the hardware interface topics that are considered of vital importance when building projects around microcontrollers. Through a scholastic examination of the critical, hardware-related practices on microcontrollers (arranged around carefully designed example codes and explanatory figures) the book has been focused on the establishment of a clear link between the software and hardware. That is, an essential area of study for building the solid background required to make the passage from *"have knowledge of"* to *"mastering"* the design of microcontroller-based projects and applications. Hereafter are some unsolved problems for the reader.

Problem 3.1 (Data Input and Output to/from the μC Using Push-Button and LED IO Units)

Write the firmware code that emulates the operation of bicycle's light. When the user presses the push-button while the state of the light is *OFF*, the LED turns (and remains) *ON*. If the user presses again the button, the light starts blinking every 0.25 seconds, until the push button is pressed again, where in this particular case the LED returns in the initial *OFF* state.

Problem 3.2 (PWM)

Write the firmware code that emulates the function of a *dimmer*. When the user presses a push-button the DC of the PWM signal, which drives a LED, is increased from 0 to 100% and all over again from the beginning (for as long as the user keeps pressing the button). If the user presses two clicks on the button (as the double click performed by the PC mouse) the dimmer changes mode of operation and decreases the DC.

Problem 3.3 (UART, SPI, I2C)

Write the firmware code that implements a UART-to-I2C bridge. In detail, connect Arduino Uno to a host PC and forward all the commands sent by the user (through a terminal console) to the BME280 sensor device (controlled by I2C), as well as transmit every response of the BME280 sensor to the host PC. Perform the same task when the sensor is controlled by SPI (i.e. implement a UART-to-SPI bridge).

4

Sensors and Data Acquisition

Data acquisition (DAQ) term refers to the process of sampling an electrical or real-world physical phenomenon (such as voltage, current, capacity, resistance, pressure, temperature, etc.) and converting the samples into digital data able to be manipulated by a computer system. The modern *DAQ* systems (aka *PC-based DAQ systems*) consist of sensors, DAQ measurement hardware, as well accompanying software running on a *personal computer* (*PC*). The DAQ measurement hardware acts as the interface between the computer and the signals from the outside world. Regularly, the DAQ measurement hardware provides a digitization of the analog signal from the outside world. However, modern sensor devices constitute complete *systems on chip (SoC)* or *system in packages (SiPs)*, as they incorporate the necessary electronics that convert the electrical or real-world physical phenomenon to digital data. Because of this particular feature of the contemporary sensors, we herein explore how to build an efficient microcontroller-based DAQ system, which is controlled by custom-designed software, toward collecting and forwarding to the computer the digital information obtained from the sensor device(s).

As explored earlier in the introductory part of the book, modern sensor devices constantly pave the wave for new solution in embedded solutions and give rise to an opportunity for creativity and innovation. Therefore, this chapter is focused on the study of some of the most interesting sensor devices, as well as the process of collecting and real-time monitoring data to a PC, using solely freeware development tools.

Environmental Measurements with Arduino Uno

This example explores the development of a DAQ system with an Arduino Uno board. The system acquires the ambient air pressure from the sensor device BME280 that was used by the previous chapter, for the exploration of the SPI and I2C serial interfaces. In consideration of the accompanying software, which runs on the PC and collects the atmospheric pressure data, the source code development is performed in *C programming language* using the free source code editor *Notepad++*. Moreover, the compilation of the source code into machine language is performed through the free tool *Minimalist GNU for Windows (MinGW)*. It is worth

Microcontroller Prototypes with Arduino and a 3D Printer: Learn, Program, Manufacture, First Edition.
Dimosthenis E. Bolanakis.

mentioning that, when exchanging data between a hardware device and software running on a PC, it is often desirable to inspect the data send out from or delivered by the computer's communication port. For that particular purpose we exploit the *Free Serial Port Monitor* tool.

Arduino Ex.4-1

The present example acquires merely air pressure data from BME280 sensor. We choose to work with air pressure measurements, because they change more rapidly than other environmental measurements (such as the temperature). Hence, the air pressure data help us to explore further the timing issues related to procedure of collecting and forwarding, to the PC, the digital information derived from the sensor. However, to obtain compensated pressure measurements, the designer should also obtain compensated temperature data from BME280 device. It should be noted that in regard to pressure and temperature control, the BME280 environmental sensor [53] is downward register compatible with the BMP280 [54] pressure sensor. Hence, to configure the BME280 for the current example, the designer should refer to both datasheets.

To control BME280 sensor, we make use of the I2C driver developed in the previous chapter along with a few amendments in the code that are explored hereafter. The most important amendment has to do the operating mode of the sensor device, based on the recommended *use cases* found in the datasheets. Table 4.1 depicts the bits of registers 0×F4 and 0×F5, which configure the pressure and temperature measurements. In regard to the "ctrl_meas" (0×F4) register we choose *Temperature oversampling × 1, Pressure oversampling × 2*, and *Normal mode* of operation (as illustrated by the bold selected text of the table). Leaving the "config" (0×F5) as is, i.e. assigned its value to 0×00 in reset state, we (i) choose inactive duration in *normal mode* of operation equal to 0.5 ms (which is the requisite waiting time before the next pressure and temperature reads), (ii) set the IIR filter of the sensor to state OFF, and (iii) deactivate the SPI interface. It should be mentioned that the IIR filter increases the resolution of pressure and temperature, but also slows down the response of the sensor. All the above initializations set the sensor device in the recommended *use case* named as *drop detection*[1] in the sensor datasheets. The latter configures a low-power operation of the sensor with a typical *output data rate* (*ODR*) equal to 125 Hz. What this means in practice is that we are able to periodically read the sensor device every (1/125 Hz=) 8 ms period so as to obtain the compensated pressure and temperature measurements. We have chosen the aforementioned configuration because the low sampling interval (of 8 ms), as it allows us to explore the timing issuers related to the DAQ system.

Figure 4.1 illustrates merely the amendments between the driver of the built-in I2C that was explored in the previous chapter (that is, the *__CH3_BME280i2c.h* given in Figure 4.1b), and the driver used by the current example (that is, the *__CH4_BME280i2c_revA.h* of Figure 4.1a). The changes in the revised version are as follows: (i) the I2C address of BME280 sensor has

---

1 A typical example of the *drop detection* use case could be a healthcare monitoring application, which, for example, could identify when a patient falls off the clinic bed and send an alert message to the mobile phone of the hospital doctor. Because such a device constitutes a battery-operated portable system (regularly attached to the body of the patient), it should be of low-power consumption (as this particular *use case* is). At this point it is worth mentioning that since the air pressure changes at a standard rate according to the variations in elevation, barometric pressure sensors are regularly utilized as barometric altimeters with numerous of creative applications (such as indoor navigation, gaming, sports applications, etc.)

Table 4.1 Setting registers "ctrl_meas" (0 × F4) and "config" (0 × F5).

Set REG 0 × F4 to BIN: 001_010_11 (reset state: 0 × 00)	Name: Value	Settings	Recommended use case (ODR)	Drop detection (125 Hz)
Bits 7,6,5	osrs_t[2 : 0]=000	Temperature oversampling skipped		
	osrs_t[2 : 0]=001	**Temperature oversampling × 1**		
	osrs_t[2 : 0]=010	Temperature oversampling × 2		
	osrs_t[2 : 0]=011	Temperature oversampling × 4		
	osrs_t[2 : 0]=100	Temperature oversampling × 8		
	osrs_t[2 : 0]= "other"	Temperature oversampling × 16		
Bits 4,3,2	osrs_p[2 : 0]=000	Pressure oversampling skipped		
	osrs_p[2 : 0]=001	Pressure oversampling × 1		
	osrs_p[2 : 0]=010	**Pressure oversampling × 2**		
	osrs_p[2 : 0]=011	Pressure oversampling × 4		
	osrs_p[2 : 0]=100	Pressure oversampling × 8		
	osrs_p[2 : 0]="other"	Pressure oversampling × 16		
Bits 1,0	mode[1 : 0]=00	Sleep mode of operation		
	mode[1 : 0]=01	Force mode of operation		
	mode[1 : 0]=10	>>		
	mode[1 : 0]=11	**Normal mode of operation**		

(*Continued*)

Table 4.1 (Continued)

Set REG 0 × F5 to BIN: 00000000 (reset state: 0 × 00)	Name: Value	Settings
Bits 7,6,5	**t_sb[2 : 0]=000**	***t* standby 0.5 ms**
	t_sb[2 : 0]=001	*t* standby 62.5 ms
	t_sb[2 : 0]=010	*t* standby 125 ms
	t_sb[2 : 0]=011	*t* standby 250 ms
	t_sb[2 : 0]=100	*t* standby 500 ms
	t_sb[2 : 0]=101	*t* standby 100 ms
	t_sb[2 : 0]=110	*t* standby 10 ms
	t_sb[2 : 0]=111	*t* standby 20 ms
Bits 4,3,2	**filter[2 : 0]=000**	**Filter off**
	filter[2 : 0]=001	Filter coefficient 2
	filter[2 : 0]=010	Filter coefficient 4
	filter[2 : 0]=011	Filter coefficient 8
	filter[2 : 0]="other"	Filter coefficient 16
Bit 0	spi3w_en[0]	Enables 3-wire SPI interface when set to 1

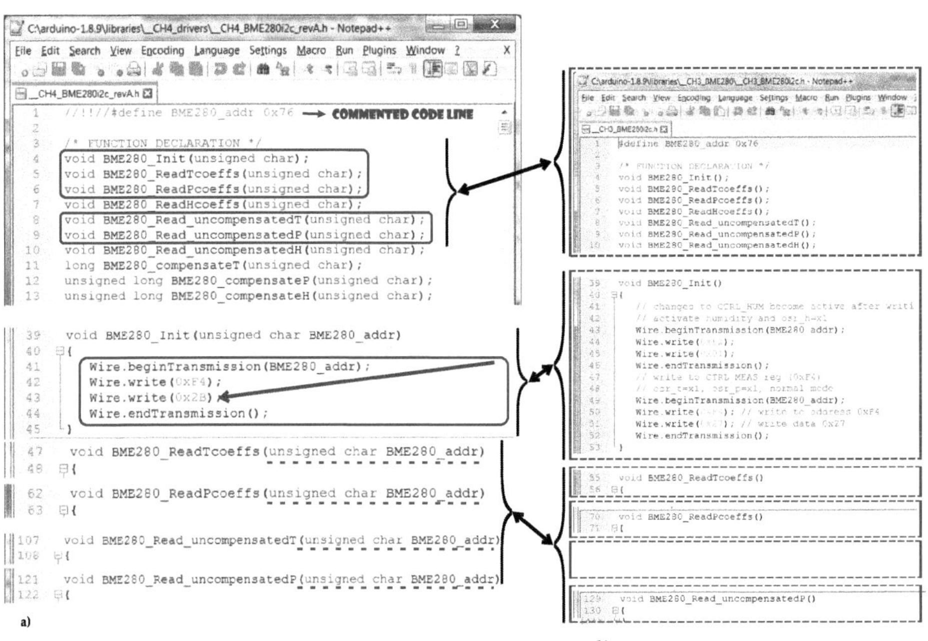

Figure 4.1 Amendments to the BME280 driver of the built-in I2C interface. (a) revision of BME280 driver of the built-in I2C interface, (b) first version of BME280 driver (presented by the previous chapter). *Source:* Arduino Software.

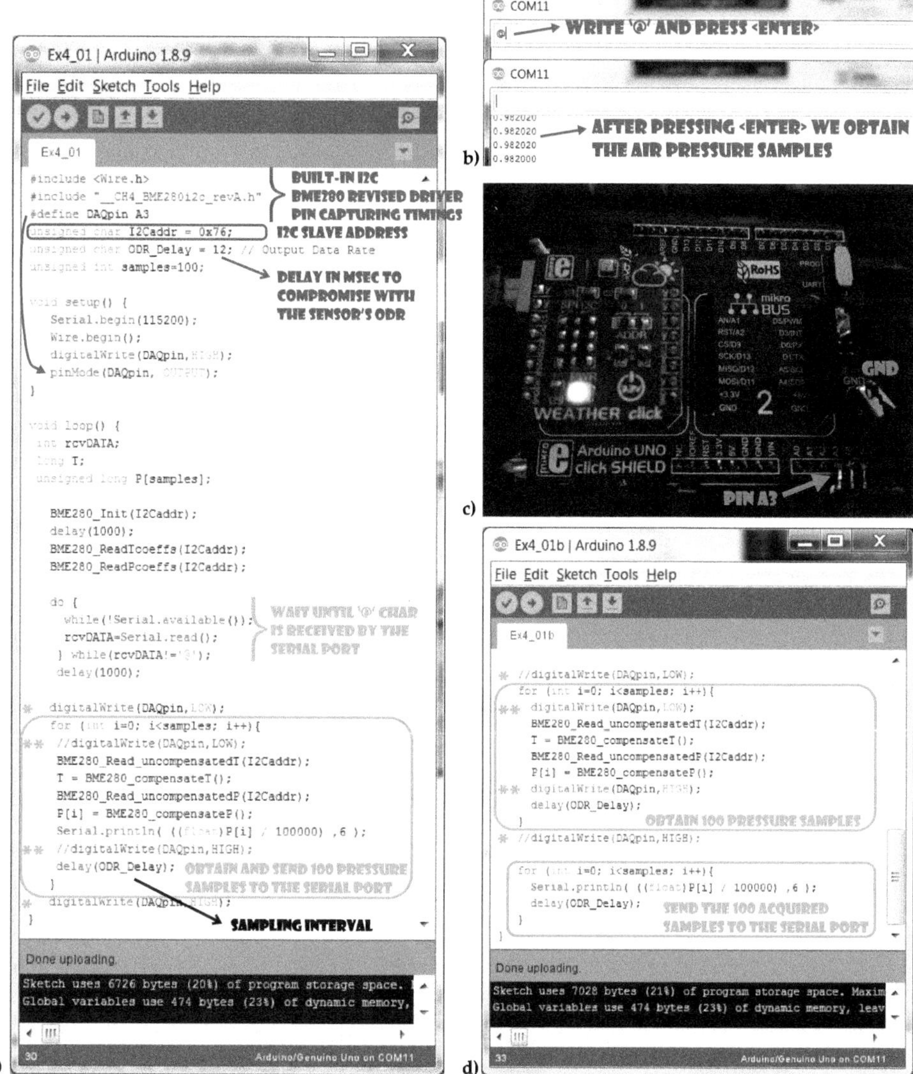

Figure 4.2 DAQ measurement hardware and firmware (air pressure with BME280 sensor). (a) application code, (b) printed results, (c) DAQ hardware, (d) a revision of the application code. *Source:* Arduino Software.

been commented, because it is going to be used by the main code, (ii) all the routines admit an argument of *unsigned char* type to be replaced by the I2C address[2] of the slave device, and (iii) the *BME_280_Init()* function incorporates only the initialization of the 0×F4 (i.e. “ctrl_meas” reg.) as the humidity measurements are not used by the current example.

Figure 4.2 illustrates the DAQ measurement hardware and firmware, which obtains air pressure measurements from the BME280 environmental sensor. The application code is

2 As we see in the upcoming examples, that revision allows us to easily control more than one identical I2C slaves by the main code.

given in Figure 4.2a, where the former three lines incorporate the Arduino I2C library as well as the revised driver of the sensor device, and they define the pin location of an output signal (name *DAQpin*) used for capturing timings of specific events. The next code line defines the I2C slave address used by the routines of the revised version of BME280 driver. The next two code lines determine the sampling interval between two successive samples as well as the number of samples to be acquired by the DAQ hardware.

At this point it is worth mentioning that the 12 ms sampling interval is selected so as to compromise with the sensor's ODR in the current mode of operation. The calculation of the *ODR_Delay* has been determined with reference to BME280 datasheet. In detail, the total cycle time of measurements in *normal mode* of operation is determined by the *active* and *standby* time. As mentioned earlier, the default value of 0×F5 ("config") register at reset state (i.e. 0×F5 = 0) initializes $t_{standby} = 0.5$ ms. According to BME280 datasheet [53], the typical and maximum active time when using temperature oversampling ×1, pressure oversampling ×4, and no humidity measurement is calculated by Formula 4.1 as, $t_{measure,typ} = 11.5$ ms and $t_{measure,max} = 13.325$ ms. Hence, the typical measurement cycle time in the current example is equal to $t_{total,typ} = t_{measure,typ} + t_{standby,typ} = 11.5 + 0.5 = 12$ ms. On the other hand, the maximum measurement cycle time is $t_{total,max} = t_{measure,max} + t_{standby,max}$[3] $= 13.325 + 0.625 = 13.95$ ms (≅14 ms).

For this example we have chosen the typical measurement cycle time. Hence, our main program should retain a measurement cycle time, no less than 12 ms, so as to assure that we typically read different successive measurement samples. We will explore in a while how the designer may apply the optimum configuration in his/her program.

Formula 4.1 Calculating typical and maximum active measurement time of sensor BME280.

$$
\begin{aligned}
t_{\text{measure,typ}} &= 1 + \left[2 \cdot T\_\text{oversampling}\right]_{\text{osrs}\_t \neq 0} \\
&\quad + \left[2 \cdot P\_\text{oversampling} + 0.5\right]_{\text{osrs}\_p \neq 0} \\
&\quad + \left[2. \cdot H\_\text{oversampling} + 0.5\right]_{\text{osrs}\_h \neq 0} \\
&= 1 + \left[2 \cdot 1\right] + \left[2 \cdot 4 + 0.5\right] + \left[0\right] = 11.5 \text{ ms}
\end{aligned}
$$

$$
\begin{aligned}
t_{\text{measure,max}} &= 1.25 + \left[2.3 \cdot T\_\text{oversampling}\right]_{\text{osrs}\_t \neq 0} \\
&\quad + \left[2.3 \cdot P\_\text{oversampling} + 0.575\right]_{\text{osrs}\_p \neq 0} \\
&\quad + \left[2.3 \cdot H\_\text{oversampling} + 0.575\right]_{\text{osrs}\_h \neq 0} \\
&= 1.25 + \left[2.3 \cdot 1\right] + \left[2.3 \cdot 4 + 0.575\right] + \left[0\right] = 13.325 \text{ ms}
\end{aligned}
$$

Next, the Arduino *setup()* function sets the baud rate of the serial port at 115 200, configures the built-in I2C library, as well as *DAQpin* as output in *HIGH* state. Subsequently, the Arduino *loop()* function calls *BME280_Init(I2Caddr)* to initialize the operating mode of BME280 sensor of I2C address 0×76 (as defined by the corresponding global variable at the

3 According to the BME280 datasheet, the max standby time accuracy is 25%. Thereby, $t_{standby,max} = 0.5 + 0.5 * (25/100) = 0.625$ ms.

beginning of the code) and, after a short delay, it obtains the coeffs, which are used later for the compensation of the raw temperature and pressure data.

Next in the *loop()* function, the *do. . .while()* statement receives data until the '@' arrives on the serial port. As soon the μC board obtains the '@' character and after a short delay, the application code starts the DAQ process within a *for()* loop. Immediately before the loop the *digitalWrite(DAQpin,LOW)* clears the state of PINA3, while immediately after the loop the *digitalWrite(DAQpin,HIGH)* sets the state of PINA3. This way we are able to measure the exact time needed to run the loop for a specified number of samples (herein, the samples are 100).

The *for()* loop code performs the following. First it reads the raw temperature data and then compensates the temperature value. Then it reads the raw pressure data and compensates the latter data (along with the utilization of the temperature value). The compensated pressure data are consecutively assigned to the *P* array of *unsigned char* type, while each acquired pressure value is transmitted to the serial port as numeric string. Before every execution of the *for()* loop, the application code calls the *delay()* routine using the *ODR_Delay* variable and, hence, it delays 12 ms.

Figure 4.2b presents the code running on the Arduino *Serial Monitor*. After opening *Serial Monitor* tool, the user types the '@' character from the PC keyboard and press <Enter> key in order to obtain 100 pressure samples (which appear on the *Serial Monitor* window as soon as the <Enter> is pressed). If we connect an oscilloscope on PINA3, as presented by DAQ measurement hardware of Figure 4.2c, we are able to measure the time of the overall run. The latter measurement is presented by Figure 4.3a, where the overall 100 samples are acquired and sent to the serial port within 1.42 s.

To measure the time required for a single sample the user should comment the *digitalWrite(DAQpin,LOW)* and *digitalWrite(DAQpin,HIGH)* instructions, which are found on the code position where the *single asterisk (*)* appears (Figure 4.2a). In addition, the user must uncomment the corresponding instructions where the *double asterisks (**)* are traced (Figure 4.2a). The latter instructions encompass the code part, which determines the timing for obtaining and transmitting to the serial port a single air pressure measurement. The pulse on PINA3 in that particular case, as acquired by the oscilloscope, is illustrated in Figure 4.3b. In detail, a single sample is acquired and sent to the serial port in 14.2 ms.

The aforementioned code example of Figure 4.2a represents a real-time monitoring system, as it acquires and transmits on the PC the air pressure measurements in real time. A revised version of the code is given in Figure 4.2d. The revised code first reads the 100 samples (within the first, in row, *for* loop) and then performs an offline transmission of the acquired samples to the PC (through the second, in row, *for* loop). That particular implementation allows us to explore air pressure measurements with a smaller sampling interval between the successive samples, as the interval is not affected by the requisite time for transmitting each sample to the PC. It is though affected merely by the sensor's ODR.

The pulse on PINA3, generated by the running code of Figure 4.2d, is presented in Figure 4.3c. The 1.4 ms duration represents the requisite time for obtaining a single compensated measurement of the air pressure. If we add the 12 ms delay determined by the *delay(ODR_Delay)* routine, which was decided by the sensor's ODR (according to the current configuration of BME280 device), we obtain an overall time of 13.4 ms for each compensated pressure measurement. Accordingly, the acquisition of the overall 100 samples

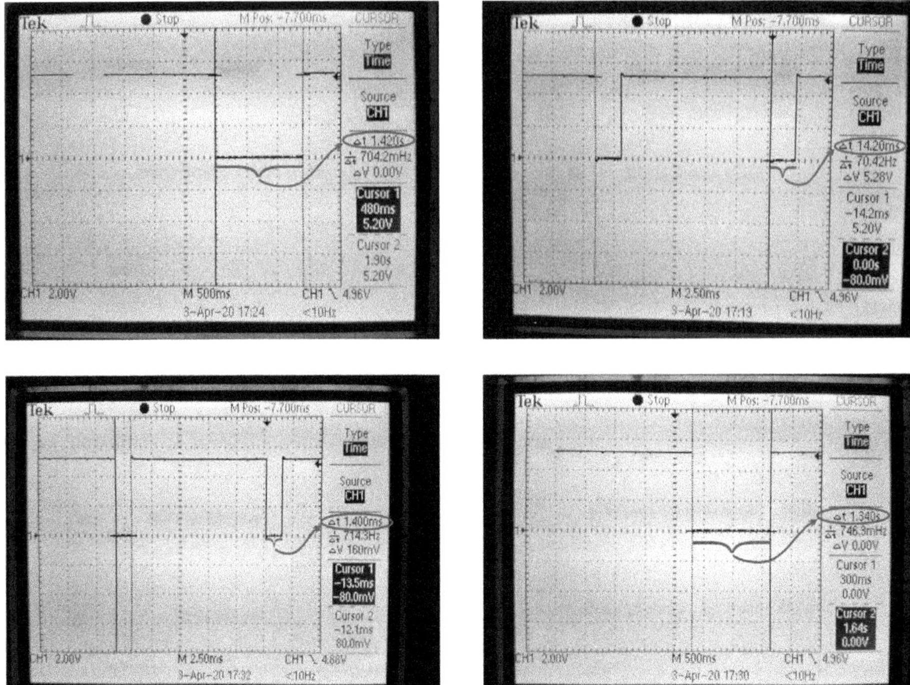

Figure 4.3 Pulse on PINA3 (*DAQpin*) for measuring the time of DAQ process. (a) pulse on PINA3 code: 100 samples acquisition @1.42s, (b) pulse on PINA3 code: single sample acquisition @14.2ms, (c) pulse on PINA3 code: single sample acquisition @1.4ms, (d) pulse on PINA3 code: 100 samples acquisition @1.34s.

is achieved in 1.34 s as illustrated by the PINA3 pulse of Figure 4.3d. The latter pulse is generated by commenting the *digitalWrite(DAQpin,LOW)* and *digitalWrite(DAQpin,HIGH)* instructions denoted by *double asterisks (**)* in Figure 4.2d, as well as uncommenting the corresponding instructions denoted by single *asterisks (*)*.

At this point it is worth mentioning that the revised code of Figure 4.2d supports the acquisition of more samples during the same time period, which might be the desirable case for some applications. This feature of the revised code relies on the fact that the DAQ of each compensated pressure sample is not overloaded by the additional time for transmitting each sample to the PC. Moreover, knowing the exact time for the acquisition a single compensated measurement (as achieved by the pulse of the so called *DAQpin*), allows us to determine more precisely the sensor's ODR within the application code. In the example presented herein, where the desired ODR is 12 ms and each compensated pressure measurement is obtained within 1.4 ms, we could call a delay of 12−1.4 = 10.6 ms (instead of 12 ms). Hence, we are able to acquire even more samples within a predefined time period. This strategy could make a significant difference in applications where the sensor's ODR is generated in a much smaller time period, compared to the requisite time period for sending a sample to the PC.

DAQ Accompanying Software of the Ex.4-1

The code explored hereafter constitutes the custom-designed DAQ software, which is used for the control and collection of the samples acquired by the DAQ hardware board. The source code development is performed in *C programming language*, while the writing process is carried out in the editor *Notepad++*. To compile the source code into machine language so as to run the executable file on a Windows operating system, we first need to install the free tool *MinGW*.

Figure 4.4 presents the first part of the header file *Serial.h*, which incorporates the functions for the access and control of the PC serial port, in the Windows operating system. The code mainly consists of the user-defined *SERIAL_Open()* function, which opens and configures the operation of the serial port. As mentioned earlier in this book, before the appearance of the USB port the most prevalent way of communication between a PC and

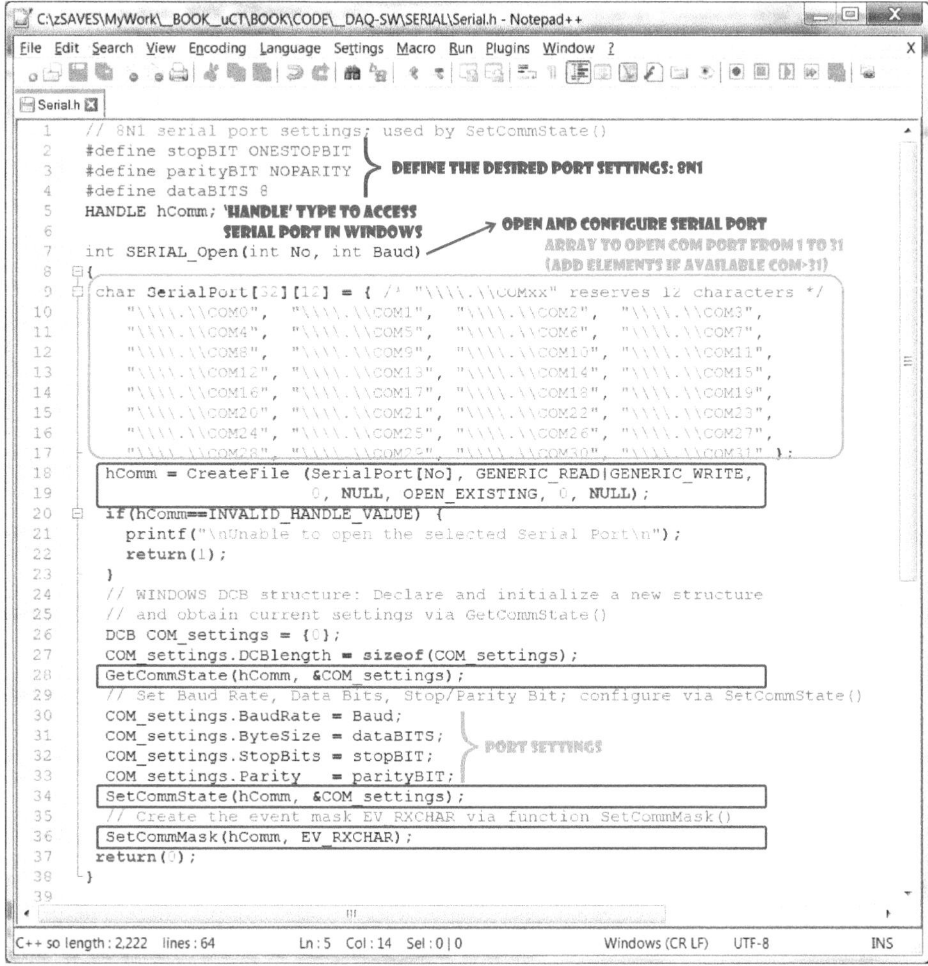

```
// 8N1 serial port settings; used by SetCommState()
#define stopBIT ONESTOPBIT
#define parityBIT NOPARITY
#define dataBITS 8
HANDLE hComm;

int SERIAL_Open(int No, int Baud)
{
 char SerialPort[32][12] = { /* "\\\\.\\COMxx" reserves 12 characters */
     "\\\\.\\COM0",  "\\\\.\\COM1",  "\\\\.\\COM2",  "\\\\.\\COM3",
     "\\\\.\\COM4",  "\\\\.\\COM5",  "\\\\.\\COM6",  "\\\\.\\COM7",
     "\\\\.\\COM8",  "\\\\.\\COM9",  "\\\\.\\COM10", "\\\\.\\COM11",
     "\\\\.\\COM12", "\\\\.\\COM13", "\\\\.\\COM14", "\\\\.\\COM15",
     "\\\\.\\COM16", "\\\\.\\COM17", "\\\\.\\COM18", "\\\\.\\COM19",
     "\\\\.\\COM20", "\\\\.\\COM21", "\\\\.\\COM22", "\\\\.\\COM23",
     "\\\\.\\COM24", "\\\\.\\COM25", "\\\\.\\COM26", "\\\\.\\COM27",
     "\\\\.\\COM28", "\\\\.\\COM29", "\\\\.\\COM30", "\\\\.\\COM31" };
  hComm = CreateFile (SerialPort[No], GENERIC_READ|GENERIC_WRITE,
                      0, NULL, OPEN_EXISTING, 0, NULL);
  if(hComm==INVALID_HANDLE_VALUE) {
    printf("\nUnable to open the selected Serial Port\n");
    return(1);
  }
  // WINDOWS DCB structure: Declare and initialize a new structure
  // and obtain current settings via GetCommState()
  DCB COM_settings = {0};
  COM_settings.DCBlength = sizeof(COM_settings);
  GetCommState(hComm, &COM_settings);
  // Set Baud Rate, Data Bits, Stop/Parity Bit; configure via SetCommState()
  COM_settings.BaudRate = Baud;
  COM_settings.ByteSize = dataBITS;
  COM_settings.StopBits = stopBIT;
  COM_settings.Parity   = parityBIT;
  SetCommState(hComm, &COM_settings);
  // Create the event mask EV_RXCHAR via function SetCommMask()
  SetCommMask(hComm, EV_RXCHAR);
 return(0);
}
```


Figure 4.4 Header file *Serial.h* for the communication with the PC serial port (1 of 2). *Source:* Arduino Software.

μC-based hardware was the UART serial port. To support downward compatibility of nowadays USB port to the older UART serial port, the modern μC motherboards incorporate USB to UART serial port converters. The accompanying driver that is installed on the PC allows the PC to exchange data with the microcontroller hardware via the USB port, while the USB port is handled as being UART serial port. What this means in practice is that the designer who develops the software code configures a UART serial communication between the PC and the μC motherboard, which is much simpler compared to the configuration and control of a true USB communication.

In consideration of the aforementioned information, the first three *#define* directives of Figure 4.4 determine the most common mode of operation of the UART serial port. That is, each frame exchanged through the serial port consists of one *stop* bit, none *parity* bits, and eight *data* bits. The subsequent code line declares a *HANDLE* type of variable, named *hComm*. In a Windows operating system, the *HANDLES* provide access to the operating system resources. Herein, the *hComm* is used by the Windows' functions to identify the serial port being utilized by the application code.

Next in the code is the declaration of the user-defined *SERIAL_Open()* function, which admits two arguments. The former defines the serial port number (aka *COM port*), while the latter determines the data rate of exchanging information (*aka* baud rate). Subsequent to the *SERIAL_Open()* declaration is a two-dimensional array (i.e. within the green frame). The array elements represent the strings in the appropriate format as being used by the corresponding Windows function, which opens and configures a COMx port (e.g. COM1, COM2). The array works for 31 serial port numbers (that is, from COM1 to COM31), but the user may add more elements in case the port number where the Arduino Uno is connected, is not found in the existing array. The rest of the *SERIAL_Open()* function makes mainly use of prototype functions for the configuration and control of the serial port, which are declared within the *Windows.h* header file. Hence, the latter file should be included either on the top of the current header file, or in the beginning of the application code.

The *CreateFile()* function is used to open and establish a connection with a COM port. The function admits the following arguments:

i) ***SerialPort[No]:*** holds the name of the serial port (herein, we make use a number from 1 to 31, which obtains the corresponding string from *SerialPort* array);
ii) ***GENERIC_READ|GENERIC_WRITE:*** constitute the most common values, which request *read* as well as *write* access to the serial port device;
iii) ***0:*** the value zero prevents the serial port device, after being opened, to be shared and opened again until the handle to the device is closed;
iv) ***NULL:*** defines a NULL pointer to a SECURITY_ATTRIBUTES structure (that is, the default security descriptor), which means that the handle returned by *CreateFile()* cannot be inherited by any child processes (which are might be created by the application code);
v) ***OPEN_EXISTING:*** constitutes the most common value, which opens the device (or file), only if it exists;
vi) ***0:*** the value zero refers to NonOverlapped IO operation (i.e. several threads cannot interact simultaneously with the port);
vii) ***Null:*** value on the device's attributes and flags means that none of the flags are set (i.e. the device or file does not match any special attributes).

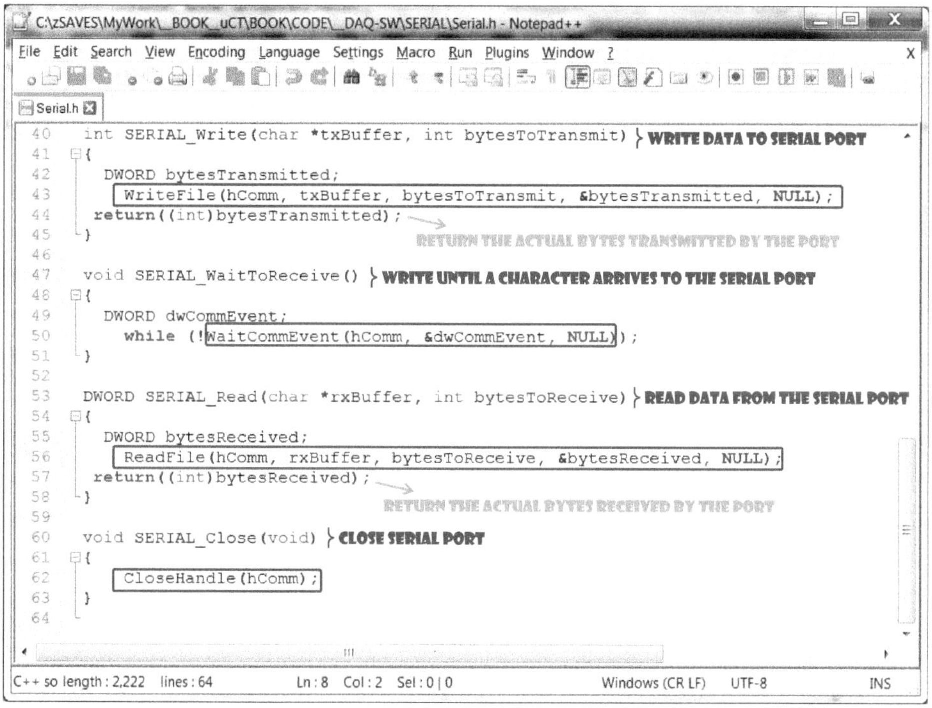

```
int SERIAL_Write(char *txBuffer, int bytesToTransmit)   WRITE DATA TO SERIAL PORT
{
  DWORD bytesTransmitted;
  WriteFile(hComm, txBuffer, bytesToTransmit, &bytesTransmitted, NULL);
 return((int)bytesTransmitted);
}
                                   RETURN THE ACTUAL BYTES TRANSMITTED BY THE PORT

void SERIAL_WaitToReceive()   WRITE UNTIL A CHARACTER ARRIVES TO THE SERIAL PORT
{
  DWORD dwCommEvent;
   while (!WaitCommEvent(hComm, &dwCommEvent, NULL));
}

DWORD SERIAL_Read(char *rxBuffer, int bytesToReceive)   READ DATA FROM THE SERIAL PORT
{
  DWORD bytesReceived;
  ReadFile(hComm, rxBuffer, bytesToReceive, &bytesReceived, NULL);
 return((int)bytesReceived);
}
                                   RETURN THE ACTUAL BYTES RECEIVED BY THE PORT

void SERIAL_Close(void)   CLOSE SERIAL PORT
{
  CloseHandle(hComm);
}
```

Figure 4.5 Header file *Serial.h* for the communication with the PC serial port (2 of 2). *Source:* Arduino Software.

The subsequent *if()* statement of the code prints the message "Unable to open the selected Serial Port" in case the selected COM device could not be opened by the *CreateFile()* function. It is worth mentioning that the current syntax of the C language *printf()* function makes use of the *new line ('\n')* character when printing the above message. Subsequent to the *if()* statement we declare and initialize a new Windows DCB structure. Next, the *GetCommState()* and *SetCommState()* functions are respectively used to obtain current setting of the serial port, and configure the desired settings of the opened device (i.e. the user-defined baud rate, 8 data bits, 1 stop bits, no parity bit). Finally, the *SetCommMask()* enables the monitoring of the *EV_RXCHAR* event, which means that we are able to inspect when a character arrives to the serial port.

Figure 4.5 presents the second part of the header file *Serial.h* for control of the PC serial port (in Windows operating system). In detail, the user-defined function *SERIAL_Write()*admits two arguments and it is used to write data to the serial port. The first argument constitutes a pointer to *char* type array, which incorporates the characters to be transmitted, while the latter incorporates an *int* type variable, which determines the number of bytes to be sent to the serial port.

The *SERIAL_Write()* function invokes the Windows *WriteFile()* function, which admits the following arguments:

i) ***hComm:*** the HANDLE type variable, which is used to identify the serial port;
ii) ***txBuffer:*** the character array incorporating the data to be sent to the serial port;

iii) ***bytestoTransmit:*** the number of bytes to be sent to the serial port;
iv) ***&bytesTransmitted:*** the number of bytes that were successfully transmitted;
v) ***NULL:*** unless the *CreateFile()* function opens the device with FILE_FLAG_OVERLAPPED, this parameter can be set to NULL.

The *SERIAL_Write()* function returns, through a *typecasting* conversion, an *int* value comprising the true bytes that were successfully transmitted.

Because the *CreateFile()* function herein enables the monitoring of the *EV_RXCHAR* event, the user-defined *SERIAL_WaitToReceive()* function waits until a character is received by the serial port. The function invokes the Windows *WaitCommEvent()* function within a *while()* statement, which admits the following arguments:

i) ***hComm:*** the HANDLE type variable, which is used to identify the serial port;
ii) ***&dwCommEvent:*** pointer to variable indicating the type of event occurred;
iii) ***NULL:*** unless the *CreateFile()* function opens the device with FILE_FLAG_OVERLAPPED, this parameter can be set to NULL.

The user-defined *SERIAL_Read()* function admits two arguments and it is used to read data from the serial port. The first argument constitutes a pointer to *char* type array, which is used to hold the received characters port, while the latter incorporates an *int* type variable, which determines the anticipated number of bytes to be received by the serial port. The *SERIAL_Read()* function invokes the Windows *ReadFile(),* which admits the following arguments:

i) ***hComm:*** the HANDLE type variable, which is used to identify the serial port;
ii) ***rxBuffer:*** the character array to hold the data received by the serial port;
iii) ***bytestoReceive:*** the anticipated number of bytes to be received;
iv) ***&bytesReceived:*** the true number of bytes that were successfully received;
v) ***NULL:*** unless the *CreateFile()* function opens the device with FILE_FLAG_OVERLAPPED, this parameter can be set to NULL.

The *SERIAL_Read()* function returns, through a *typecasting* conversion, an *int* value comprising the true bytes that were successfully received by the serial port.

The user-defined *SERIAL_Close()* function invokes the Windows *CloseHandle(),* which is used to close the serial port, otherwise the opened device will become unavailable to other applications.

Figure 4.6 presents the application code, in C programming language, which uses the serial port toward acquiring data from Arduino Uno board. The *stdio.h* header file holds the C language definitions and functions for inputting/outputting data to/from the PC; e.g. the *printf()* function which prints data on the PC monitor. The *Windows.h* file incorporates the Windows functions, which are exploited by the operating system resources toward controlling and accessing the computer's serial port. The *serial.h* file encompasses the user-defined functions (explored before), which are addressed to facilitate the configuration and accessing process of the serial port through the application code.

The application code consists of a single function, which implements the DAQ process. The so-called *SERIAL()* function admits five arguments of *int* data type, which determine (i) the *COM* port number that will be used by the application code, (ii) the *baud rate* of the

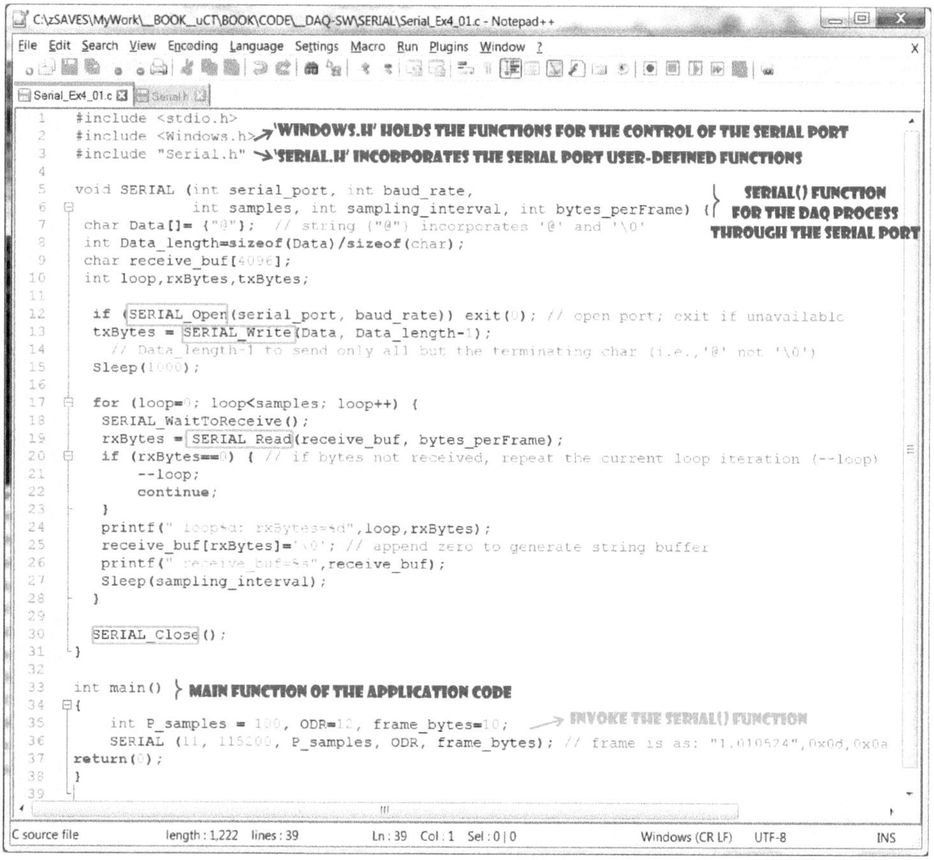

C:\zSAVES\MyWork_BOOK_uCT\BOOK\CODE_DAQ-SW\SERIAL\Serial_Ex4_01.c - Notepad++

```c
#include <stdio.h>
#include <Windows.h>
#include "Serial.h"

void SERIAL (int serial_port, int baud_rate,
             int samples, int sampling_interval, int bytes_perFrame) {
 char Data[]= {"@"};  // string {"@"} incorporates '@' and '\0'
 int Data_length=sizeof(Data)/sizeof(char);
 char receive_buf[4096];
 int loop,rxBytes,txBytes;

  if (SERIAL_Open(serial_port, baud_rate)) exit(0); // open port; exit if unavailable
  txBytes = SERIAL_Write(Data, Data_length-1);
    // Data_length-1 to send only all but the terminating char (i.e.,'@' not '\0')
  Sleep(1000);

  for (loop=0; loop<samples; loop++) {
   SERIAL_WaitToReceive();
   rxBytes = SERIAL_Read(receive_buf, bytes_perFrame);
   if (rxBytes==0) { // if bytes not received, repeat the current loop iteration (--loop)
       --loop;
       continue;
   }
   printf(" loop%d: rxBytes=%d",loop,rxBytes);
   receive_buf[rxBytes]='\0'; // append zero to generate string buffer
   printf(" receive_buf=%s",receive_buf);
   Sleep(sampling_interval);
  }

  SERIAL_Close();
}

int main()
{
    int P_samples = 100, ODR=12, frame_bytes=10;
    SERIAL (11, 115200, P_samples, ODR, frame_bytes); // frame is as: "1.010524",0x0d,0x0a
return(0);
}
```

'WINDOWS.H' HOLDS THE FUNCTIONS FOR THE CONTROL OF THE SERIAL PORT
'SERIAL.H' INCORPORATES THE SERIAL PORT USER-DEFINED FUNCTIONS
SERIAL() FUNCTION FOR THE DAQ PROCESS THROUGH THE SERIAL PORT
MAIN FUNCTION OF THE APPLICATION CODE
INVOKE THE SERIAL() FUNCTION

C source file length : 1,222 lines : 39 Ln : 39 Col : 1 Sel : 0 | 0 Windows (CR LF) UTF-8 INS

Figure 4.6 DAQ software for the communication with the PC serial port (*Serial_Ex4_01.c*). *Source:* Arduino Software.

serial port, (iii) the number of samples that will be obtained from the DAQ hardware upon a single run of the software code, (iv) the delay between adjacent samples in terms of the sensor's ODR, and (v) the bytes held by each individual sample.

Immediately below the definition of the *SERIAL()* function are the variables used by the function. In detail, the *char* datatype array named*DATA*, holds a string, as determined by the *double quotes ("...")*. That is, a *null ('\0')* terminated character array and, hence, the *DATA* array holds the characters '@' and '\0'. It is here noted that the latter array is used for writing data to serial port. The next *int* datatype variable named *Data_length* exploits the *sizeof()* operator so as to return the number of bytes reserved by the *DATA* array. The *receive_buf* character array of 4096 memory locations is used to store the data arrived from the serial port. The integer variable *loop* is used within a *for()* statement so as to count the number of samples acquired by the serial port. The integer variables *rxBytes*, *txBytes* are respectively used for holding the actual bytes that were successfully received and transmitted by the PC serial port.

Subsequent to the variable definitions is the user-defined function *SERIAL_Open()*, encompassed within an *if()* statement. Thereby, if the selected COM port is unavailable and cannot be opened, the *exit(0)* function terminates the execution of the application code. If the COM port is successfully opened, the user-defined function *SERIAL_Write()* sends to the available COM port all the elements of *Data* array, except the very last element (a choice made by the directive *Data_Length-1*). Since the current *Data* array holds merely two elements, the *SERIAL_Write()* function sends to the serial port the '@' character. After that, the code delays 1 s.

Next, the *for()* loop obtains the number of samples determined by the user definitions. The first instruction within the loop waits until a character arrives to the PC serial port. As soon as that character arrives, the user-defined function *SERIAL_Read()* obtains from the COM port the available data, where the maximum number of bytes read by the *SERIAL_Read()* is determined by the *bytes_perFrame* variable. The received data are assigned to the *receive_buf* array. The subsequent *if()* statement is addressed to confirm that data has arrived to the serial port. Otherwise, the *loop* counter is unary decreased and the *for()* execution resumes from the beginning, as determined by the execution of the *continue* statement. The unary decrement operation along with the execution of the *continue* statement, causes the current loop iteration to be executed again in case none data arrived to the serial port.

If data have successfully arrived to the serial port, the subsequent *printf()* function prints on the PC screen the current decimal value (as defined by the *%d* directive) of the loop counter, as well as the number of bytes received (e.g. *loop17: rxBytes=10*). It is worth noting that the current example obtains ten ASCII characters, where each one of them represents a decimal digit (or the decimal point) of a number. To simplify the printing process of the received characters, the *receive_buf[rxBytes]='\0';* instruction appends a zero value ('\0') to the *receive_buf* array after the last available character. Hence, each received frame is arranged as a *null ('\0')* terminated numeric string (e.g "0.996470") and the subsequent *printf()* function prints that string (as defined by the *%s* directive) on the PC screen. The very last instruction of the *for()* loop delays the program for a period that is in agreement with the sensor's ODR, as determined by DAQ firmware.

Immediately below the *for()* loop is the user-defined function *SERIAL_Close(),* which closes the opened COM port, as soon as the acquisition of the overall samples has been concluded. Inside the *main()* function of the C code, the user may modify the COM number, the baud rate, the total number of samples, the sampling interval (i.e. delay between adjacent samples), as well as the number of bytes of each received frame. The present example establishes the following configuration for the DAQ software: COM11, 115 200 baud rate, 100 samples per run, 12 ms sampling interval, 10 bytes for each received samples.

At this point is worth mentioning that the user may identify the available COMx port number (where the μC board is attached) by the Windows Device Manager. In the example of Figure 4.7, we confirm that the Arduino Uno is attached to COM11 port.

Figure 4.8 presents the command lines for the compilation and execution of the DAQ software, as well as the 100 samples printed on the *command prompt* console. First of all, the user should run a Windows *command prompt* console[4] and enter into the working folder where the source code is located. To perform the latter task the user types command

---

4 One way to open a *command prompt console* is to press the *Windows* along with the *R* key (i.e. *Win* + *R*), type *cmd* in the *Run dialog box*, and press <Enter>.

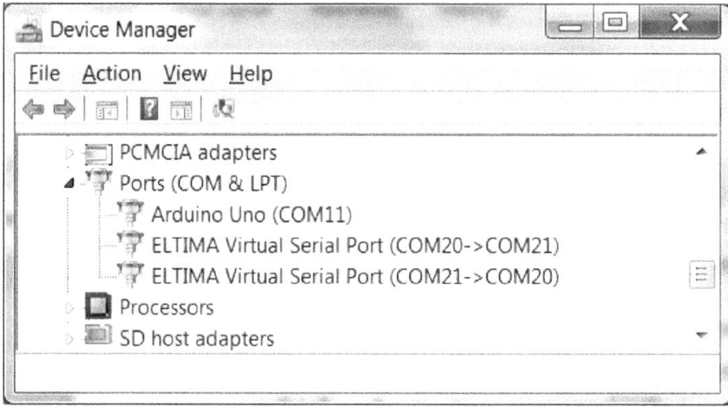

Figure 4.7 Arduino Uno connected to COM11 serial port (Windows Device Manager). *Source:* Arduino Software.

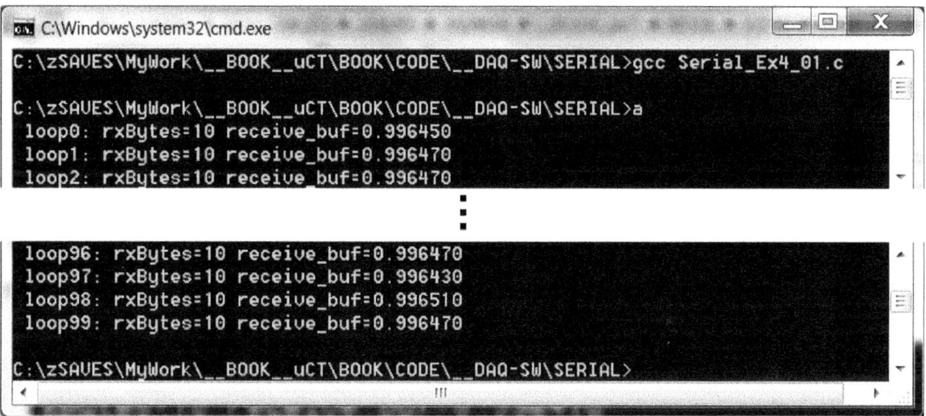

C:\Windows\system32\cmd.exe
C:\zSAVES\MyWork__BOOK__uCT\BOOK\CODE__DAQ-SW\SERIAL>gcc Serial_Ex4_01.c
C:\zSAVES\MyWork__BOOK__uCT\BOOK\CODE__DAQ-SW\SERIAL>a
loop0: rxBytes=10 receive_buf=0.996450
loop1: rxBytes=10 receive_buf=0.996470
loop2: rxBytes=10 receive_buf=0.996470
loop96: rxBytes=10 receive_buf=0.996470
loop97: rxBytes=10 receive_buf=0.996430
loop98: rxBytes=10 receive_buf=0.996510
loop99: rxBytes=10 receive_buf=0.996470
C:\zSAVES\MyWork__BOOK__uCT\BOOK\CODE__DAQ-SW\SERIAL>

Figure 4.8 Compilation and execution of the DAQ software. *Source:* Arduino Software.

cd, presses <Space> key, types the path of the working folder (e.g. *cd C:\zSAVES\MyWork__BOOK__uCT\BOOK\CODE__DAQ-SW\SERIAL*) and presses <Enter>. To compile the C language code into machine language, the user types the *gcc* command for the compiler, presses the <Space>, types the filename (e.g. *gcc Serial_Ex4_01.c*) and presses <Enter>. The latter command line generates an executable file of name *a.exe*. To run the executable file, the user types either *a.exe* or just *a* and presses <Enter>. To generate an executable of name identical to the source code file, the user types the command *gcc<Space>filename.c<Space>-o<Space>filename.exe* (e.g. *gcc Serial_Ex4_01.c -o Serial_Ex4_01.exe*). The latter command line generates the executable file *Serial_Ex4_01.exe*, which can be executed by typing either *Serial_Ex4_01.exe* or just *Serial_Ex4_01*.

In Figure 4.9 the source code *Serial_Ex4_01.c* is enriched by the requisite instructions for the storage of the acquired measurements to a *text (.txt)* file on the PC hard disk. The additional instructions are enclosed within the red rectangles, as presented in Figure 4.9a. In detail, the code first declares a pointer of *FILE* type (i.e. via the instruction *FILE *fp;*).

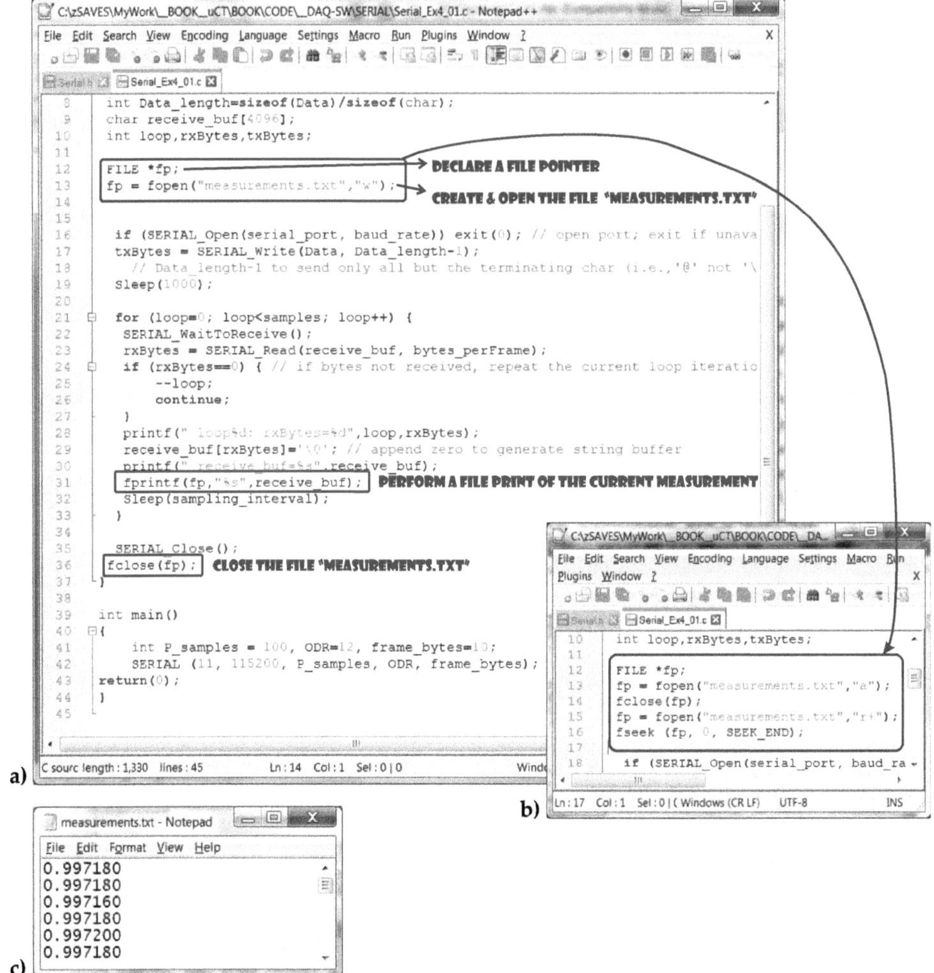

Figure 4.9 Enrichment of the DAQ software *Serial_Ex4_01.c* to store data to a text file. (a) code instructions to create a new file and write data to it, (b) code instructions to keep the existing file and append new data to it, (c) data written to the file measurements.txt. *Source:* Arduino Software.

The data type *FILE* is defined within the header file *stdio.h* and is used to handle files. The so called *file pointer* (which is herein given the identical name **fp*) is used to handle IO operations on the opened file. The file is created, or opened (if already created), during the execution of the *fopen()* function. The first argument of the latter function holds the name of the file, while the second argument holds the mode of accessing as well as the privileges of the opened file. Herein, the "*w*" directive creates the file *measurement.txt* in case the latter does not exist, or truncate the file to zero length (if it already exists).

To append data to an opened file we use the function *fprintf()*. In this example, the *fprintf(fp,"%s",receive_buf);* instruction appends to the file the current string found on the *receive_buf* register (that is, the current measurement sample). Before exiting the DAQ software, the *fclose(fp);* instruction closes *measurement.txt* file. Otherwise, we would not be able to open the file in our PC, so as to see or analyze the measurements.

Every time the code example of Figure 4.9a is executed, it deletes any previous data found on *measurement.txt* file and writes the new acquired samples. To keep the previous data and append new samples to the file, we can create/open the file through the instructions presented in Figure 4.9b. The *"a"* directive, currently used by the *fopen()* function, creates a new file and if the file already exists, it does not delete its content. In each case the file is closed and reopened using the directive *"r+"*, which allows us to update the file. Thereafter, the *fseek()* function identifies the end of file position via the *SEEK_END* value (i.e. third argument), while also assigning a *zero (0)* offset value from the identified position (i.e. second argument). As a result, the new measurements are appended at the end of the file (without destroying the previous data). The assigned data on *measurements.txt* file are presented in Figure 4.9c.

Figure 4.10 presents a new tool, the *Free Serial Port Monitor*, which allows us to inspect an open COM port and, hence, explore the data sent out from, or delivered by the DAQ software. This program is particular useful when developing software code that controls the PC serial port. Figure 4.10a,b presents how to configure a new session on the *Serial Port*

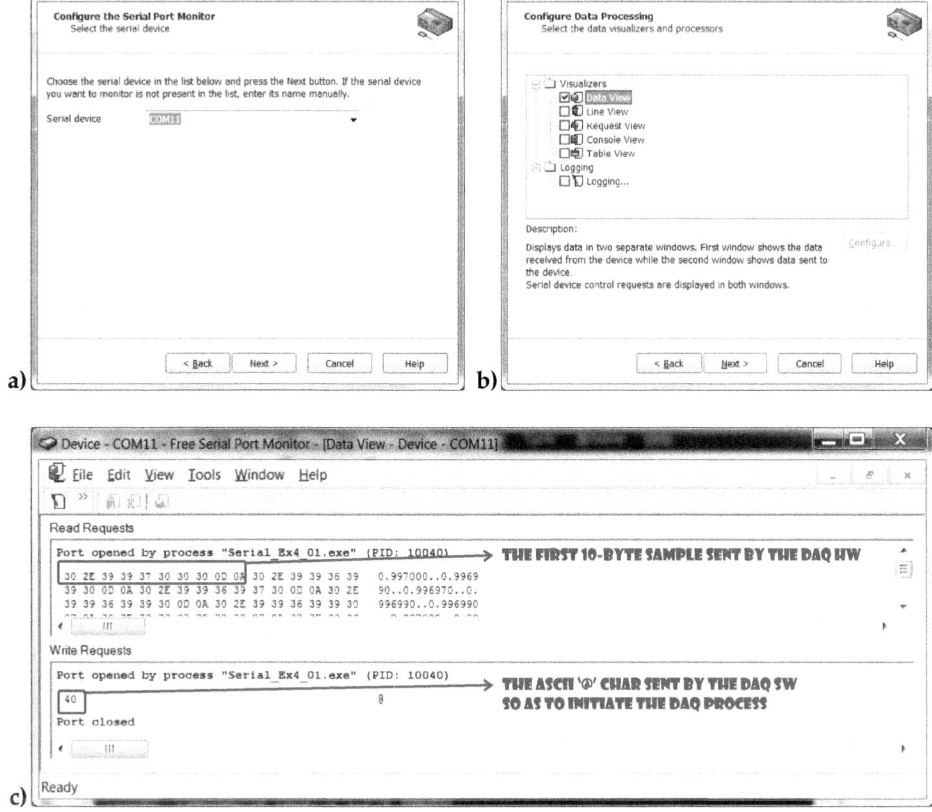

Figure 4.10 Free Serial Port Monitor tool for inspecting the data exchanged by the COM port. (a) create new session: select *serial device,* (b) create new session: select *data view* mode, (c) data captured by the software. *Source:* Arduino Software.

Monitor, while Figure 4.10c presents the read and write requests captured by this software. The *read requests* window incorporates the data arrived to the host PC, while the *write requests* window incorporates the data sent out from the PC.

DAQ Accompanying Software with Graphical Monitoring Feature Via gnuplot

The example presented hereafter is enriched by the instructions able to provide a graphical monitoring of the acquired pressure samples. To run the example, the designer should install the free *gnuplot* software available at www.gnuplot.info/. The *gnuplot* constitutes a command-line program that can generate two- or three-dimensional plots of data, and can be invoked by C language code. The software code is designed to run with the previously explored firmware **Ex4-01**, but to make the *gnuplot* graphing utility working with the C software code the designer should change the baud rate to 9600. Moreover, while the originally decided sensor's ODR of 12 ms works fine, it has herein changed to 100 ms so as to provide a smoother graphical representation of the real-time monitoring process. Those two amendments of the code **Ex4-01** are highlighted in Figure 4.11b and outline the minor revision of the original code, currently named **Ex4-01c**.

Figure 4.11a presents the DAQ software enriched by a graphical monitoring feature. All new instructions that generate the graphing utility are enclosed within the frames appeared on the *SERIAL()* function. The red and green frames respectively illustrate the new initialization as well as the new part of the main code, while the blue frame depicts the required amendments to the previous code example. As in the previous DAQ software addressed for storing data within a *.txt* file, the current example performs IO operations that are also associated with functions and directives of the *C Standard Input and Output library* (aka *stdio.h*). Hence, no other header files are included in the code.

In terms of the variable definitions, the code utilizes a *char* type array of 10 elements, defined as *X_Axis* (i.e. code line 6). It is here noted that the example generates a two-dimensional plot of data, where in *x* axis is the number of the currently acquired measurement (obtained from the loop counter), and in *y* axis is the value of the current pressure sample held by *receive_buf* variable. Accordingly, both plotted variables incorporate a numeric string. An additional variable defined by the code is the *int* type *gnuplotWindow,* which is assigned to the value 25. We will explore later how the *gnuplotWindow* can be used to determine the number of samples appeared on the gnuplot window.

Below the variable definitions, the *fclose()* function (code line 8) terminates the *stderr* IO device. The latter device is mapped to the Windows terminal console, that is, *the command prompt*, and generates the error messages. Because the opened *gnuplot* program might send warning messages while running, the *fclose()* function terminates the *stderr* IO device. Thereby, all warning messages are suppressed during the code execution.

An external program can be invoked by the C language code through the execution of the *popen()* function. Accordingly, the code line 14 creates a *pipe* between the *gluplot* external program and the executed command and returns the *pipe pointer* (herein given the identical name **pipe*). The pointer is used either for reading or writing to the pipe. The function *popen()* encompasses two directives, that is, the "*w*" directive, which establishes a writing process to the open pipe with the *gnuplot* program, as well as the directive *-persist,* which keeps the plot open, even after the termination of the executable C code. If the latter

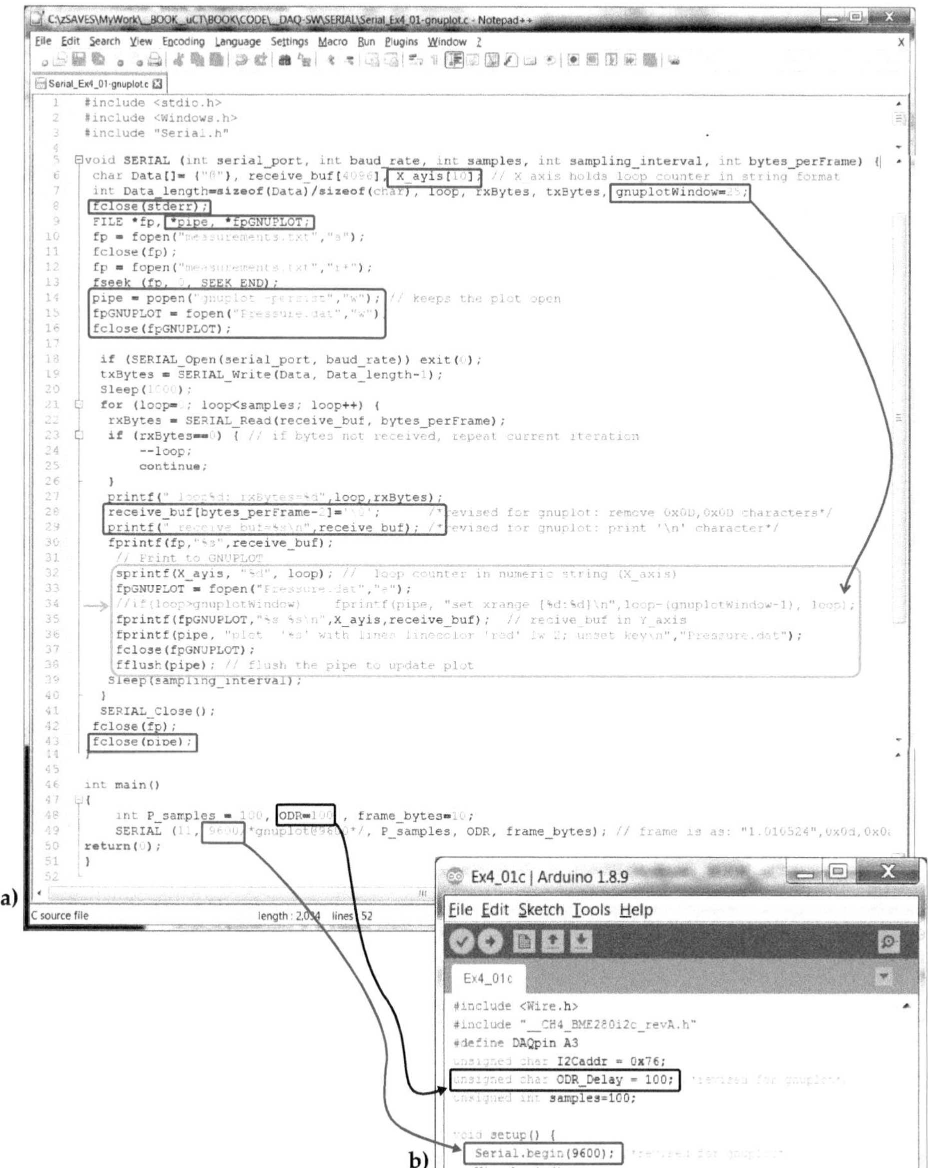

Figure 4.11 DAQ software with graphical monitoring feature (*Serial_Ex4_01-gnuplot.c*). (a) application software code, (b) application firmware code. *Source:* Arduino Software.

directive is not used, the plot appears on the PC screen for as long the executable DAQ programs runs on the PC. Then the termination of the executable code closes the plot graph.

Code lines 15 and 16 are equivalent to those used for generating the *measurement.txt* file. Because the *gnuplot* program plots the data found within a file, code line 15 generates the data file *Pressure.dat* using the "*w*" directive (i.e. if the file already exists from a previous run, it deletes its content). The *pressure.dat* file is handled by the file pointer

**fpGNUPLOT*. Code line 16 closes the generated file *Pressure.dat*, which is later reopened through the *for()* loop so as to append the acquired samples.

The code inside the blue frame incorporates no new instructions, but revision to the previous DAQ software appropriate for the plotting process. It is here reminded that each sample sent by the firmware consists of 10 ASCII characters representing a number, such as "1.031 713", followed by *carriage return* $(0 \times 0D)$ and *new line* $(0 \times 0A)$ characters. To arrange the samples appropriate for the *pressure.dat* file (used by the *gnuplot* program), code line 28 appends a *null* (0×00) character immediately after the number and, hence, it forms a numeric string such as, "1.031 713<null>". Accordingly, the code line 29 prints on the *command prompt* window the content of the revised *receive_buf* variable ("1.031 713<null>") along with a '\n' character so that each sample reserves a separate line on the window.

The main code that generates the plot inside the *for()* loop is enclosed with the green frame. In detail, the *sprint()* function in code line 32, reforms the integer value of the *loop* counter into string, and assigns the result into the *x_axis* character array. Next, the code line 33 reopens *Pressure.dat* file using the *"a"* directive in order to append data to the file on each single execution of the *for()* loop. The commented *if()* statement of code line 34 decides on the number of the acquired samples appeared on the plot, as determined by the *set xrange* command in the direction of the *gnuplot* program. In detail, the first 25 samples are plotted on the *gnuplot* window and, as soon as the *loop* counter exceeds the number of *gnuplotWindow* variable (herein set to 25), the x axis is recalculated to uphold values from *loop-(gnuplotWindow-1)* till *loop*. For instance, if the *loop* counter value is equal to 26, the x axis of the plot incorporates the samples: 26-(25-1) to 26 (i.e. the samples 2–26). Accordingly, the plot initially prints the samples 1–25 and, thereafter, it successively plots the samples 2–26, 3–27, 4–28, and so forth. If we comment the code line 34 (as illustrated in the example), the *gnuplot* window is forced to plot all the acquired samples on the graph (herein, it plots 100 samples).

The *fprint()* function of code line 35 prints to the *Pressure.dat* file (as decided by the *fpGNUPLOT* file pointer), two strings per line with a <Space> character in between them (as defined by the directives *%s %s\n*). The first string is reserved in the *x_axis* variable and holds the *loop* counter value, while the second is found in the *receive_buf* variable and holds the current air pressure measurement.

The *fprint()* function of code line 36 prints to the opened pipe (which is connected to the *gnuplot* program) the stings (as defined by the *%s* directive) found in the *"Pressure.dat"* file. The samples are connected with a line of red color and line width 2. The latter task is determined by a directive addressed for the *gnuplot* program, that is, *with line color 'red' lw 2*. The *unset key* directive is also meant for the *gnuplot* program and tells to the latter program not to print any key (i.e. legend) to the generated plot.

The *fclose()* function of code line 37 closes the *Pressure.dat* file, while the *fflush()* function of code line 38 moves the buffered data of the previous *plot* task to the console. Hence, the data are delivered to the host environment and update the plot. Outside the *for()* loop the *fclose()* function of code line 43 closes the opened pipe of the software code along with the *gnuplot* program. It is worth mentioning two revisions made to the *main()* function of the software code, which are addressed to provide compatibility with the firmware. That is, (i) the ODR value (in code line 48) is revised to 100 (i.e. 100 ms between successive

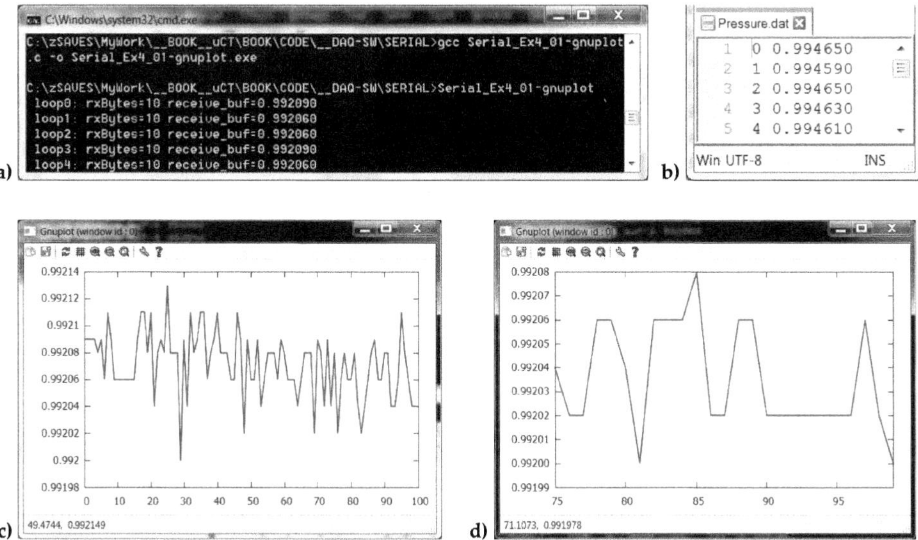

Figure 4.12 Compilation and execution of the *Serial_Ex4_01-gnuplot.c* DAQ software. (a) Windows command prompt console, (b) data stored to *Pressure.dat*, (c) plot of measurements with *gnuplot* (example 1), (d) plot of measurements with *gnuplot* (example 2). *Source:* Arduino Software.

samples) so as to provide a smoother graphical representation of the real-time monitoring process, and (ii) the baud rate is revised to 9600 so as to make gnuplot graphing utility working properly with the C software code.

Figure 4.12a illustrates the compilation command of the software DAQ code and the execution of the generated *.exe* file, as well as the printed results on the terminal console. Figure 4.12b depicts the content of the generated *Pressure.dat* file, which holds the data to be plotted by the *gnuplot* program. Finally, Figure 4.12c,d depicts two plots generated by two different runs of the executable code. The former incorporates the overall 100 samples acquired by the software code, while the latter incorporates merely the last 25 samples. The latter plot is generated by uncommenting code line 34 of the DAQ software given in Figure 4.11a.

Arduino Ex.4–2

The code explored herein is similar to the previous example, except that now the DAQ hardware and accompanying software are designed to acquire samples from two different sensor devices. Since the air pressure measurements could be rearranged into barometric altitude data [55], a potential use of such a system could be the estimation of the absolute height in between the sensor devices. Typical applications can be found in indoor navigation projects for, perhaps, the detection of floor changes, the elevator detection, etc. Figure 4.13 presents the DAQ measurement firmware (Figure 4.13a) and hardware (Figure 4.13b), which obtains data from two identical BME280 sensor devices, as well as a plot of the acquired samples using the *Serial Plotter* program of Arduino IDE (Figure 4.13c).

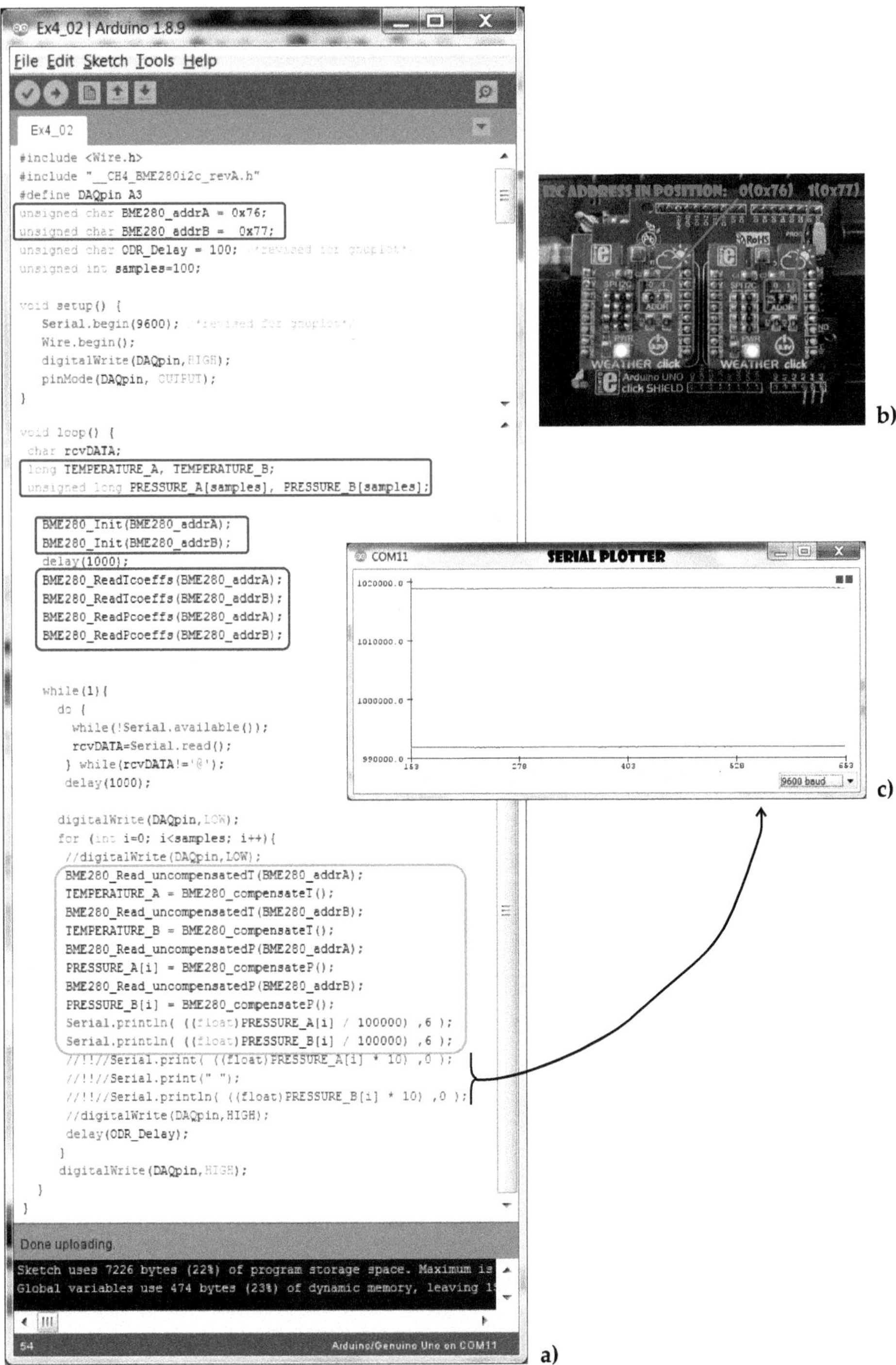

Figure 4.13 DAQ measurement hardware and firmware (reading data from two sensors). (a) DAQ firmware, (b) DAQ hardware, (c) plot of measurements with *serial plotter*. *Source:* Arduino Software.

In regard to the DAQ measurement hardware the designer should change the I2C address in one of the two *Weather clicks*. As presented within the red frames on the top of the *Weather clicks* in Figure 4.13b, by applying short circuit to the two contacts of the board on the *position* named *0*, the I2C address of the sensor device is set to 0×76. Similarly, by applying short circuit to the contacts of *position 1*, the I2C address of the sensor device changes to 0×77.

In regard to the DAQ measurement firmware in Figure 4.13a, the frames highlight the differences between the current, and the previous example code, which obtains data from a single sensor device. The red frames highlight the variable definitions, which are twice as the variables used by the single-sensor setup. For instance, there are two *long* type variables (i.e. *TEMPERATURE_A and TEMPERATURE_B*) to hold the raw temperature data. Similarly, the blue frames highlight the sensors' initialization as well as the acquisition of pressure and temperature coeffs (which are performed for two sensors instead of one). Finally, the DAQ process of the two sensors is performed by the code found inside of the green frame.

To generate a quick plot of the acquired samples using the *Serial Plotter* tool, as illustrated by Figure 4.13c, the designer should comment the two *Serial.println()* functions within the green frame, and uncomment the three subsequent commented *Serial.println()* functions. The revision reforms the pressure samples in *μbar* units (i.e. $1\,\mu\text{bar} = 10^{-6}$ bar) so as to provide a more clear plot compared to the one generated when using *bar* units. It is worth noting that identical sensor devices feature a deviation to the output signal when acquiring the exact same pressure, under identical environmental conditions.

Figure 4.14 presents the DAQ software that obtains data from two identical sensor devices (Figure 4.14a), as well as the generated results when executing the code (Figure 4.14b,c). In consideration of the DAQ software code of Figure 4.14a, the red frames illustrate the differences between the current and the previous example (which was used for the acquisition of data from a single sensor device). The code addresses two buffers for the acquisition of pressure samples from the serial port (that is, the *receive_bufA* and *receive_bufB*), and hence, all the corresponding processes are twice as the original tasks of the single-sensor setup. It is worth noting that if we comment the *fclose()* function of code line 8, warning messages from the *gnuplot* program may appear on the *command prompt* console during the code execution (as highlighted by the green frame of Figure 4.14b). In consideration of the plot of Figure 4.14c generated by the running code, the spikes on the two plots have been occurred by blowing closely to the corresponding sensor device.

Orientation, Motion, and Gesture Detection with Teensy 3.2

The advancement of the modern μC devices is, in fact, outstanding. While the previous generation of μC technology mainly focused on 8-bit microcomputer systems, it has nowadays made the passage to the fabrication of 32-bit systems with tremendous computing power and additional embedded features. The additional effect of maker culture on microcontroller applications, along with the progress in IC fabrication in respect of making electronic chips smaller (with more technology inside the chip), has resulted in today's

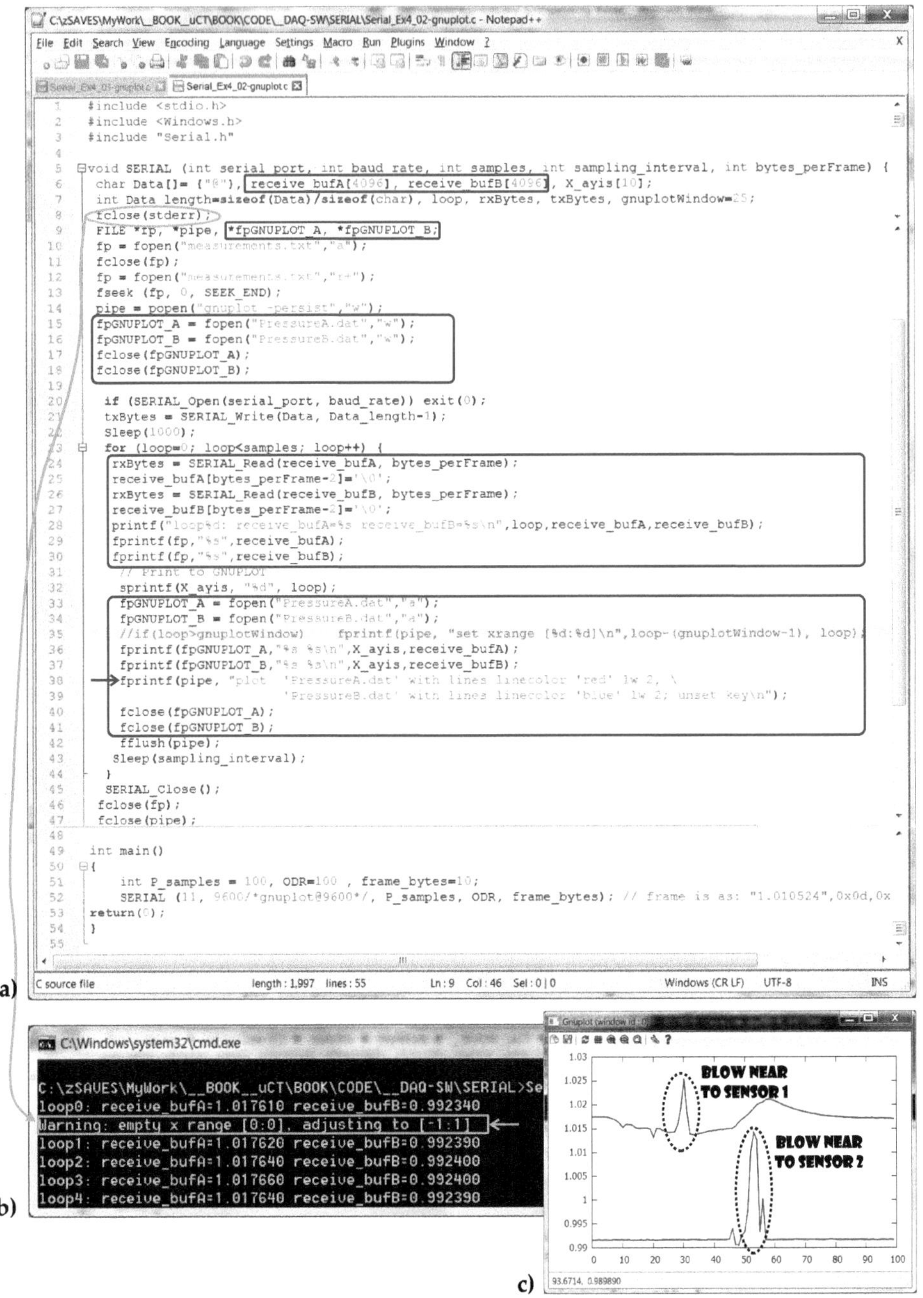

Figure 4.14 DAQ with data monitoring acquired by 2 sensors (*Serial_Ex4_02-gnuplot.c*). (a) application software code, (b) data output with Windows command prompt console, (c) plot of measurements with *gnuplot*. *Source:* Arduino Software.

availability of particularly sophisticated motherboards and daughterboards for microcontroller projects.

Herein, we explore an Arduino-compatible 32-bit μC motherboard of very small footprint; that is the Teensy 3.2 delivered by Electronic Project Components (www.pjrc.com/teensy/). The Teensy USB development board is particularly popular in the maker community. In addition, we explore notable daughterboards made by makers, which can be found at www.tindie.com; that is, a website to buy electronics directly by makers.

Arduino Ex.4–3

This example attempts to help the designer in getting started with Teensy 3.2 board. The Teensy 3.2 is a 32-bit USB-based microcontroller development system, in a very small footprint of standard header dimensions, that is, "0.100" (2.54 mm). Table 4.2 summarizes the difference between the 32-bit Teensy 3.2 and the 8-bit Arduino Uno board utilized by the previous chapters. Despite the fact that Teensy 3.2 is able to manipulate 32-bit data and consists of considerably larger memory, as well as IO pins and buses, it also features clock speed of approximately 4.5 times greater than the Arduino Uno clock speed.

Moreover, Teensy 3.2 features other embedded subsystems not available in the Arduino Uno. That is, the (i) *touch sense* capability of some pins, (ii) *controller area network* (*CAN*)

Table 4.2 Specs of Teensy 3.2 (32-bit μC) and the Arduino Uno (8-bit μC) development boards.

Specs	**Teensy 3.2 board**	**Arduino Uno board**
Processor	MK20DX256 (32-bit ARM Cortex-M4)	ATmega328 (8-bit AVR)
Clock speed	72 MHz	16 MHz
Flash memory	256 kB	32 kB (0.5 kB is used by the bootloader)
RAM memory	64 kB	2 kB
EEPROM	2 kB	1 kB
Operating voltage	3V3 (with 5 V pin tolerance)	5 V
Digital IO pins	34	20
Analog pins	21	6
Touch sense pins	12	0
PWM	12	6
UART bus	3	1
I2C bus	2	1
SPI bus	1	1
CAN bus	1	0
DAC	1	0
RTC	1	0

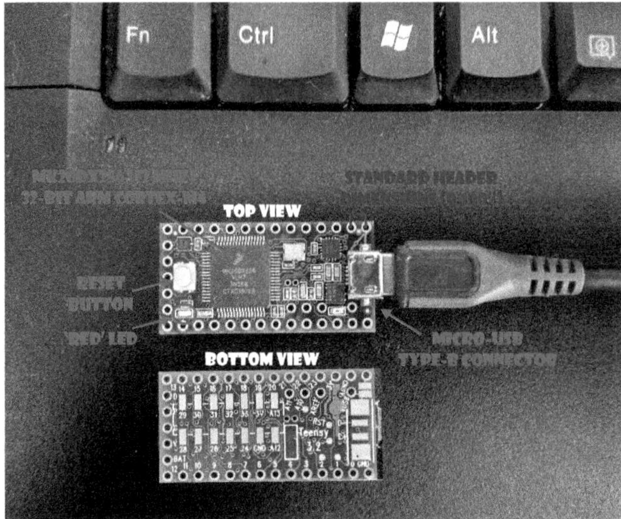

Figure 4.15 Teensy 3.2 USB development board (*Top* and *Bottom* view).

bus, i.e. a robust vehicle bus standard to allow microcontrollers to communicate with each other, (iii) DAC module, and (iv) *real-time clock* (*RTC*). Figure 4.15 depicts top and bottom side of the Teensy 3.2 board. The board is powered and programmed through a *Micro-USB type-B* connector.

To make the board working with Arduino IDE, the user must install the add-on *Teensyduino* software, available from www.pjrc.com/teensy. Figure 4.16 illustrates how to configure Arduino IDE after the aforementioned installation from the *Tools* menu (Figure 4.16b) in order to upload the example firmware (Figure 4.16a). The firmware sends to the serial port the "Hello World" string and blinks the LED (attached to Teensy board) with time delay 100 ms (between the ON/OFF states of the LED). During uploading the firmware code to microcontroller's memory, the *Teensyduino* software pop ups the messages given in Figure 4.16c–e. If required, the user should press the reset button (depicted in Figure 4.15) in order to enter into the programming mode of operation. Figure 4.16f presents the printed result on an Arduino IDE *Serial Monitor* tool.

Arduino Ex.4–4

The present example explores the control of a daughterboard designed for the Teensy 3.2 (or other conventional Teensy board). That is, the *BNO055+BMP280* designed by *Pesky Products* in USA. The daughterboard can be purchased at the website (https://www.tindie.com/). Pesky's *BNO055+BMP280* module employs two sensors, that is, the (i) BNO055 motion orientation sensor [56], and (ii) BMP280 barometric pressure sensor. Figure 4.17a presents the Pesky's daughterboard next to Teensy 3.2 motherboard and highlights the four pins on the latter board that are used to power and control, over the I2C, the daughterboard (that is, the 3V3, GND, SLC0, and SDA0 pins).

At this point it is worth mentioning that, the designer should pay particular attention to the interconnection between the μC motherboard and third-party daughterboards.

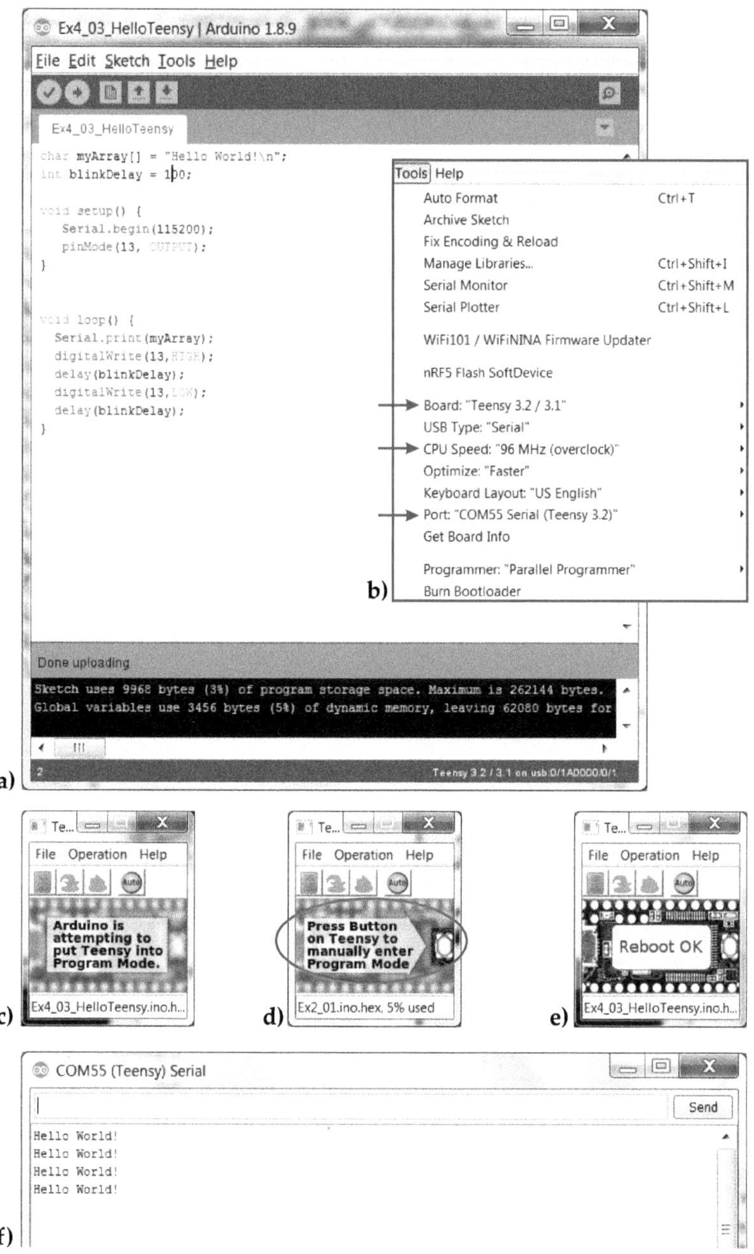

Figure 4.16 Getting started with Teensy 3.2 board (example firmware). (a) firmware code, (b) configuring Teensy board in Arduino IDE, (c) download code to Teensy board (step 1), (d) download code to Teensy board (step 2), (e) download code to Teensy board (step 3), (f) printed results. *Source:* Arduino Software.

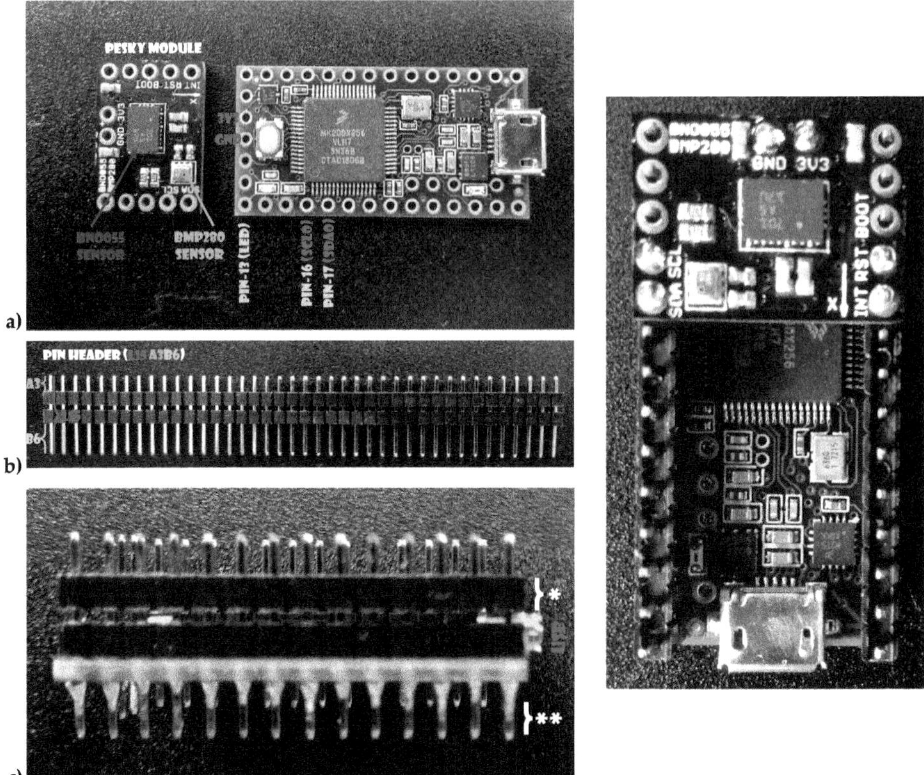

Figure 4.17 BNO055+BMP280 Pesky module for Teensy 3.2 motherboard. (a) the Pesky module (left side) and the Teensy board (right side), (b) pin header for attaching together the two modules, (c) pin header soldered to the Teensy board, (d) the Pesky module soldered on the top of Teensy board.

In this example we have selected a standard 2.54 pin-header of overall height 15 mm (i.e. L15), as presented in Figure 4.17b. The metal top and bottom dimension of each pin is 3 and 6 mm, accordingly, and thereby they leave an additional space of 6 mm in between them. That is the space of plastic contacts that separates the daughterboard from the motherboard.

The soldering task the pin header, as presented by Figure 4.17c, allows us to leave the appropriate distance to the daughterboard from taller material found on the motherboard (that is, the micro-USB type-B connector). In addition, placing the long metallic pin (i.e. B6) at the bottom of Teensy board, supports the interconnection of more daughterboards at the bottom area of this setup, or perhaps the interconnection of the overall setup to a different PCB. Figure 4.17d illustrated the setup of the Teensy 3.2 motherboard holding the *BNO055+BMP280* Pesky daughterboard.

To get started with the I2C control of the daughterboard we exploit the compatibility between BME280 environmental sensor and the BMP280 barometric pressure sensor. Accordingly, we address the previous example, which was used to read data from BME280 device. In consideration of the requisite amendments made to the firmware code, as they are presented in Figure 4.18a, the I2C address of BMP280 sensor (attached to *BNO055+BMP280* Pesky module) is set to 0 × 77. Moreover, the pin location of the I2C lines of the Teensy 3.2 board (which are different from the Arduino Uno board) is configured inside the blue frame, through the *Wire.setSDA()* and *Wire.setSCL()* functions. Finally, the instructions for the control of the *DAQpin* (used for capturing timings of specific events) have been removed by the code.

A few amendments of the previous DAQ software are given in Figure 4.18b,c. Because the Teensy 3.2 board has been attached to COM55 serial port, we have added more string values to the corresponding array found in the *Serial.h* file (Figure 4.18b). Consequently, the *main()* function of the DAQ software has been changed to match the number (i.e. 55) of the reserved (by Teensy) COMx port (Figure 4.18c). The compilation and execution of the code DAQ software is given in Figure 4.18d, while the generated plot is presented in Figure 4.18d.

Arduino Ex.4–5

The current example exploits BNO055 sensor found on Pesky's daughterboard, and applies to *absolute orientation* detection expressed with *Euler angles*. Before exploring the application hardware and software code, it is worth making a reference to BNO055 sensor. The BNO055 device incorporates three different sensors, that is, (i) triaxial *accelerometer*, (ii) triaxial *gyroscope*, and (iii) triaxial *geomagnetic sensor* (aka *magnetometer*). Moreover, it incorporates a 32-bit *cortex M0+* microcontroller running *sensor fusion* algorithms and, hence, BNO055 constitutes a complete *SiP*. The *accelerometer*, *gyroscope*, and *magnetometer* are suitable sensors for motion detection. Particular examples are the motion sensing games, indoor navigation systems, gesture recognition implementations, virtual and augmented reality applications, and so forth.

The *accelerometer* has a mass attached to a spring, which is confined to move along one direction so as to measure acceleration. When at rest, the moving mass of the accelerometer can also sense the gravitation force. What a triaxial accelerometer means, is that there are three available mass-spring configurations to move along *x*, *y*, and *z* axis, respectively. The combination of the three mass-spring configurations along the three axes provides three measureable data points, thereby portraying a sensor of 3 *degrees of freedom (DOF)*.

The *gyroscope* measures angular velocity (i.e. the rate of rotation) around a particular axis, by means of the *Coriolis Effect*. The gyroscope features a constantly oscillating mass and when the device rotates around a particular axis, a displacement to the mass is realized and measured. Similarly to the triaxial accelerometer, the triaxial gyroscope portrays a 3DOF sensor device.

The *magnetometer* detects and measures any kind of magnetic field by means of the *Hall Effect*. According to that effect, a magnetic field near to a conductive plate with current to flow through it, generates a voltage across the electrical conductor which depends to the strength of the magnetic field. Similarly to the triaxial accelerometer and gyroscope, the triaxial magnetometer portrays a 3DOF sensor.

When the *accelerometer*, *gyroscope*, and *magnetometer* are utilized individually, they feature severe limitations that affect precision and response. More accurate results as well as

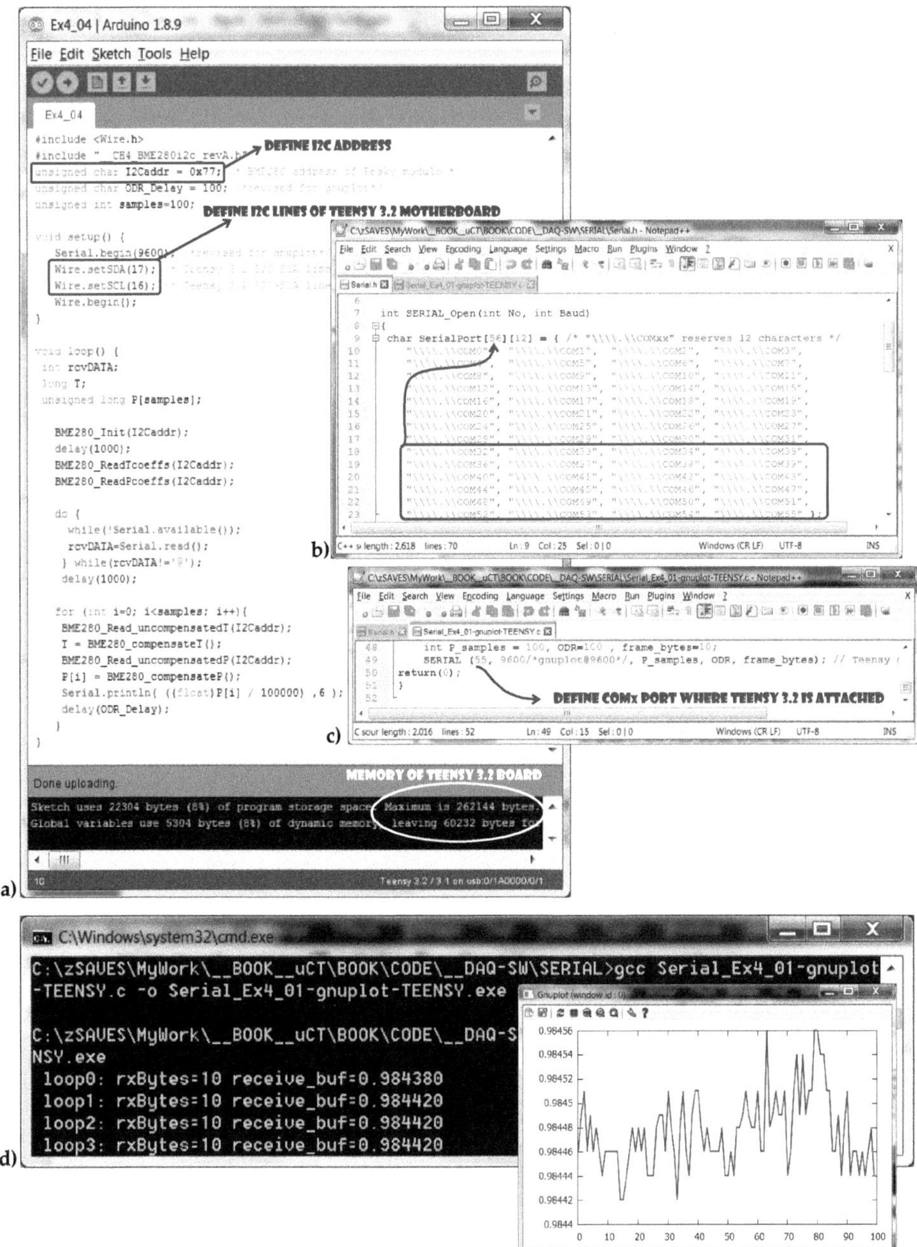

Figure 4.18 Reading air pressure from BMP280 sensor of Pesky's module (with Teensy 3.2). (a) application firmware code, (b) application software code (header file), (c) application software code (source code), (d) data output with Windows command prompt console, (e) plot of measurements with *gnuplot*. *Source:* Arduino Software.

faster response is feasible when combining data derived from the three sensors (of overall 9DOF), a process that is generally known as *sensor fusion*. The BNO055 SiP is able to provide raw data from the aforementioned sensors in the *non-fusion* modes of operation as well as fused data. In the fusion mode, the BNO055 device provides orientation of the device expressed in either *Euler* angles (i.e. *roll*, *pitch*, and *yaw* angle) or as *quaternion*. Moreover, sensor fusion tasks can separate gravity from acceleration and, hence, BNO055 device can provide, separately, *linear acceleration* (i.e. acceleration applied by the device' movement) as well as *gravity vector*.

The present example explores how to sense orientation of BNO055 device when orientation is expressed with *Euler* angles. The so called *roll*, *pitch*, and *yaw* angles provide a combination of three rotations with respect to a fixed coordinate system toward an intuitive 3D representation of an object. Figure 4.19a presents the default *x*, *y*, and *z* reference axes of BNO055 device, when holding the Teensy 3.2 board (which employs BNO055 sensor) in the way illustrated by Figure 4.19b. It is here noted that BNO055 device can be reconfigured to arrange a setup of different reference axes.

Based on the coordinate system of Figure 4.19 (determined by the default configuration of BNO055 of the upcoming code examples), the *roll* angle represents a rotation of the Teensy 3.2 board around *y* axis (Figure 4.19d). Accordingly, the *pitch* angle represents a rotation of the Teensy 3.2 around *x* axis (Figure 4.19c), and the *yaw* angle represents a rotation of the Teensy 3.2 around *z* axis (Figure 4.19e). Rotating the device clockwise when of looking toward the positive direction of the axis (i.e. looking in the direction pointed by the

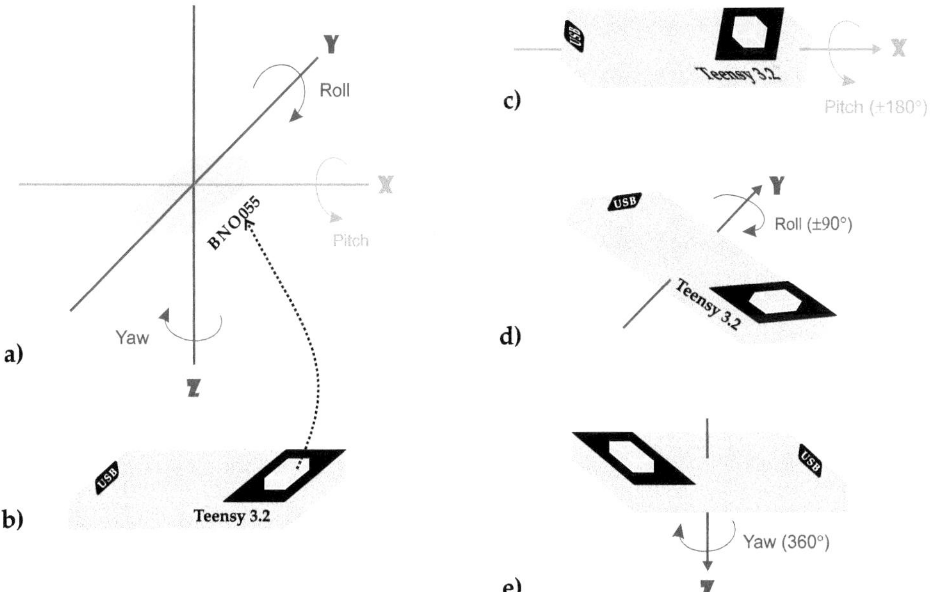

Figure 4.19 3D orientation of BNO055 device using *Euler* angles. (a) reference x,y,z axes of BNO055 device, (b) reference position of BNO055 device on the top of Teensy board, (c) rotation of the hardware setup around x axis: *pitch* angle, (d) rotation of the hardware setup around y axis: *roll* angle, (e) rotation of the hardware setup around z axis: *yaw* angle.

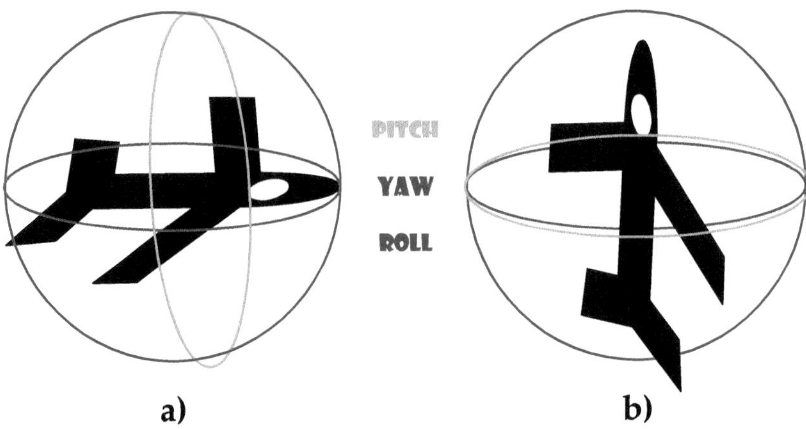

Figure 4.20 *Gimbal lock* phenomenon in *Eule*r angles. (a) independent gimbals.

arrow as illustrated by Figure 4.19c–e), the corresponding *pitch*, *roll*, or *yaw* angle increases. The opposite rotation (i.e. counter-clockwise) decreases the value of the *Euler* angles. The *pitch* angle admits values from −180° to +180°, the *roll* admits values from −90° to +90°, while the *yaw* admits values 0 to +360°. At this point it is worth mentioning that *Euler* angles feature the unwanted *gimbal lock* phenomenon, which, in a three-dimensional (i.e. three gimbal) scheme, results in the loss of one DOF.

Figure 4.20a illustrates three independent gimbals, while Figure 4.20b illustrates a gimbal-locked airplane. Because the aircraft pitches 90°, the *pitch* and *yaw* gimbals become aligned. Hence, changes to the *roll* and *yaw* apply the same rotation of the aircraft. A *gimbal lock*–free solution is accomplished by the orientation expression with *quaternion*; however, the latter approach constitutes not so intuitive 3D representation of an object.

Table 4.3 presents the BNO055 registers used by the book examples. In memory address 0×00 is the *CHIP_ID* register, which holds the value 0×A0. The *CHIP_ID* is useful when the designer assesses whether or not the device responds to the μC's transaction.[5] The subsequent 18 registers hold the calibrated (i.e. fused) data, which are available in case the device operates in fusion mode. In detail, addresses 0×1A–0×1F hold the *Euler* angles, addresses 0×28–0×2D hold the *Linear Acceleration* data, and addresses 0×2E–0×33 hold the *Gravity Vector* data. The examples make use of the default units in each case and, hence, the *Euler* angles are expressed in *degrees*, while both the *Linear Acceleration* and *Gravity Vector* are expressed in ms^{-2}.

According to the sensor's datasheet, for each one of the *x,y,z* axes the corresponding parameter holds a signed value of 2 bytes (such, the *EUL_HEADING* parameter). Table 4.4 provides the data representation of the *Euler* angles as well as the *Linear Acceleration/Gravity Vector* data. For instance, to reform the 16-bit variable *EUL_HEADING* into *degrees*, we divide its content to the number 16.

The initialization of the sensor in the desired mode of operation is accomplished through the *OPR_MODE* register found in 0×3D memory address. To enter into fusion mode and,

5 It is here noted that the BNO055 sensor supports both I2C and UART interfaces to communication with the master device. However, the book examples apply to I2C interface.

Table 4.3 BNO055 registers used by the book examples.

ADDR	NAME	RST	bit7	bit6	bit5	bit4	bit3	bit2	bit1	bit0
0×00	CHIP_ID	0×A0	1	0	1	0	0	0	0	0
0×1A	EUL_HEADING_LSB	0×00	Yaw Data [7..0]							
0×1B	EUL_HEADING_MSB	0×00	Yaw Data [15..8]							
0×1C	EUL_ROLL_LSB	0×00	Roll Data [7..0]							
0×1D	EUL_ROLL_MSB	0×00	Roll Data [15..8]							
0×1E	EUL_PITCH_LSB	0×00	Pitch Data [7..0]							
0×1F	EUL_PITCH_MSB	0×00	Pitch Data [15..8]							
0×28	LIA_Data_X_LSB	0×00	Linear Acceleration Data X [7..0]							
0×29	LIA_Data_X_MSB	0×00	Linear Acceleration Data X [15..8]							
0×2A	LIA_Data_Y_LSB	0×00	Linear Acceleration Data Y [7..0]							
0×2B	LIA_Data_Y_MSB	0×00	Linear Acceleration Data Y [15..8]							
0×2C	LIA_Data_Z_LSB	0×00	Linear Acceleration Data Z [7..0]							
0×2D	LIA_Data_Z_MSB	0×00	Linear Acceleration Data Z [15..8]							
0×2E	GRV_Data_X_LSB	0×00	Gravity Vector Data X [7..0]							
0×2F	GRV_Data_X_MSB	0×00	Gravity Vector Data X [15..8]							
0×30	GRV_Data_Y_LSB	0×00	Gravity Vector Data Y [7..0]							
0×31	GRV_Data_Y_MSB	0×00	Gravity Vector Data Y [15..8]							
0×32	GRV_Data_Z_LSB	0×00	Gravity Vector Data Z [7..0]							
0×33	GRV_Data_Z_MSB	0×00	Gravity Vector Data Z [15..8]							
0x3D	OPR_MODE	0×1C	Operating Mode [3..0]							

Table 4.4 Data representation of *Euler* angles and *Linear Acceleration/Gravity Vector* data.

Unit	Representation
Degrees (Euler angles)	$1° = 16$ LSB
$m\,s^{-2}$ (Linear acceleration)	$1\,m\,s^{-2} = 100$ LSB
$m\,s^{-2}$ (Gravity vector)	>>

in particular, into *NDOF* operating mode (as in the examples presented herein) the designer should set *OPR_MODE* register to the value 0×0C. The default value of the *OPR_MODE* register is 0×00, which means that after *power-on* or *reset*, BNO055 enters into the so-called *CONFIGMODE* of operation. Accordingly, the application code should first switch from *CONFIGMODE* to *NDOF* operating mode, where the requisite switching time in that case is 7 ms (and 19 ms to switch from any operating mode to *CONFIGMODE*). The *NDOF* mode of 9DOF provides a fused *absolute orientation*. That is, the orientation of the

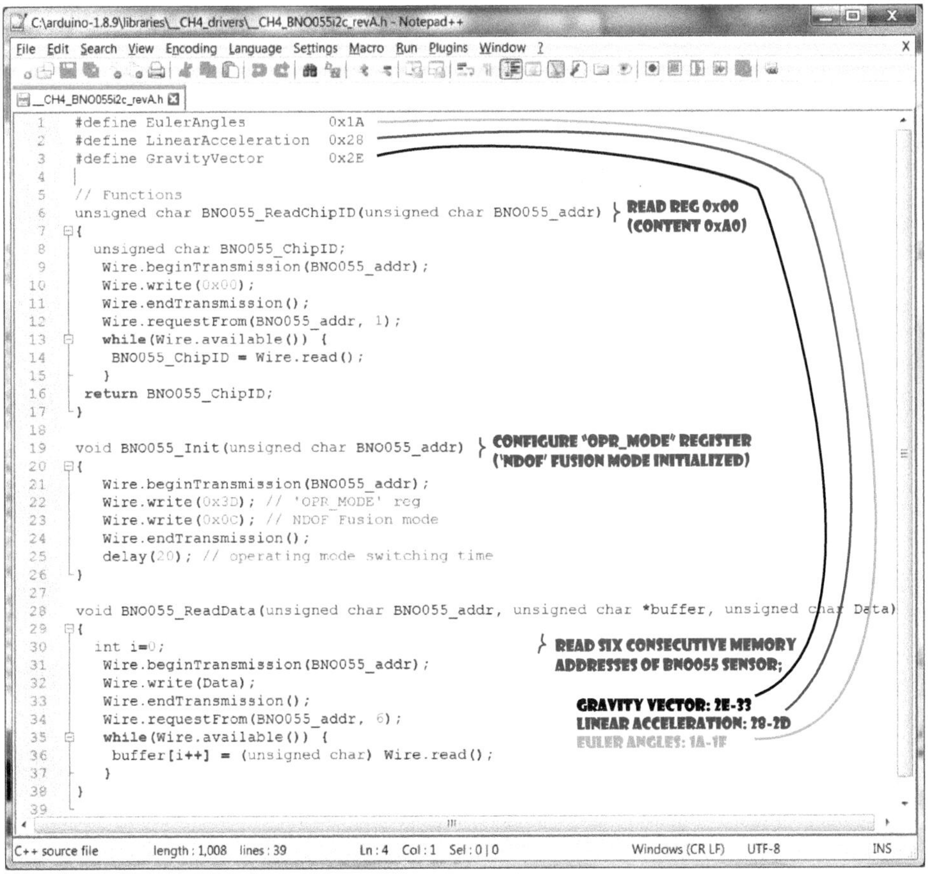

```cpp
#define EulerAngles          0x1A
#define LinearAcceleration   0x28
#define GravityVector        0x2E

// Functions
unsigned char BNO055_ReadChipID(unsigned char BNO055_addr)
{
  unsigned char BNO055_ChipID;
   Wire.beginTransmission(BNO055_addr);
   Wire.write(0x00);
   Wire.endTransmission();
   Wire.requestFrom(BNO055_addr, 1);
   while(Wire.available()) {
    BNO055_ChipID = Wire.read();
   }
 return BNO055_ChipID;
}

void BNO055_Init(unsigned char BNO055_addr)
{
   Wire.beginTransmission(BNO055_addr);
   Wire.write(0x3D); // 'OPR_MODE' reg
   Wire.write(0x0C); // NDOF Fusion mode
   Wire.endTransmission();
   delay(20); // operating mode switching time
}

void BNO055_ReadData(unsigned char BNO055_addr, unsigned char *buffer, unsigned char Data)
{
  int i=0;
   Wire.beginTransmission(BNO055_addr);
   Wire.write(Data);
   Wire.endTransmission();
   Wire.requestFrom(BNO055_addr, 6);
   while(Wire.available()) {
    buffer[i++] = (unsigned char) Wire.read();
   }
}
```

READ REG 0x00 (CONTENT 0xA0)
CONFIGURE "OPR_MODE" REGISTER ('NDOF' FUSION MODE INITIALIZED)
READ SIX CONSECUTIVE MEMORY ADDRESSES OF BNO055 SENSOR;
GRAVITY VECTOR: 2E-33
LINEAR ACCELERATION: 28-2D
EULER ANGLES: 1A-1F

Figure 4.21 BNO055 driver (using the built-in I2C).

sensor with respect to the earth and its magnetic field (i.e. the sensor fusion modes calculate the direction of the magnetic north pole). The ODR of the *NDOF* mode is equal to 100 Hz and, hence, the designer is able to obtain data with sampling interval of at least 10 ms period.

Figure 4.21 illustrates the driver of BNO055 device, which is used by the subsequent examples so as to sense orientation, motion, as well as gesture recognition. As mentioned earlier, orientation is expressed in *Euler* angles, while motion and gesture recognition is respectively identified through the readings of *Linear Acceleration* and *Gravity Vector*.

The driver consists of three different functions. *BNO055_ReadChipID()* is used for testing the I2C communication, as it obtains the content of 0 × 00 is the *CHIP_ID* register (holding 0 × A0 value). *BNO055_Init()* inserts the device into the *NDOF* fusion operating mode by setting *OPR_MODE* register to 0 × 0C value. Before returning from the function, *BNO055_Init()* delays 20 ms to comply with the time needed for configuring the device. Finally, *BNO055_ReadData()* obtains the content of six consecutive memory locations of BNO055 device. The driver is currently designed to obtain the memory locations associated with

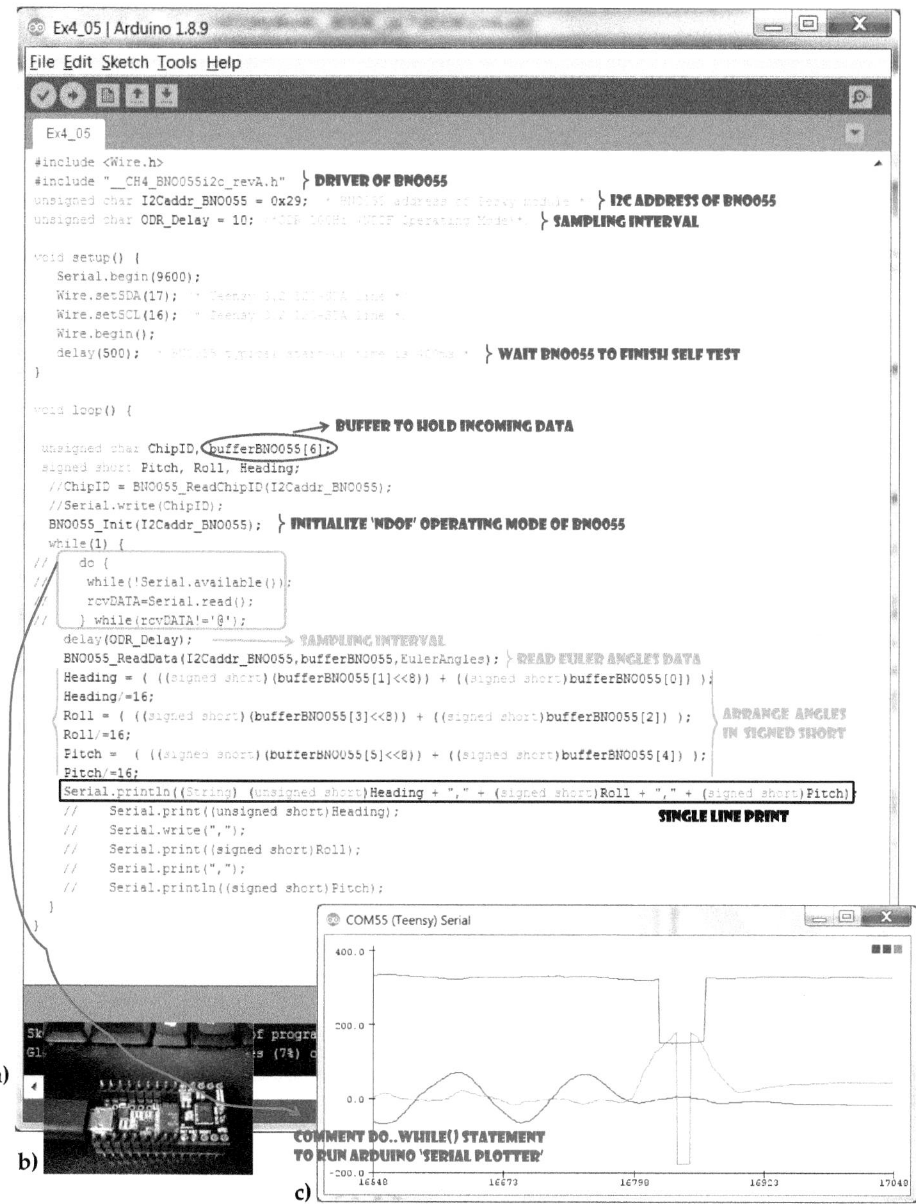

Figure 4.22 Orientation detection expressed with *Euler* angles. (a) application firmware, (b) hardware setup, (c) plot of measurements with *serial plotter*. *Source:* Arduino Software.

Euler angles (i.e. 0×1A–0×1F), *Linear Acceleration* (i.e. 0×28–0×2D) and *Gravity Vector* (i.e. 0×2E–0×33), presented earlier in Table 4.3.

Figure 4.22a presents the firmware code for obtaining Euler angles so as to detect the orientation of BNO055. The example addresses the hardware setup used before (as presented by Figure 4.22b). It is here noted that (despite the accompanying DAQ software that will be explored next in the chapter), the functionality of the firmware can be explored by the *Serial Plotter* built-in tool of Arduino IDE. That is why the *do..while()* statement within the green frame has been commented. Running the *Serial Plotter* tool, the user obtains the results given in Figure 4.22c. Any kind of displacement of the hardware setup (Figure 4.22b) in respect to the scheme presented earlier in Figure 4.19, modifies the corresponding *yaw* (blue graph), *pitch* (green graph), and *roll* (red graph) angle (Figure 4.22c).

In consideration of the firmware code of Figure 4.22a, the first two lines include the built-in I2C driver of Arduino as well the custom-designed driver of BNO055 SiP. The subsequent two lines determine the I2C address of BNO055 device and the sampling interval of the DAQ process. Next, the *setup()* function configures the serial as well as the I2C port of Teensy board and, then, it performs a call to delay of 0.5 s. The delay is required because (as defined by the datasheet of BNO055) the start-up time of the sensor device from *off* to *configuration mode* is typically 400 ms. Inside the *loop()* function the *unsigned char* type array named *bufferBNO055[6]* (which reserves six memory location) is used hold the incoming data, that is, two bytes for each *Euler* angle. Each angle should be converted to a 16-bit signed value and, hence, the *signed short* type is used to hold the conversion result by each angle.

The main code starts with two commented lines that may be used to test the I2C transaction, in case some communication problem arises. The code actually starts with the configuration of BNO055 device to *NDOF* mode, through the call of *BNO055_Init()* function. Subsequent to the function goes an endless loop, which (as soon as receiving '@' char from serial port and generating a sampling interval) repeatedly obtains the *Euler* angles from BNO055 device, and reforms each angle to 16-bit signed format divided by 16 (as 1° corresponds to 16 LSB). Finally, the *Euler* angles are sent to the serial port through a single *Serial.println()* instruction. The three angles are arranged into a *String* object (defined by the Arduino *String* library) and separated by *comma (',')* character. The single-line print corresponds to the commented lines found below.

Arduino Ex.4–6

Apart from *Euler* angles, the fusion algorithm of the previous example can also provide *Gravity Vector* data for *x*, *y*, and *z* axis. Readings of *Gravity Vector* data can be applied to mobile gaming applications, such as steering a vehicle by changing orientation of the mobile device. Figure 4.23 presents the revised firmware for reading *Gravity Vector. Linear Acceleration* data can also be obtained with just a minor modification. That is, by changing the final argument of the *BNO055_ReadData()* from '*GravityVector*' to '*LinearAcceleration.*' The latter amendment directs the firmware code to read memory locations 0×28–0×2D (i.e. *Linear Acceleration*) instead of 0×2E–0×33 (*Gravity Vector*). As illustrated by Figure 4.23, the current example has minor amendments compared to the previous example applied to *Euler* angles. The revisions are found within the blue and red frames.

Figure 4.24 presents a draft setup to demonstrate gesture recognition through *Gravity Vector* readings. The attachment of BNO055 device (employed by Teensy board) to a sleeve, as illustrated by Figure 4.24c, can transform the sleeve to a sophisticated one-handed product (i.e. a product for people with permanent disability who have one hand to use). The sleeve-based setup can be used to send commands to several systems.

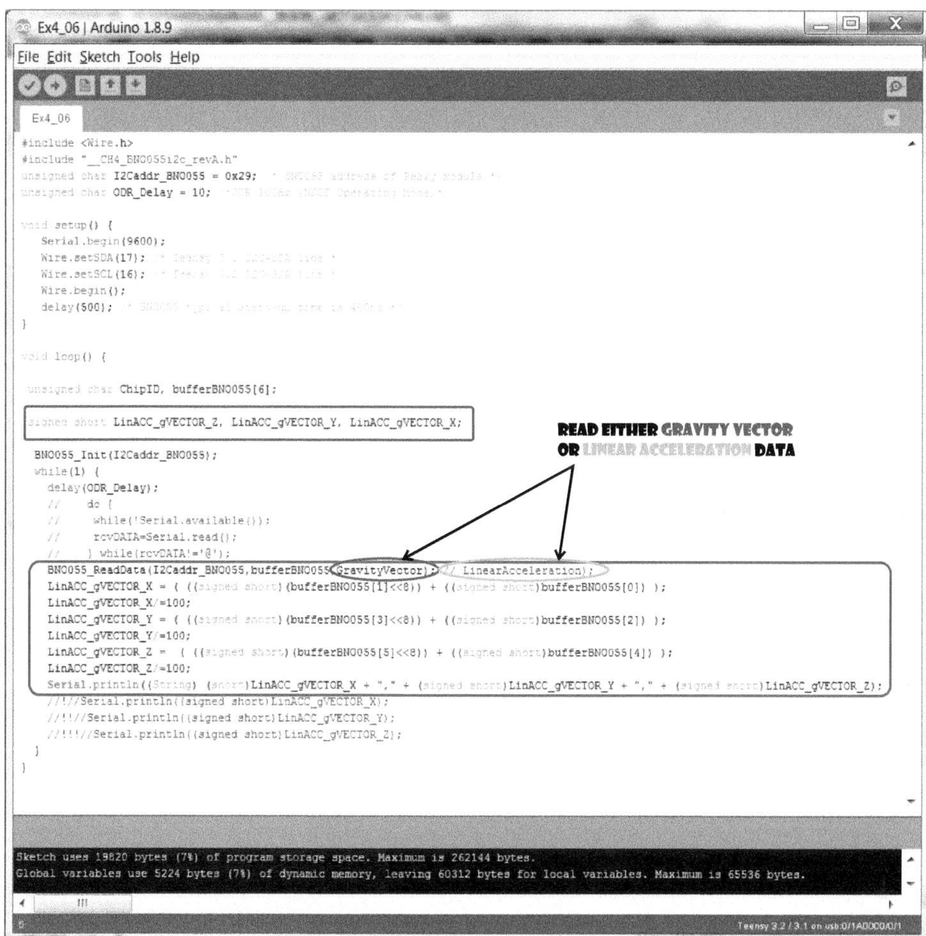

Figure 4.23 Gesture and motion detection firmware. *Source:* Arduino Software.

The current example exploits gravity sensed by the accelerometer device. The *accelerometer (ACCLRM)* senses both gravity and acceleration, but as mentioned earlier, the data fusion algorithm of BNO055 separates gravity from acceleration. A sensor that detects only gravity can be easily attached to gesture recognition applications, in case each gesture is aligned with the gravity. To illustrate that kind of gesture recognition, we consider the scheme in Figure 4.24b depicting a single-axis ACCLRM in rest. When the device is placed horizontally in relation to the Earth's surface (as illustrated by the blue frame of Figure 4.24b), it indicates 0 g. When the ACCLRM is rotated 90° clockwise (as depicted by the green arrow) or counterclockwise (as depicted by the red arrow) it indicates −1 and +1 g, respectively (i.e. the gravity acts on the proof mass, and as a result, the ACCLRM spring is either pulled or pushed). Accordingly, the single-axis sensor of Figure 4.24b can identify three different gestures (e.g. gesture-1, gesture-3, and gesture-4 of Figure 4.24a).

A three-axis ACCLRM gives us more choices in gesture recognition. Figure 4.24d illustrates what happens to the three-axis ACCLRM of BNO055 SiP when its coordinate system

GESTURES AS OBSERVED WHEN LOOKING DIRECTLY AT USER'S HAND

GESTURE-1 GESTURE-6

GESTURE-2 GESTURE-4 GESTURE-7

GESTURE-3 GESTURE-5 GESTURE-8

a)

GESTURES AS OBSERVED WHEN LOOKING OVER THE USER'S HAND

GESTURE-1

GESTURE-3

c)

GRAVITY APPLIED TO THE MASS OF ACCLRM

-1G 0G +1G SPRING MASS +1G 0G -1G

b)

ROTATION OF A 3-AXIS ACCLRM (ALIGNED WITH THE DIRECTION OF GRAVITY)

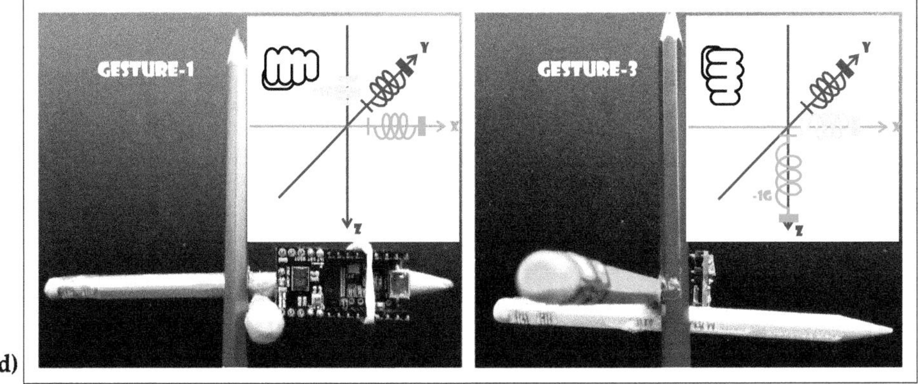

d)

Figure 4.24 Gesture detecting sleeve and gesture types. (a) the various types of gesture of the current application, (b) gravity applied to the mass of a single-axis accelerometer, (c) hardware setup and gesture demonstration examples, (d) gravity applied to the mass of a three-axis accelerometer.

(represented by the *cyan*, *magenta*, and *yellow* pencils) rotates 90° clockwise relative to the fixed coordinate system of the Earth (represented by the X, Y, Z axes of *green*, *red*, and *blue* color, respectively). In detail, the rotation has an effect on two of the three accelerometers: (i) the yellow ACCLRM indication changes from +1 to 0 g, (ii) the cyan ACCLRM indication changes from 0 to −1 g. As a general rule we could state that on each 90° clockwise or counterclockwise rotation of the device, we expect one ACCLRM to be pulled or pushed by the gravity, one ACCLRM to be released (from a previous pushed/pulled condition), and one ACCLRM to remain unchanged to 0 g indication.

Figure 4.25 explores the gravity effect on the three-axis accelerometer of BNO055 device, in consideration the aforementioned gestures (i.e. Figure 4.24a). To reach a sense of realization of the patterns in Figure 4.25, the reader may utilize the *cyan-*, *magenta-*, *yellow-*pencil setup of Figure 4.24d. The graphs for each individual gesture are presented in Figure 4.26, where the gravity is expressed in $m\,s^{-2}$ (i.e. $g \approx 9.81\,m\,s^{-2}$). Because the code stores the gravity vector to *signed short* variable, the value of each graph varies from −9 to +9, that is, the integral part of the gravity value (for instance, the integral part of 9.81 is equal to 9).

It is worth noting that different gestures can produce the same graph pattern, as gesture-2 with gesture-8 (i.e. 0g, +1g, 0g), and gesture-5 with gesture-7 (0g, −1g, 0g). However, we are able to distinguish two gestures if we consider the pattern of the preceding one. To this end, the designer could reserve a *neutral* gesture, which should always be used as an intermediate step before switching from one gesture to another.

The example of Figure 4.27 aims in helping the reader realize how to achieve *simplicity in design*. Based on the aforementioned information, we present how to use the sleeve as remote controller of an RC car, using six gestures of unique pattern each. That is, *forward*(0g, 0g, +1g), *backward*(0g, 0g, −1g), *forward+right*(+1g, 0g, 0g), *forward+left*(−1g, 0g, 0g), *backward+left*(0g, +1g, 0g), and *backward+right*(0g, −1g, 0g). This way, the designer does not have to consider the preceding (neutral) gesture in order to identify the command sent to the RC car.

Motion detection with BNO055 device can be performed by reading *Linear Acceleration*. Figure 4.28 presents how to identify motion by accelerating BNO055 sensor (employed by Teensy board) along x axis (Figure 4.28b), y axis (Figure 4.28c), or z axis (Figure 4.28d).

Figure 4.29a presents the amendments (i.e. instructions within the frame) required to read *Linear Acceleration* instead of *Gravity Vector*. Uncommenting all lines within the back frame results in the graph of Figure 4.29b. Commenting all but one code lines (i.e. setting to zero two of the three variables), we obtain the graphs for device's displacement along x axis (Figure 4.29c), along y axis (Figure 4.29d), and along z axis (Figure 4.29e). The circled part of the last three graphs depicts the sensor's response when device moves toward the x, y, or z positive infinity.

Up until this point in this chapter, we have explored BNO055 and BME280 sensor devices. The association of the two sensors (taking into account their possibilities explored by the examples) can direct readers in developing *critical thinking* in consideration of contemporary μC-based applications. For instance, the air pressure readings (obtained from BME280) can be rearranged into barometric altitude data. As explored earlier in the chapter, air pressure varies radically with the passage of time and, hence, the absolute height data cannot be accurate with a single barometer. However fusing *Linear Acceleration* and *Barometric Altitude* data may result in an accurate estimation of the absolute height variations (in particular applications). For example, it could be used to determine changes in the absolute height of a drone system (Figure 4.30), by reading the difference of barometric altitude when identifying linear acceleration on the z axis. Data fusion of *Barometric Altitude* (obtained from BME280) and *Linear Acceleration* (obtained from BNO055) toward

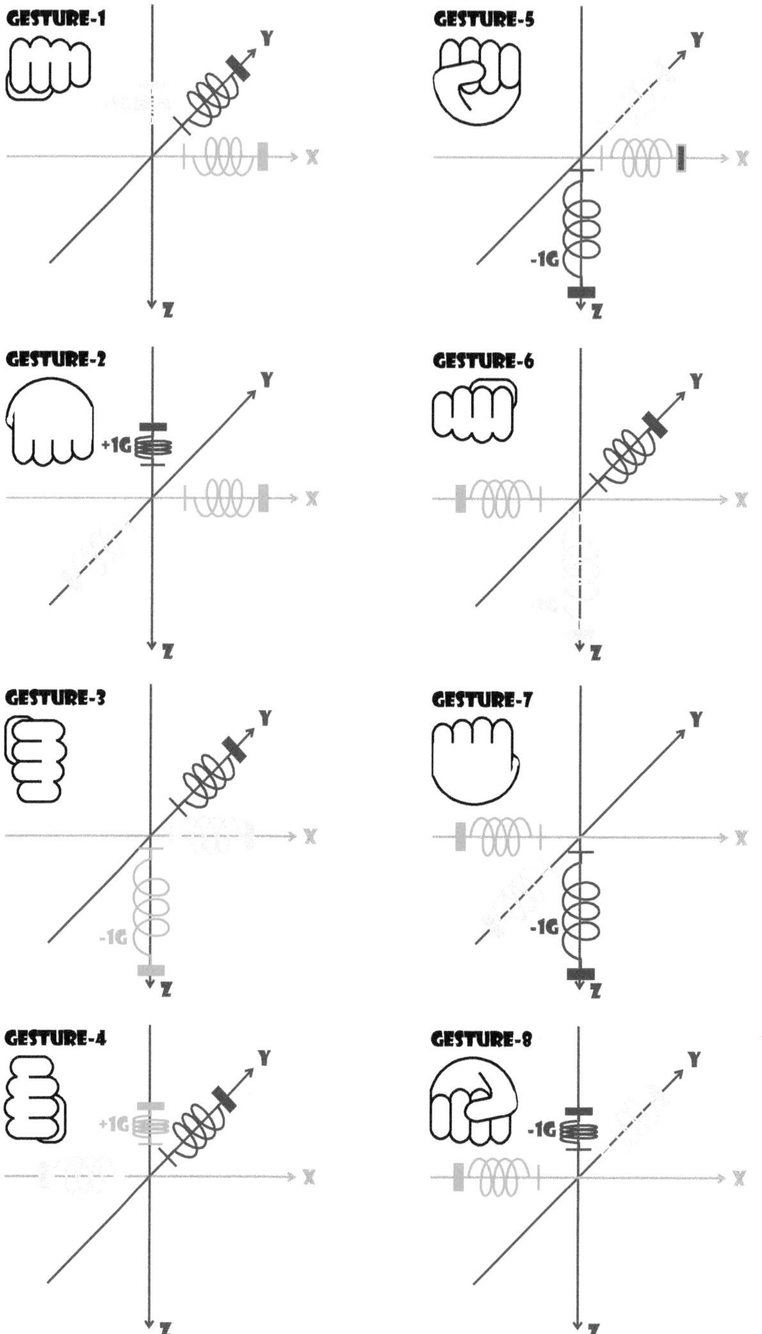

Figure 4.25 Gesture detection via *Gravity Vector* (gestures aligned with the direction of gravity).

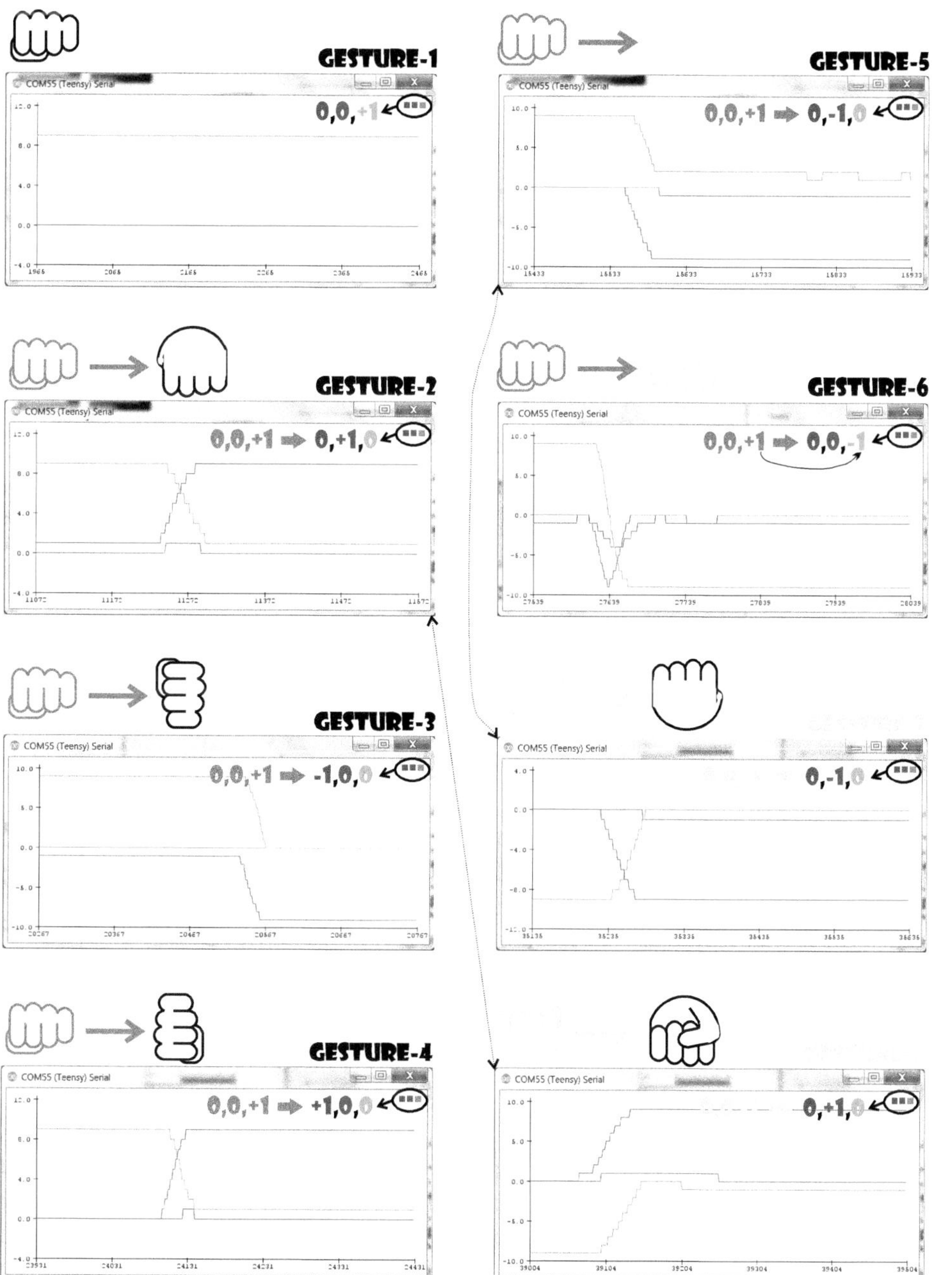

Figure 4.26 Gesture recognition graphs. *Source:* Arduino Software.

estimating the vertical displacement of an object, is left as an exercise to the reader. An empirical equation to rearrange atmospheric pressure into altitude data (aka *international barometric formula*) is given in Formula 4.2a. Based on that formula, the *absolute height* estimation between two vertical positions is given by Formula 4.2b.

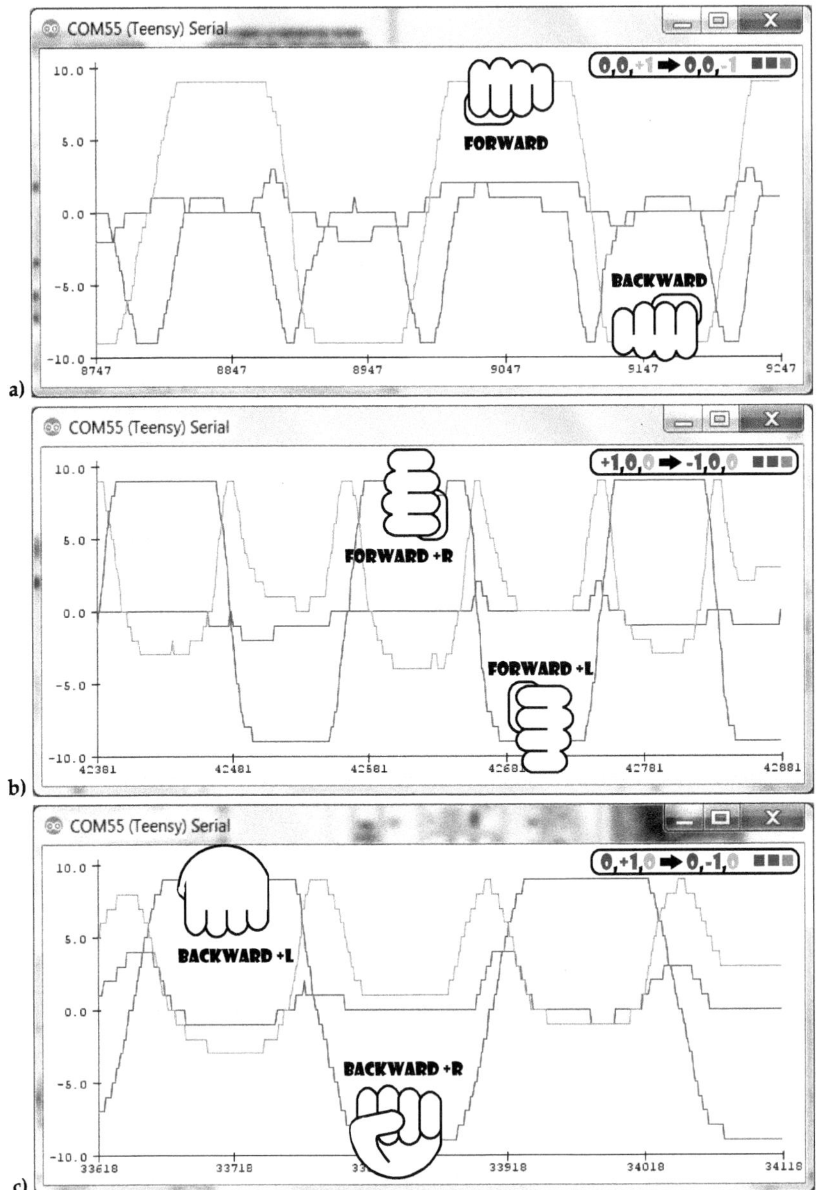

Figure 4.27 Gesture recognition example toward steering an RC car. (a) plot of gravity measurements of a particular gesture (example 1), (b) plot of gravity measurements of a particular gesture (example 2), (c) plot of gravity measurements of a particular gesture (example 3). *Source:* Arduino Software.

Formula 4.2 Absolute height estimation through the *International Barometric Formula*.

$$z(\text{meters}) = 44\ 330 - 4935.125 \cdot P_{(z)}^{0.1903} \quad \text{(a) International Barometric Formula}$$

$$h = z_A - z_B = 4935.125 \cdot \left(P_{(zB)}^{0.1903} - P_{(zA)}^{0.1903}\right) \quad \text{(b) Absolute Height}$$

where: $z_A > z_B$ and $P_{(zA)} < P_{(zB)}$

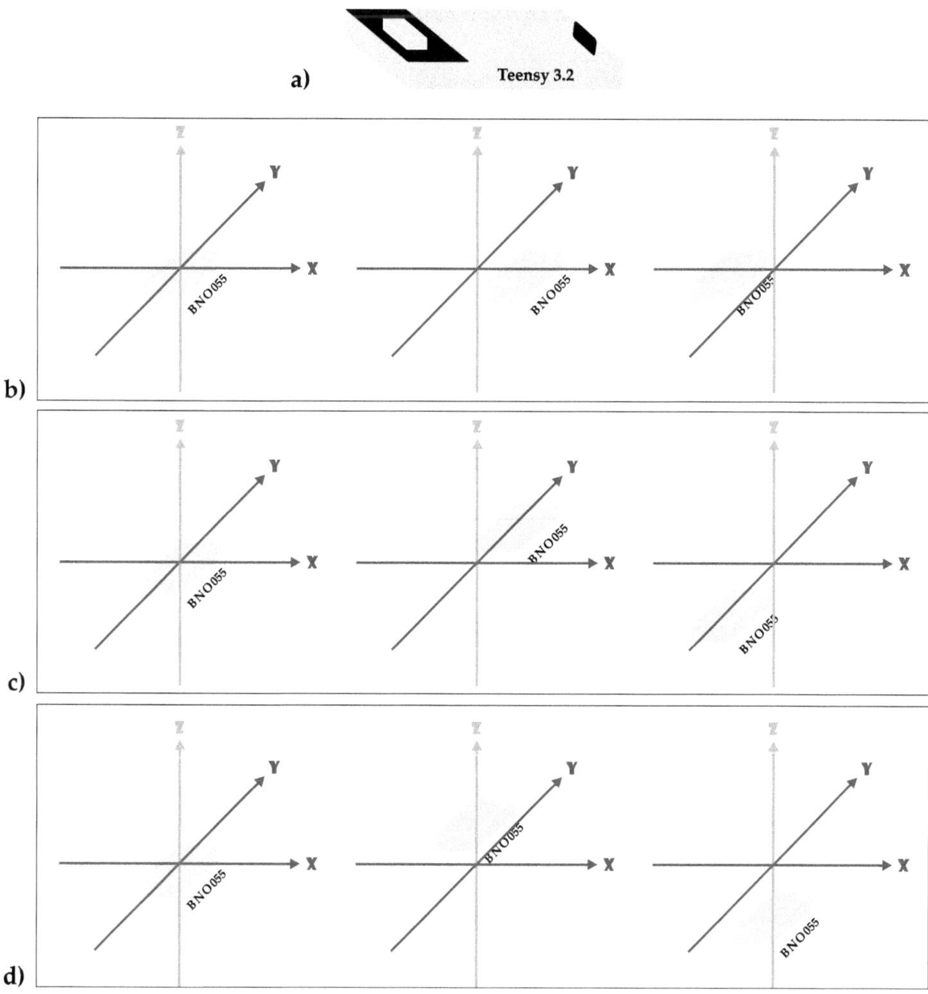

Figure 4.28 Motion detection via *Linear Acceleration*. (a) reference position of BNO055 device on the top of Teensy board, (b) accelerating BNO055 sensor along x axis, (c) accelerating BNO055 sensor along y axis, (d) accelerating BNO055 sensor along z axis.

DAQ Accompanying Software for Orientation, Motion, and Gesture Detection with gnuplot

The example presented herein applies to the accompanying software, designed to obtain data (i.e. *Euler* angles, *Gravity Vector*, or *Linear Acceleration*) from BNO055 SiP. To run the DAQ software, the designer should perform some changes to the firmware examples presented before. The changes (which are used to synchronize the communication between the host PC and the microcontroller) are found within the blue frame of Figure 4.31a,b. In detail, the microcontroller waits until '@' character is received by the serial port before sending the acquired data to the host PC. *Euler* angles are obtained from the code example given in Figure 4.31a, while *Gravity Vector* data are obtained from the example of

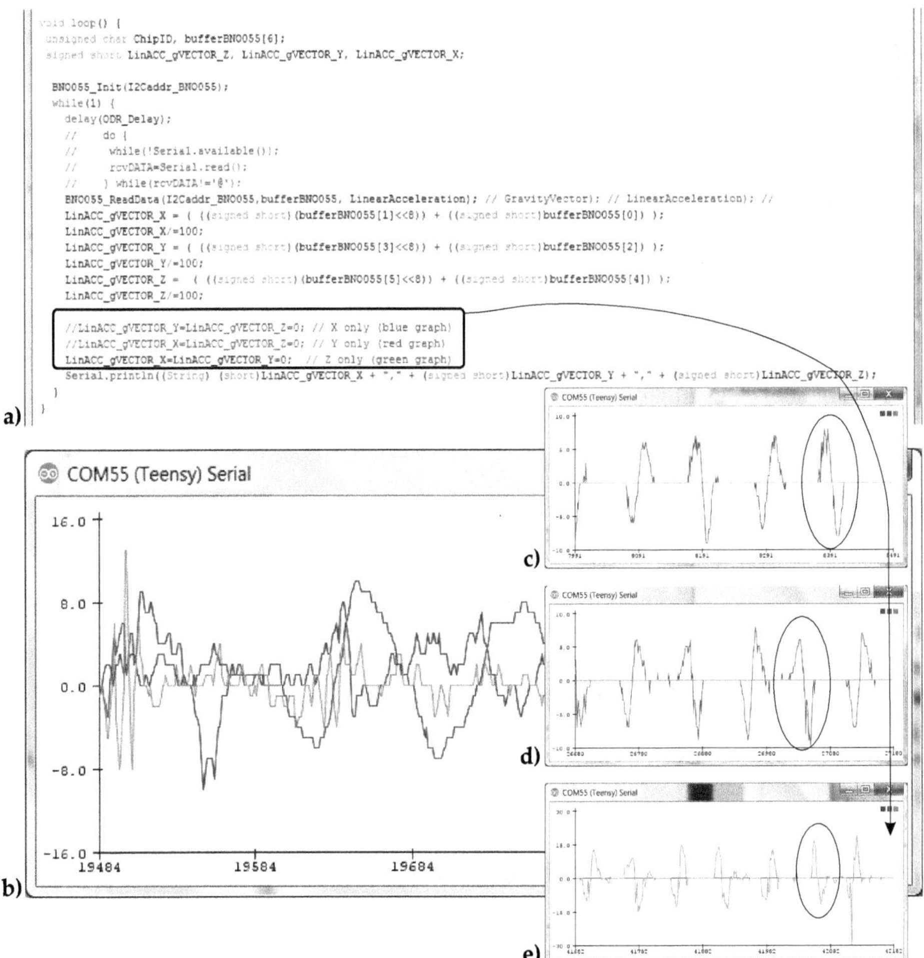

Figure 4.29 Motion detection (*Linear Acceleration*) graphs and revised firmware. (a) application firmware: revisions for reading *gravity* and *linear acceleration,* (b) plot of BNO055 displacement on x,y,z axis, (c) plot of BNO055 displacement along x axis, (d) plot of BNO055 displacement along y axis, (e) plot of BNO055 displacement along z axis. *Source:* Arduino Software.

Figure 4.31b. To read *Linear Acceleration* data, the designer should replace the function argument named *GravityVector* to the one called *LinearAcceleration*. Figure 4.31c depicts the compilation and execution[6] of the DAQ software, where Figure 4.31d–f illustrate the plotted data of *Euler* angles, *Gravity Vector*, and *Linear Acceleration*, respectively.

The DAQ software (which looks pretty much the same to the preceding DAQ example explored before) is given in Figure 4.32. The green rectangle encompasses the new elements of the code. Because *Yaw*, *Roll*, and *Pitch* data (or *x*, *y*, *z* for *Gravity* and *Linear Acceleration*) are transmitted all together through serial port (separated by ';' and appended with '\r' and

6 It is here noted that the user may need to run (i.e. just open and close) the Arduino *Serial Monitor* tool before the foremost run of the current DAQ software.

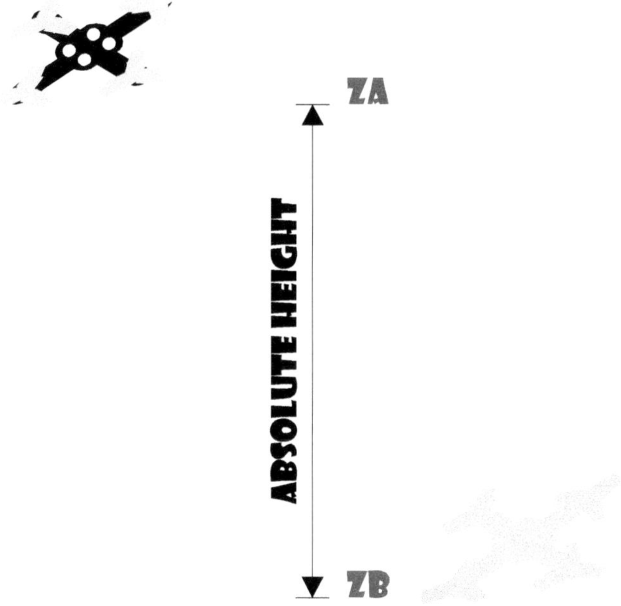

Figure 4.30 Fusing *Linear Acceleration* and *Barometric Altitude* toward absolute height calculation.

'*\n*' characters), the *while()* statement separates each individual information and converts it into *float* format. The latter task is performed through the *atof()* built-in function of C language, which interprets a *numeric string* value into *floating point* number (code lines 39–41).

The *while()* statement runs three consecutive times, until the terminating ('\n') character of the received data is traced. The *do..while()* statement (code lines 35–37) copies to the *strEuler* character array the content of *receive_buf* array until the *','* or '*\n*' character is found. The '*\n*' character is traced in the final run of the *do..while()* statement. The *k* variable is used to identify each run of the loop, and during the first run it arranges the *Yaw* (or *x*) data, during the second run arranges the *Roll* (or *y*) data, and during the third and final run it arranges the *Pitch* (or *y*) data.

The blue frame encompasses the instructions that print *Yaw*, *Roll*, and *Pitch* data (or *x*, *y*, *z* for *Gravity* and *Linear Acceleration*). It is here noted that the code in its current form runs for *Euler* angles, as well as *Gravity* and *Linear Acceleration*. However, the user could replace the information referred to as *Yaw*, *Roll*, and *Pitch* to *x*, *y*, and *z*, in order to arrange a more readable code when acquiring *Gravity Vector* or *Linear Acceleration* data.

Real Time Monitoring with Open GL

The software example presented herein applies to *open graphics library (OpenGL)* cross-language (i.e. independent of the utilized programming language) as well as cross-platform API (i.e. independent of any windowing system), which is used to portray 2D and 3D vector graphics. Because of its independency to windowing systems, OpenGL requires a library to work on particular operating system. GLUT library constitutes a regular OpenGL utility toolkit for developing code on MS-Windows using MinGW.

Figure 4.31 Updated firmware and DAQ running for Euler angles, *Gravity*, *Linear Acceleration*. (a) revised firmware code to work with a custom-designed software *(Euler angles)*, (b) updates of the latest firmware code for obtaining *gravity* and *linear acceleration*, (c) compilation and execution of the custom-designed software, (d) plot of *Euler angles* obtained from the custom-designed software, (e) plot of *gravity vector* obtained from the custom-designed software, (f) plot of *linear acceleration* obtained from the custom-designed software. *Source:* Arduino Software.

C:\zSAVES\MyWork_BOOK_uCT\BOOK\CODE_DAQ-SW\SERIAL\BNO055-gnuplot.c - Notepad++

File Edit Search View Encoding Language Settings Macro Run Plugins Window ?

BNO055-gnuplot.c

```
#include <stdio.h>
#include <windows.h>
#include "Serial.h"

float Yaw,Roll,Pitch;

void SERIAL (int serial_port, int baud_rate, int samples, int sampling_interval, int bytes_perFrame) {
  char Data[]= {"3"}, receive_buf[4096], X_ayis[10];
  int Data_length=sizeof(Data)/sizeof(char), loop, rxBytes, txBytes, gnuplotWindow=50;
  char strEuler[8]; //### Yaw: 0,360 ### Roll: -90,90 ### Pitch:-180,180
  int i,j,k;
  fclose(stderr);
  FILE *fp, *pipe, *fpGP_Yaw, *fpGP_Roll, *fpGP_Pitch;
  fp = fopen("measurements.txt","a");
  fclose(fp);
  fp = fopen("measurements.txt","r+");
  fseek (fp, 0, SEEK_END);
  pipe = popen("gnuplot -persist","w");
  fpGP_Yaw = fopen("Yaw.dat","w");
  fpGP_Roll = fopen("Roll.dat","w");
  fpGP_Pitch = fopen("Pitch.dat","w");
  fclose(fpGP_Yaw);
  fclose(fpGP_Roll);
  fclose(fpGP_Pitch);

   if (SERIAL_Open(serial_port, baud_rate)) exit(0);

   for (loop=0; loop<samples; loop++) {
    txBytes = SERIAL_Write(Data, Data_length-1);
    Sleep(sampling_interval);
    rxBytes = SERIAL_Read(receive_buf, bytes_perFrame);
    receive_buf[rxBytes]='\0';
    i=j=k=0; //### Yaw: 0,360 ### Roll: -90,90 ### Pitch:-180,180 ###//
    while( receive_buf[j] != ('\n') ) {
     do{
        strEuler[i++] = receive_buf[j++];
     } while ( (receive_buf[j] != ',') & (receive_buf[j] != '\r') );
     strEuler[i]=0;
     if(k==0) Yaw = atof(strEuler);
     if(k==1) Roll = atof(strEuler);
     if(k==2) Pitch = atof(strEuler);
     i=0;
     j++;
     k++;
    }
    printf ("%.0f, %.0f, %.0f            \r",Yaw, Roll, Pitch);
    fprintf(fp,"%.0f\t %.0f\t %.0f\n",Yaw, Roll, Pitch);
    // Print to GNUPLOT
    sprintf(X_ayis, "%d", loop);
    fpGP_Yaw = fopen("Yaw.dat","a");
    fpGP_Roll = fopen("Roll.dat","a");
    fpGP_Pitch = fopen("Pitch.dat","a");
    if(loop>gnuplotWindow)   fprintf(pipe, "set xrange [%d:%d]\n",loop-(gnuplotWindow-1), loop);
    fprintf(fpGP_Yaw,"%s %.0f\n",X_ayis,Yaw);
    fprintf(fpGP_Roll,"%s %.0f\n",X_ayis,Roll);
    fprintf(fpGP_Pitch,"%s %.0f\n",X_ayis,Pitch);
    fprintf(pipe, "plot  'Yaw.dat' with lines linecolor 'blue' lw 2, \
                      'Roll.dat' with lines linecolor 'red' lw 2, \
                      'Pitch.dat' with lines linecolor 'green' lw 2; unset key\n");
    fclose(fpGP_Yaw);
    fclose(fpGP_Roll);
    fclose(fpGP_Pitch);
    fflush(pipe);
   }

   SERIAL_Close();
  fclose(fp);
  fclose(pipe);

}

int main()
{
    int P_samples = 10000, ODR=10 , frame_bytes=32; /*frame_bytes max is 14*/
    SERIAL (55, 9600/*gnuplot@9600*/, P_samples, ODR, frame_bytes);
 return(0);
}
```

SEPARATE DATA OF EACH AXIS

C source file length : 2,563 lines : 77 Ln : 46 Col : 25 Sel : 0 | 0 Windows (CR LF) UTF-8 INS

Figure 4.32 DAQ software for acquiring *Euler* angles, *Gravity*, and *Linear Acceleration*. *Source:* Arduino Software.

To install GLUT, download the corresponding toolkit from www.opengl.org/ and copy *GL* directory to *C:\MinGW\include* path. Inside the *C:\MinGW\include\GL* the designer should find the *glut.h* file. Next, the designer should copy *glut32.lib* file within the working folder where the source code file (i.e. *<file-name>.c*) is found and the executable (i.e. *<file-name>.exe*) is created. The compilation command of the source code *<file-name>.c*, when the latter invokes GLUT functions, is as follows:

```
gcc -o <file-name>.exe -Wall <file-name>.c glut32.lib
-lopengl32 -lglu32 -w
```

Figure 4.33a presents the compilation and execution of the code, which rotates a 3D box printed on the generated *BNO055 Euler Angles* window. The rotation is in agreement with the *Euler* data obtained from BNO055. Figure 4.33b,c depicts how the 3D box responses (in real time) as the user rotates Teensy board (with BNO055 on it).

The source code of the example is given in Figures 4.34 and 4.35. Starting from the basic elements of the program, code line 4 encompasses the *glut.h* header file found inside *GL* folder. Before exploring the preceding function prototypes, we examine the *main()* function in code lines 116–128 (Figure 4.35) where GLUT execution starts from.

The *glutInit()* in code line 117 initializes the GLUT library. The subsequent *glutInitDisplayMode()* enables *double buffering* technique, which eliminates the *flickering effect* in graphics programs. That is, the display must be erased before drawing new information on it, which causes the *flickering effect*. The next three glut functions of code lines 119–121 generate the window on the PC screen. In detail, the window is initialized to (i) $width \times height$ equal to 640×480 pixels, (ii) it appears on the $x \times y$ location of the monitor equal to 50×50 pixels (where for the top-left corner of the screen $x \times y = 0 \times 0$), and (iii) it admits window title *BNO055 Euler Angles* (as defined by the *string* of code line 7).

The *glutDisplayFunc()* of code line 122 belongs to the group of GLUT functions aka *callback functions*, which admit a function as argument. Thus, the *glutDisplayFunc()* invokes the *display()* function (i.e. code lines 49–100), which draws new information to the current window. The *glutReshapeFunc()* callback function invokes the *reshape()* function when the window first appears, and whenever the window is re-sized (e.g. when the user changes the window size manually). The *initGL()* is invoked only once and is addressed to carry out initialization tasks, such as clearing the background color via *glClearColor()* function (of code line 40).

The *glutTimerFunc()* callback function in code line 125 invokes the *timer()* function prototype. The *glutPostRedisplay*() within the latter function (code line 103), forces the *display()* to redraw the window. Then the *glutTimerFunc()* determines 10 ms interval (as defined by the *refreshPeriod* variable in code line 8) before redrawing the window again. Finally, the *glutMainLoop()* function of code line 126, passes the control to GLUT so as to repeatedly process events. The *glutPostRedisplay*() called by *glutTimerFunc()* callback function constitutes the essence of our code, as the application constantly reads the *Euler* angles from BNO055 device and updates orientation of the 3D box in our window. Hence, we need to periodically redraw the window (herein, every 10 ms) in order to update orientation of the 3D object. On the other hand, if we try (for example) to change the size of *BNO055 Euler Angles* window (as in every window running on our computer) the *glutMainLoop()* will invoke *glutReshapeFunc()* to respond to the resize task.

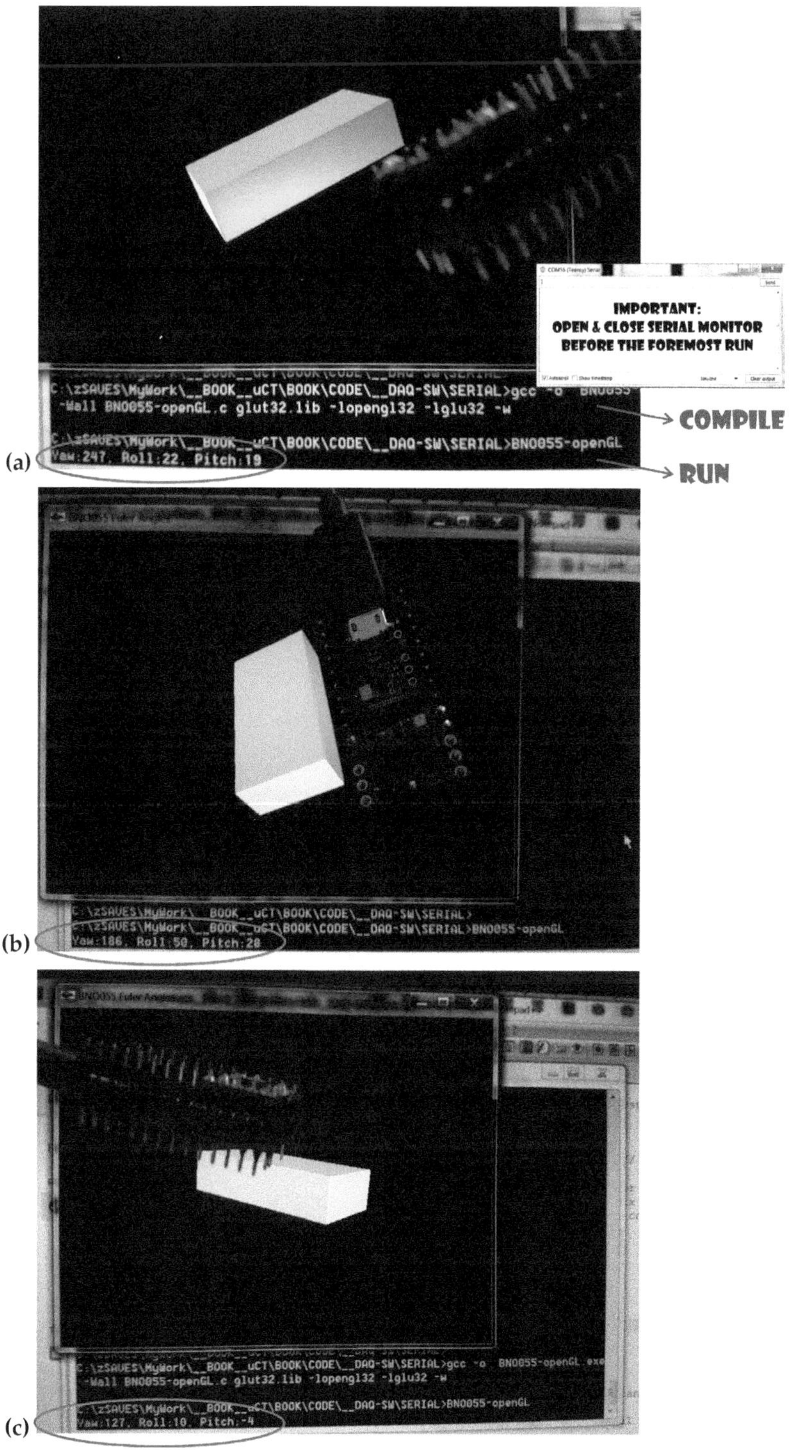

Figure 4.33 Compilation and execution of the Open GL example code. (a) real-time response of a 3D box (printed on the PC screen) according to the *Euler angles* obtained from BNO055 sensor (demonstration example 1), (b) real-time response of the 3D box (demonstration example 2), (c) real-time response of the 3D box (demonstration example 3). *Source:* Arduino Software.

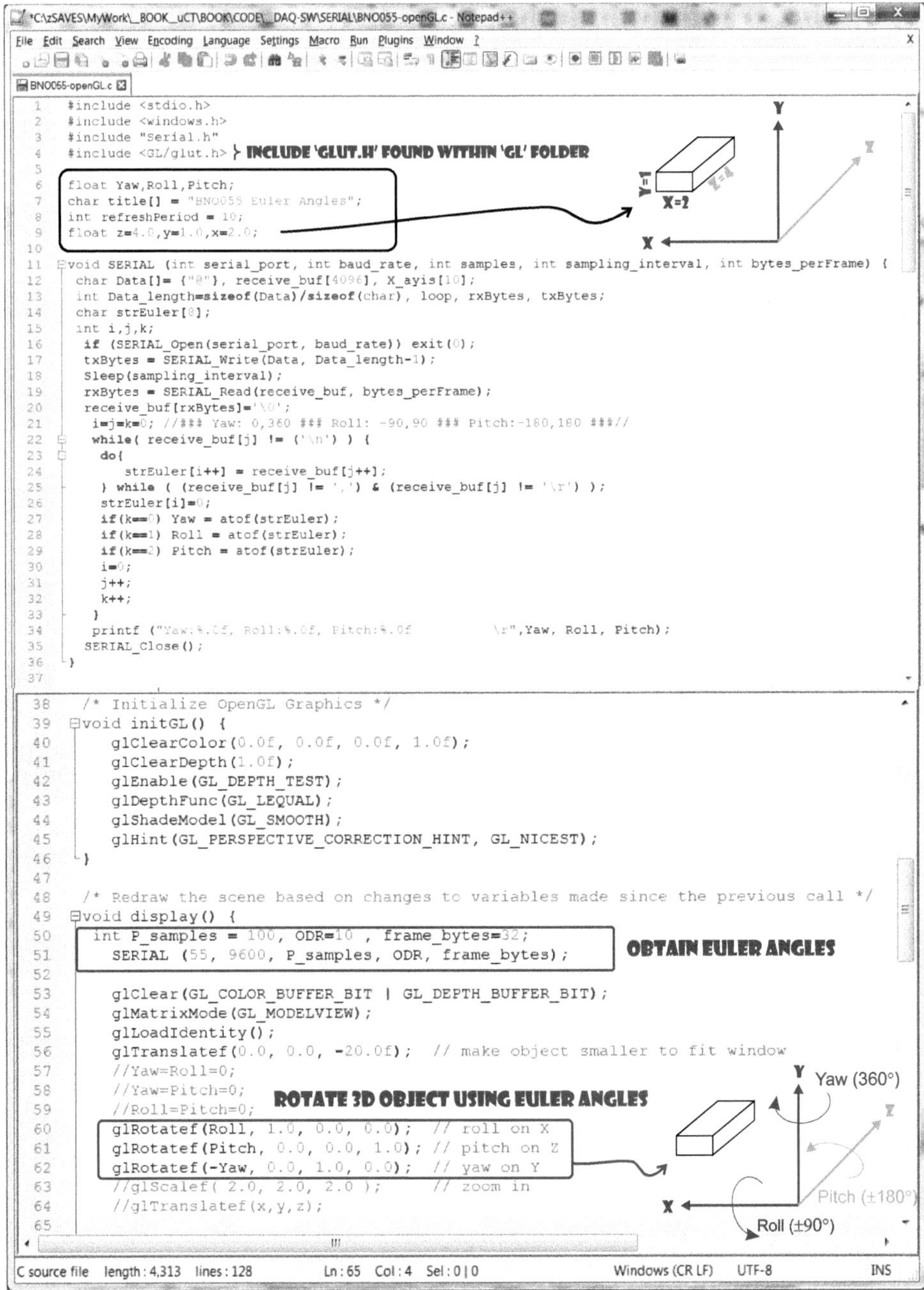

*C:\zSAVES\MyWork_BOOK_uCT\BOOK\CODE_DAQ-SW\SERIAL\BNO055-openGL.c - Notepad++
File Edit Search View Encoding Language Settings Macro Run Plugins Window ?
BNO055-openGL.c
#include <stdio.h>
#include <windows.h>
#include "Serial.h"
#include <GL/glut.h>
INCLUDE 'GLUT.H' FOUND WITHIN 'GL' FOLDER
float Yaw,Roll,Pitch;
char title[] = "BNO055 Euler Angles";
int refreshPeriod = 10;
float z=4.0,y=1.0,x=2.0;
Y
X
Z
Y=1
X=2
Z=4
void SERIAL (int serial_port, int baud_rate, int samples, int sampling_interval, int bytes_perFrame) {
char Data[]= {"0"}, receive_buf[4096], X_ayis[10];
int Data_length=sizeof(Data)/sizeof(char), loop, rxBytes, txBytes;
char strEuler[8];
int i,j,k;
if (SERIAL_Open(serial_port, baud_rate)) exit(0);
txBytes = SERIAL_Write(Data, Data_length-1);
Sleep(sampling_interval);
rxBytes = SERIAL_Read(receive_buf, bytes_perFrame);
receive_buf[rxBytes]='\0';
i=j=k=0; //### Yaw: 0,360 ### Roll: -90,90 ### Pitch:-180,180 ###//
while(receive_buf[j] != ('\n')) {
do{
strEuler[i++] = receive_buf[j++];
} while ((receive_buf[j] != ',') & (receive_buf[j] != '\r'));
strEuler[i]=0;
if(k==0) Yaw = atof(strEuler);
if(k==1) Roll = atof(strEuler);
if(k==2) Pitch = atof(strEuler);
i=0;
j++;
k++;
}
printf ("Yaw:%.0f, Roll:%.0f, Pitch:%.0f \r",Yaw, Roll, Pitch);
SERIAL_Close();
}
/* Initialize OpenGL Graphics */
void initGL() {
glClearColor(0.0f, 0.0f, 0.0f, 1.0f);
glClearDepth(1.0f);
glEnable(GL_DEPTH_TEST);
glDepthFunc(GL_LEQUAL);
glShadeModel(GL_SMOOTH);
glHint(GL_PERSPECTIVE_CORRECTION_HINT, GL_NICEST);
}
/* Redraw the scene based on changes to variables made since the previous call */
void display() {
int P_samples = 100, ODR=10 , frame_bytes=32;
SERIAL (55, 9600, P_samples, ODR, frame_bytes);
OBTAIN EULER ANGLES
glClear(GL_COLOR_BUFFER_BIT | GL_DEPTH_BUFFER_BIT);
glMatrixMode(GL_MODELVIEW);
glLoadIdentity();
glTranslatef(0.0, 0.0, -20.0f); // make object smaller to fit window
//Yaw=Roll=0;
//Yaw=Pitch=0;
//Roll=Pitch=0;
ROTATE 3D OBJECT USING EULER ANGLES
glRotatef(Roll, 1.0, 0.0, 0.0); // roll on X
glRotatef(Pitch, 0.0, 0.0, 1.0); // pitch on Z
glRotatef(-Yaw, 0.0, 1.0, 0.0); // yaw on Y
//glScalef(2.0, 2.0, 2.0); // zoom in
//glTranslatef(x,y,z);
Y
Yaw (360°)
Z
X
Pitch (±180°)
Roll (±90°)
C source file length : 4,313 lines : 128 Ln : 65 Col : 4 Sel : 0 | 0 Windows (CR LF) UTF-8 INS

Figure 4.34 Open GL example applying to *Euler* angles (1 of 2). *Source:* Arduino Software.

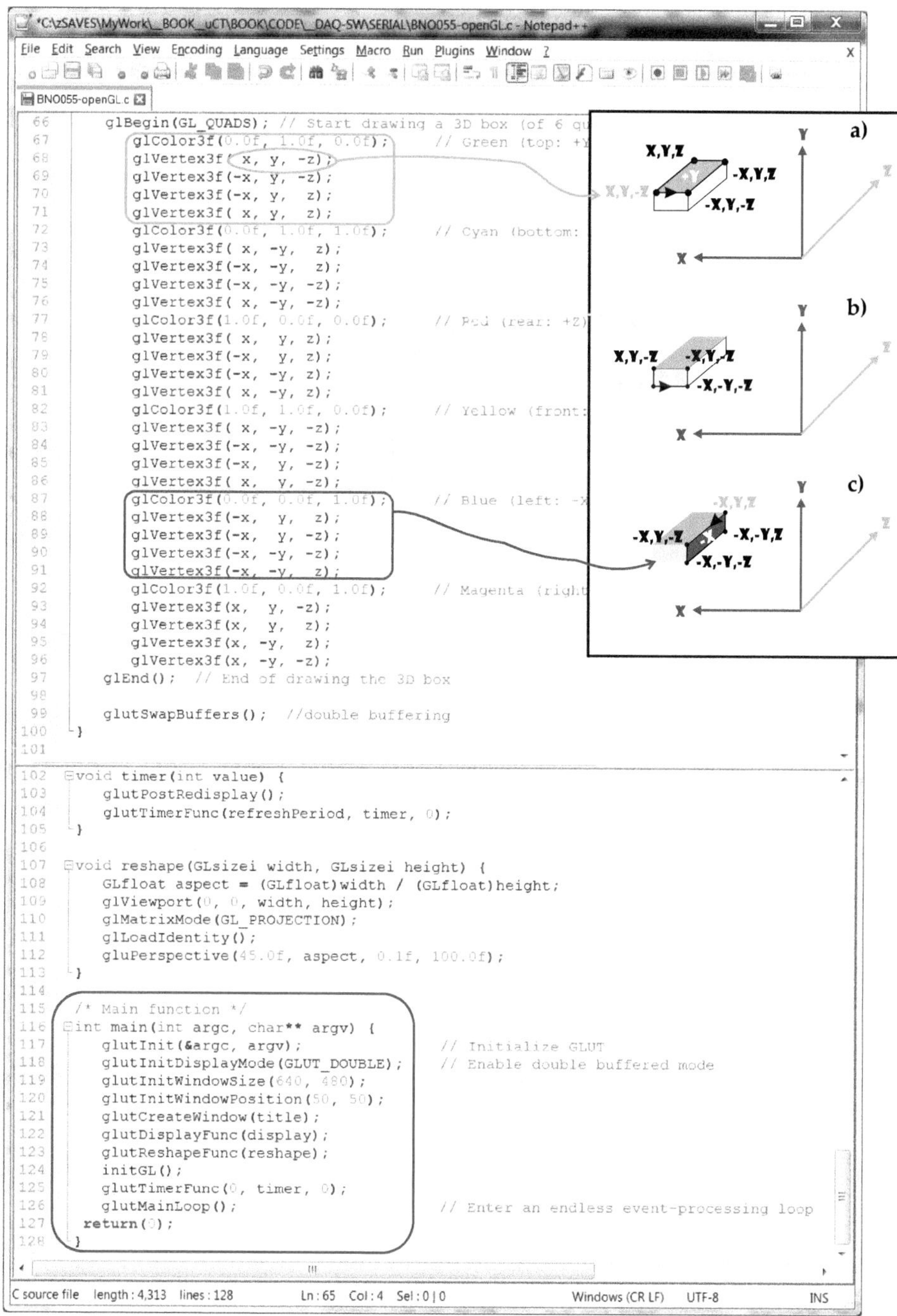

```
    glBegin(GL_QUADS); // Start drawing a 3D box (of 6 qu
      glColor3f(0.0f, 1.0f, 0.0f);     // Green (top: +Y
      glVertex3f( x, y, -z);
      glVertex3f(-x, y, -z);
      glVertex3f(-x, y,  z);
      glVertex3f( x, y,  z);
      glColor3f(0.0f, 1.0f, 1.0f);     // Cyan (bottom:
      glVertex3f( x, -y,  z);
      glVertex3f(-x, -y,  z);
      glVertex3f(-x, -y, -z);
      glVertex3f( x, -y, -z);
      glColor3f(1.0f, 0.0f, 0.0f);     // Red (rear: +Z)
      glVertex3f( x,  y, z);
      glVertex3f(-x,  y, z);
      glVertex3f(-x, -y, z);
      glVertex3f( x, -y, z);
      glColor3f(1.0f, 1.0f, 0.0f);     // Yellow (front:
      glVertex3f( x, -y, -z);
      glVertex3f(-x, -y, -z);
      glVertex3f(-x,  y, -z);
      glVertex3f( x,  y, -z);
      glColor3f(0.0f, 0.0f, 1.0f);     // Blue (left: -X
      glVertex3f(-x,  y,  z);
      glVertex3f(-x,  y, -z);
      glVertex3f(-x, -y, -z);
      glVertex3f(-x, -y,  z);
      glColor3f(1.0f, 0.0f, 1.0f);     // Magenta (right
      glVertex3f(x,  y, -z);
      glVertex3f(x,  y,  z);
      glVertex3f(x, -y,  z);
      glVertex3f(x, -y, -z);
    glEnd();  // End of drawing the 3D box

    glutSwapBuffers(); //double buffering
}

void timer(int value) {
    glutPostRedisplay();
    glutTimerFunc(refreshPeriod, timer, 0);
}

void reshape(GLsizei width, GLsizei height) {
    GLfloat aspect = (GLfloat)width / (GLfloat)height;
    glViewport(0, 0, width, height);
    glMatrixMode(GL_PROJECTION);
    glLoadIdentity();
    gluPerspective(45.0f, aspect, 0.1f, 100.0f);
}

/* Main function */
int main(int argc, char** argv) {
    glutInit(&argc, argv);                // Initialize GLUT
    glutInitDisplayMode(GLUT_DOUBLE);     // Enable double buffered mode
    glutInitWindowSize(640, 480);
    glutInitWindowPosition(50, 50);
    glutCreateWindow(title);
    glutDisplayFunc(display);
    glutReshapeFunc(reshape);
    initGL();
    glutTimerFunc(0, timer, 0);
    glutMainLoop();                       // Enter an endless event-processing loop
  return(0);
}
```


Figure 4.35 Open GL example applying to *Euler* angles (2 of 2). *Source:* Arduino Software.

The dimensions of the 3D box are defined in code line 9 and in order to realize its shape, we consider the fixed *X,Y,Z* coordinate system illustrated in the top-right corner of Figure 4.34. Holding the Teensy board in such orientation so that the front side of the 3D box (i.e. yellow side) appears on our window as presented by Figure 4.33b, we are able to test and realize *roll*, *pitch*, and *yaw* angles (following the direction of scheme on the bottom right corner of Figure 4.34). Code lines 60–62 (within the green frame) change the orientation of the 3D object through *glRotatef()* function, and using the *Euler* data obtained from BNO055 SiP in code lines 50,51 (within the blue frame).

It should be noted that if we change the *yaw* angle 90° to the left or right (i.e. as presented in Figure 4.33a,c, respectively) the coordinate system of the Teensy+BNO055 setup does not much anymore to the fixed X,Y,Z coordinate system illustrated in the right side of Figure 4.34. In detail, the *roll* angle of Teensy+BNO055 setup is originally orientated around *x* axis of the fixed X,Y,Z coordinate system, while *pitch* angle is orientated around *z* axis. Changing the *yaw* angle 90° to the left or right, *roll* angle of Teensy+BNO055 is orientated around *z* axis and the *pitch* angle is orientated around *x* axis. Thereby, *pitch* becomes *roll* and vice versa. One possible solution for the designer is to apply on-the-fly switching of the *Euler* angles so that the Teensy+BNO055 coordinate system is always adjusted to the fixed X,Y,Z coordinate system of the 3D box.

The drawing task of the 3D object is performed in code lines 66–97, within the *glBegin()* and *glEnd()* functions. As an example, the top-right scheme of Figure 4.35 illustrates how each side of the box is drawn via *glVertex3f()*, and colored by *glColorf()* function. For instance, the generation of the top side (i.e. green side) of the 3D box, where all four vertices are found in the positive area of *y* axis (i.e. +*y*), the four consecutive *glVertex3f()* functions draw the vertices {*x*,*y*,−*z*}, {*x*,*y*,−*z*), {−*x*,*y*,*z*}, and {*x*,*y*,*z*}, respectively.

It is worth mentioning that *glRotatef()* function produces a rotation of the 3D object around *x*,*y*,*z* vector. If we uncomment *glTranslatef()* in code line 64 we cause a rotation of the scene (i.e. the 3D object) around −*x*,−*y*,−*z* vertex. Moreover we could zoom in the 3D object by uncommenting code line 63, while a change to the arguments of *glScale()* function from 2.0 to 0.5, causes a zoom out task. Moreover, we could uncomment one of the three code lines 57–59 so as to observe one *Euler* angel at the time.

Distance Detection and 1D Gesture Recognition with TinyZero

In this subchapter we explore another remarkable 32-bit μC motherboard which is compatible with Arduino IDE. That is, the TinyZero board manufactured by TinyCircuits (www.tinycircuits.com/). As notified by its brand name, TinyCircuits is focused on the development on really tiny motherboards and daughterboards for microcontroller applications. Compared to other corporations that develop stackable boards systems for *makers*, TinyCircuits has been innovated in the way boards are stacked to each other. While industries regularly make use of the standard 2.54 pin-header, TinyCircuits takes advantage of a *surface-mount technology* (*SMT*) board-to-board connector of really low profile. Hence, the useless space between two stackable boards is eliminated, but also the number of available pins is increased.

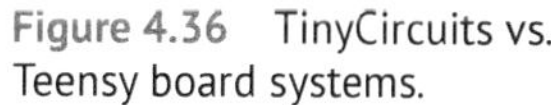

Figure 4.36 TinyCircuits vs. Teensy board systems.

Figure 4.36 depicts a stackable setup of three TinyCircuits board systems next to the Teensy+BNO055 setup of the previous subchapter. We observe that despite the tiny area of the 3-board setup (TinyCircuits), its height is less than the height of the 2-board (Teensy) setup. It should be noted that the Teensy and BNO055 boards have been directly soldered to each other (i.e. not attached to each other with male–female headers) so as to additionally reduce the unused space in between, and hence, the latter setup is not (anymore) straightforwardly detachable as the TinyCircuits setup. The examples presented hereafter apply to the TinyZero processor board of TinyCircuits Corporation, which based on the SAMD21 32-bit ARM Cortex M0+ processor (that is, the same processor used in the Arduino Zero board delivered by Arduino Corporation).

Arduino Ex.4–7

The current example is addressed to help the designer in getting started with TinyZero. Table 4.5 illustrates the TinyZero board specs compared to Teensy board. In general, we could assume that Teensy board provides more features to the designer, which mainly results from the different processor employed by each board. At this point it is worth mentioning that, while there is a plethora of different μC boards available for *makers*, the designer should select the proper board according to the specs of the custom-designed application that is meant for. Another issue the designer should keep in mind is that availability of the daughterboards (by either the same, or third-party manufacturer), which are compatible with the selected μC motherboard. Minimizing the hardware involvement during the development process of a μC-based system is a critical design principle of simplicity.

For instance, if we need to work with a daughterboard that provides 5V output signals, Teensy could a possible option as the processor features 5V tolerance on its IO pins. On the other hand, if we are developing an autonomous battery-operated system TinyZero could

Table 4.5 Specs of TinyZero and Teensy 32-bit μC development boards.

Specs	**TinyZero board**	**Teensy 3.2 board**
Processor	ATSAMD21G18A (32-bit ARM Cortex-M0+)	MK20DX256 (32-bit ARM Cortex-M4)
Clock speed	48 MHz (32.728 KHz clock crystal on board)	72 MHz
Flash memory	256 kB	256 kB
RAM memory	32 kB	64 kB
Operating voltage	3V3	3V3 (with 5 V pin tolerance)
Digital IO pins	14	34
Analog pins	6	21
PWM	8	12
UART bus	1	3
I2C bus	1	2
SPI bus	1	1
DAC	1	1
RTC	1	1

Figure 4.37 Battery operated system with TinyZero and Teensy (with Adafruit Feather). (a) *TinyZero* board, (b) *Teensy 3.X Feather Adapter* along *Teensy3.2* board.

be a straightforward solution as the board employs Li–Ion battery connector (Figure 4.37a). On the other hand, if we need both features, then Teensy board could be used along with the Teensy 3.X Feather Adapter by Adafruit Industries (Figure 4.37b). Accordingly, it would be wise to make a market research before deciding what μC motherboard(s) and daughterboard(s) to use in our custom-designed application, as minimizing hardware involvement is a wanted feature toward achieving *simplicity in design*.

The installation of TinyZero board in Windows requires some effort and, therefore, we give below a detailed description of the steps that should be followed:

1) Delete **.tmp* files (if any) from the path:
   C:\Users\<Current-User>\AppData\Local\Arduino15
2) In the Arduino IDE, select *File→Preferences* and enter the following URL to the position *Additional Boards Manager URLs*:

   http://files.tinycircuits.com/ArduinoBoards/package_tinycircuits_index.json

   (If there are other URLs, insert a *coma (,)* before writing the new URL.)
3) Connect to the internet, and in the Arduino IDE, select *Tools→Board→Boards Manager...*
4) Then install the following package:
   Arduino SAMD Boards (32-bits ARM Cortex-M0+) by Arduino
5) Through the same selection, scroll down to select and install the following package (downloaded by the internet):
   TinyCircuits SAMD Boards by TinyCircuits
   (After the installation you should be able to see the result of Figure 4.38.)
6) Connect the TinyZero to your computer using a *microUSB* cable and make sure the TinyZero power (slide) switch is turned ON.
7) In the Arduino IDE, select *Tools→Board→TinyZero*.
8) The TinyZero uses the driver of Teensy board that was installed in the previous subchapter. If the driver is not automatically found for TinyZero, open the Windows *Device Manager* and, by pressing *right click* on the *Unknown device*, select the option *Update Driver Software...*
9) In the pop-up window select *Browse my computer for Driver software*.
10) In the updated screen select *Let me pick from a list of device drivers on my computer* and click *Next* button.

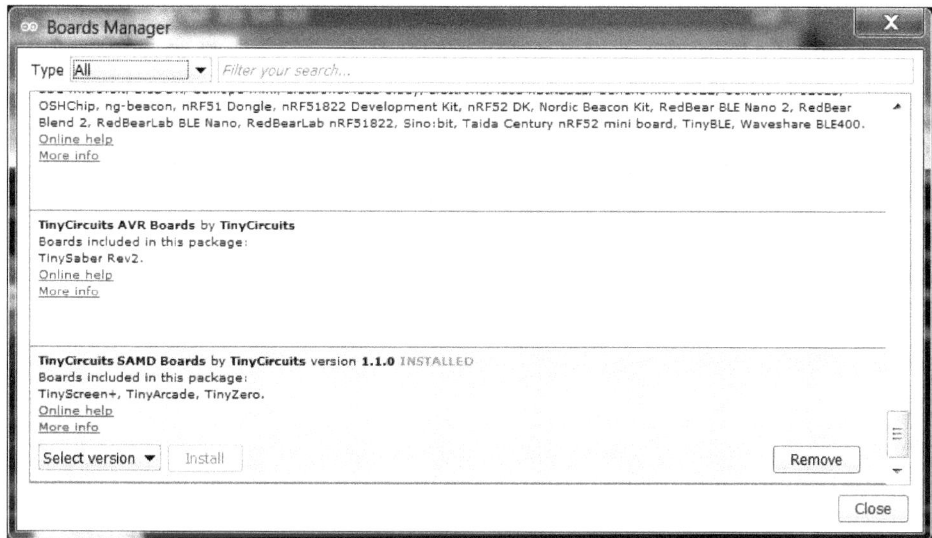

Figure 4.38 *TinyCircuits SAMD Boards* by *TinyCircuits* installed. *Source:* Arduino Software.

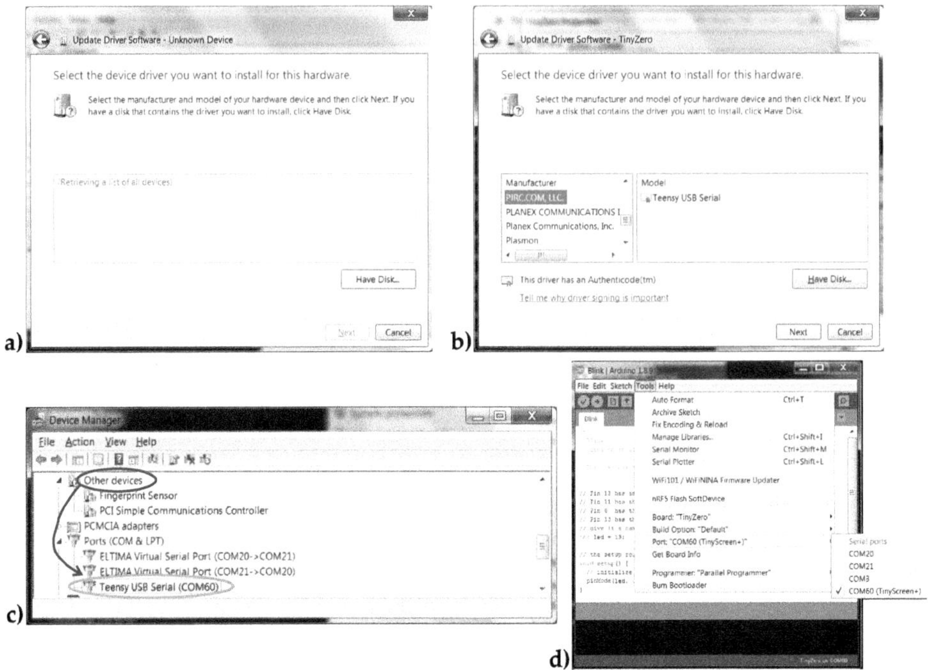

Figure 4.39 Teensy USB Serial driver installed for TinyZero board. (a) driver installation process (step 1), (b) driver installation process (step 2), (c) driver installation process (step 3), (d) configuring TinyZero board in Arduino IDE. *Source:* Arduino Software.

11) In the updated screen select *Show All Devices* and click *Next* button.
12) Wait for a while in the screen of Figure 4.39a, till the drivers are listed, and then, select the *Teensy USB Serial* driver presented in Figure 4.39b. Then click *Next* button and press *Yes* button in the pop-up window *Update Driver Warning*.
13) After the installation, the *Unknown device* (that was previously found on *Other devices list*, as described in step 8), should be available by the *(COM & LPT) Ports'* list as highlighted by the green frame of Figure 4.39c.
14) Return to the Arduino IDE and select the available COM Port from the menu *Tools→Port→COMXX (TinyScreen+)* as presented in Figure 4.39d.
15) The same installation (from step 8) might be needed when the Arduino IDE is trying to upload code to TinyZero board for the first time. To force TinyZero into bootloader code (if not automatically inserted when pressing the *Upload* button in the Arduino IDE), press the push-button at the bottom side of the board and, then, slide ON the switch found at the top side (Figure 4.40).

A simple firmware to get started with TinyZero board is given in Figure 4.41a, which blinks a LED every 1 s and sends to the serial port the "Hello World!" string (Figure 4.41d). The LED in TinyZero (Figure 4.41b) is driven by PIN13 (as in the Arduino Uno). To exchange data with the UART device, one should make use of *SerialUSB* object (i.e. blue frames of Figure 4.41a) instead of *Serial* used by the previous examples. To maintain compatibility among different µC motherboards we could "cheat" the code by the *#define* directive of Figure 4.41c (blue frame).

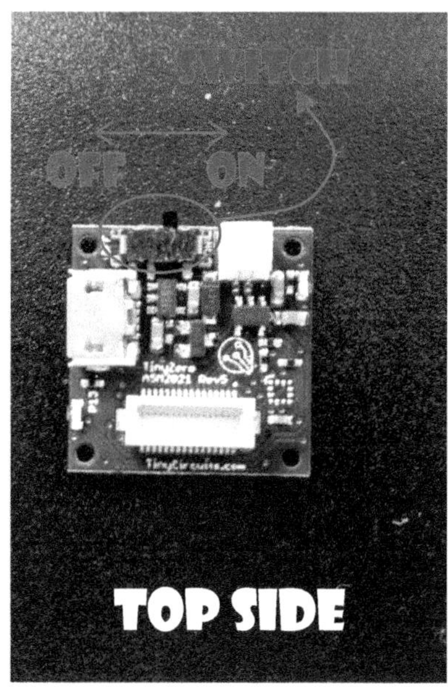

Figure 4.40 Force TinyZero board into bootloader code (if not automatically inserted).

Arduino Ex.4–8

The current example applies to VL53L0X Time-of-Flight sensor [57] by STMicroelectronics. The sensor is addressed for accurate absolute distance measurements up to 2 m. Particular applications of such a sensor are in agreement with obstacle detection in robotics, hand detection in faucets and soap dispensers, etc. However, the utilization of two (or more) VL53L0X sensors can lead us in 1D gesture recognition. The example presented hereafter, applies to both distance measurements (acquired by one sensor) and 1D gesture recognition (via two sensors), by merely modifying a constant value (i.e. 1 or 2, respectively) before compiling and uploading the firmware code to TinyZero board.

It is worth mentioning that, up until this point, we have worked out thoroughly on the development of firmware code from scratch. We've paid attention on that particular trend as it is exactly what is required for the learner to be directed to a deep understanding of microcontroller programming and application development. However, the DIY and maker cultures of our age have established a wave of, not only ready-to-use board systems, but also free and shareable libraries available over the internet. Accordingly, to keep pace with the current trends of our age it is important to understand what kind of third-party code to re-use, in order to avoid putting extra effort on reinventing the wheel when not really worth it.

In this example we will need two different ready-to-use libraries developed by TinyCircuits. One of them is the library that is used for the configuration and control of VL53L0X device, employed by *TOF Sensor Wireling* board (of TinyCircuits). It should be noted that the configuration of VL53L0X sensor is not trivial and it would be wise to start with a driver delivered by trusted suppliers. The VL53L0X driver can be downloaded by the https://tinycircuits.com/products/tof-distance-sensor-wireling-vl53l0x. It is worth mentioning that the learner should now be able to modify the driver, if needed. The second (ready-to-use) library is

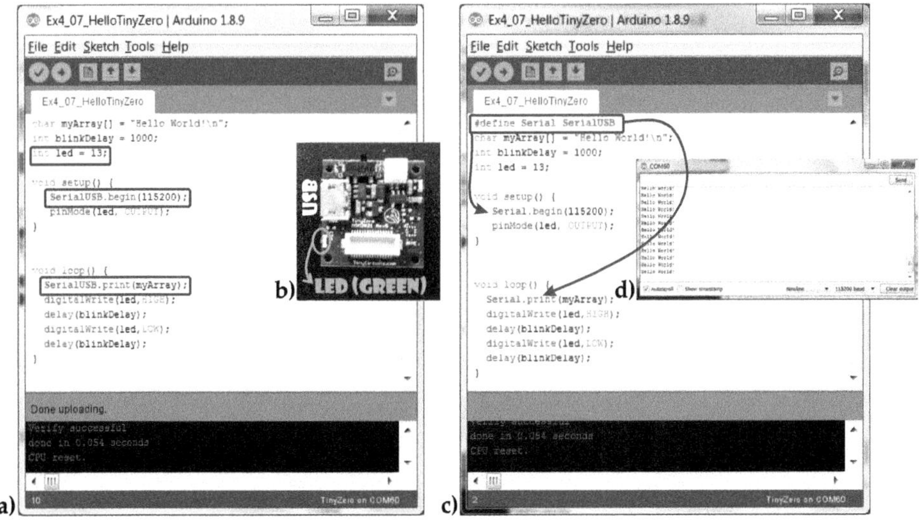

Figure 4.41 Getting started with "Hello World!" and blinking LED firmware (TinyZero board). (a) example firmware for TinyZero, (b) LED on TinyZero board, (c) updated firmware to maintain code compatibility with different hardware platforms, (d) printed results of the example firmware. *Source:* Arduino Software.

addressed for the control of the *Wireling Adapter TinyShield*, also delivered by TinyCircuits. The latter adapter is used as an intermediate board for the connection of the *TOF Sensor Wireling* daughterboard to TinyZero motherboard. The adapter features a 4-channel Multiplexer, which allows us to connect up to four different I2C device of identical address. This option serves the feature of our 1D gesture recognition example by reading two identical sensors (i.e. 2×VL53L0X). The latter driver can be found at www.arduinolibraries.info/libraries/wireling.

For the development of the current example, we install the latter driver into the root directory of the Arduino IDE and inside the *libraries* folder (e.g. *C:\arduino-1.8.9\libraries\Wireling-0.1.1*), while the VL53L0X driver (i.e. the files *VL53L0X.h* and *VL53L0X.cpp*) is kept inside the working folder (herein, the *Ex4_08*). Figure 4.42a presents the hardware for reading distance measurements (i.e. from sensor-1) as well as applying 1D gesture recognition (i.e. from two sensors). The hardware incorporates the *Wireling Adapter* attached on the top of *TinyZero* processor as well as 2x*TOF Sensor Wireling* boards (holding the sensors). The firmware code is given in Figure 4.42b.

The firmware applies to both distance detection (using sensor-1 attached to Port-0) and 1D gesture recognition (using both sensors attached to Port-0 and Port-3) by merely changing the *sensors constant* value (inside the red circle); that is, from 1 to 2, respectively. Inside the black frame at the top of the firmware code, the *#include* directive invokes the driver for VL53L0X sensor. The *header* file VL53L0X.h is placed within *double quotes ("…")* because it is located inside the working folder (i.e. where the source code *Ex4_08.ino* is found). This syntax directs the Arduino IDE to open the involved files VL53L0X.h and VL53L0X.cpp, as presented at top of Figure 4.42b.

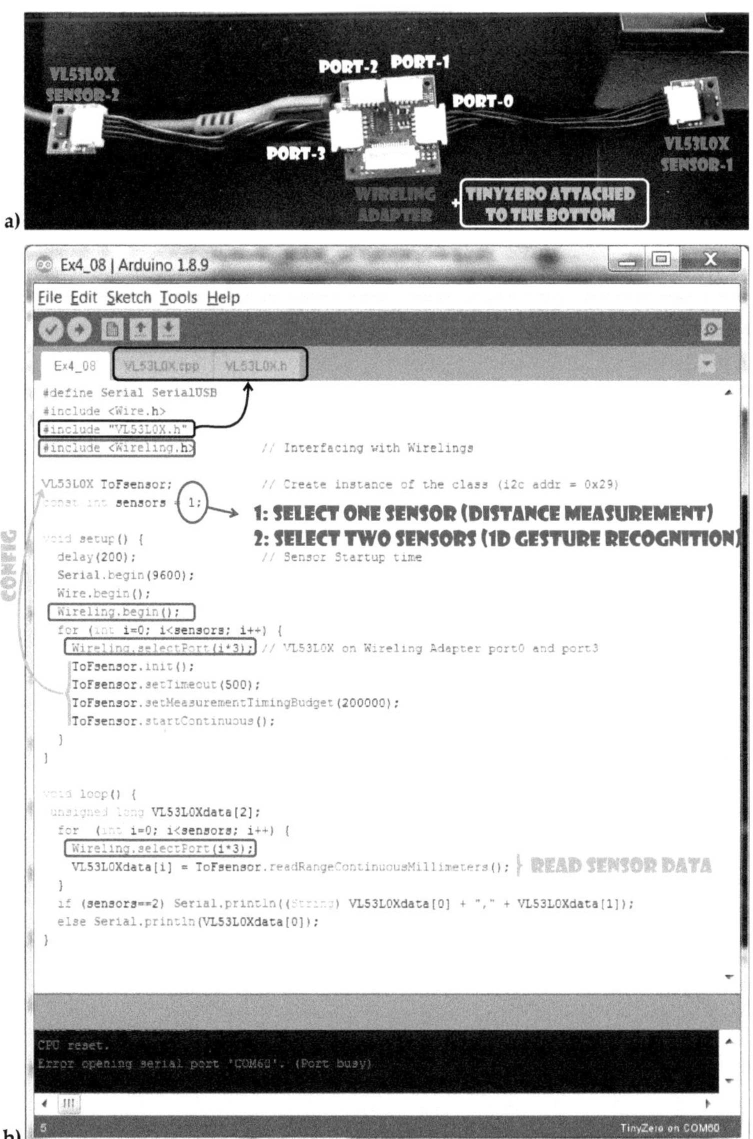

Figure 4.42 Distance measurements and gesture recognition setup & firmware. (a) application hardware, (b) application firmware.

The subsequent *#include* directive within the blue frame invokes the driver for the *Wireling Adapter*, while the rest of the blue frames invoke functions of that particular driver. In detail, the *Wireling.begin()* powers the adapter, while the *Wireling.selectPort()* determines which port is going to be used so as to access the I2C device. To render the firmware code parameterizable, we have incorporated the *Wireling.selectPort()* function within a *for* loop. The loop runs either once or twice, as being configured by the *const int sensors* value at the beginning of the code. In case it runs only once, the loop counter *i* admits zero value and, hence, the argument *i*×*3* is equal to zero. Thereby, only Port-0 of the *Wireling Adapter* is accessed. In case the loop executes two iterations, the loop counter *i* admits the value 1 during the second iteration. Hence, the argument *i*×*3* is equal to 3 during the second run of the *for* loop and, thereby, Port-3 of the *Wireling Adapter* is accessed during the second loop iteration.

The code instructions highlighted by green color, refer to the VL53L0X driver. In detail, after creating an instance of the class VL53L0X (using the name *ToFsensor*) the corresponding instructions inside the *setup()*, which are highlighted by the green *curly brace ({)*, initialize the VL53L0X sensor device(s). Inside the *loop()*, the function *readRangeContinuousMillimeters()* acquires the detected distance in *millimeters (mm)*. If the sensor is out of range (i.e. greater than 2 m), it returns the value 8190. In case the sensor is not responding (e.g. not found) on the selected port of the *Wireling Adapter*, it returns the value 65 535. Figure 4.43 illustrates the distance measurements with one VL53L0X sensor using the *Serial Monitor* (Figure 4.43a) and *Serial Plotter* (Figure 4.43b). The data have been acquired by placing our

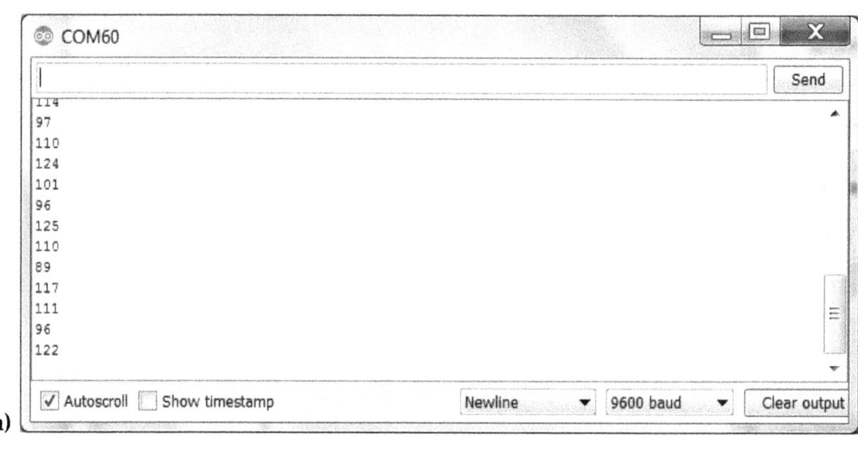

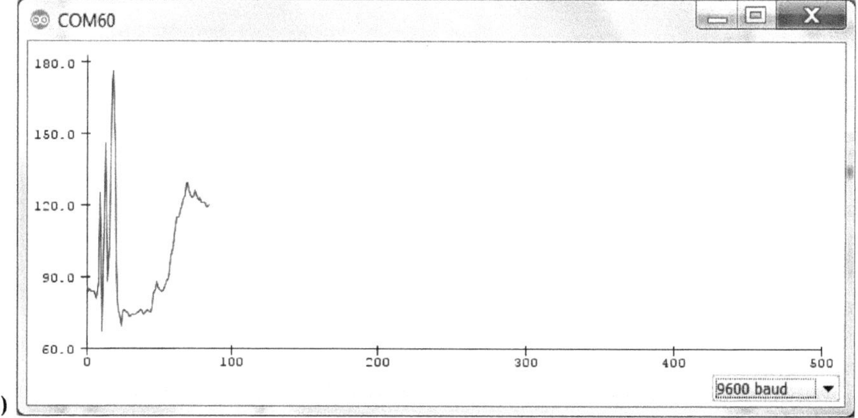

Figure 4.43 Distance measurements with one VL53L0X: (a) serial monitor, (b) serial plotter. *Source:* Arduino Software.

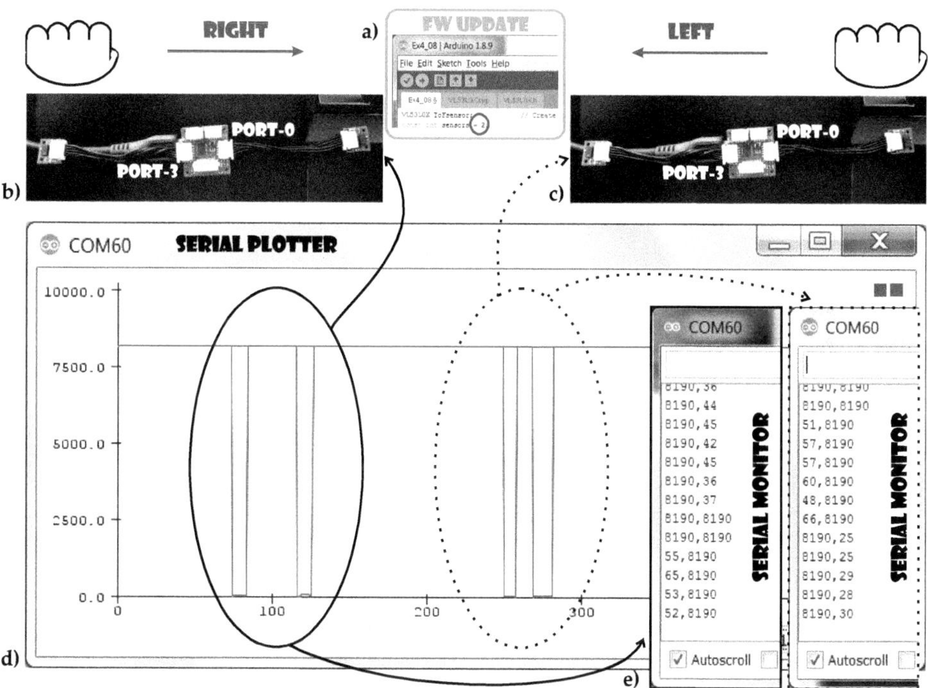

Figure 4.44 1D gesture recognition by reading two identical VL53L0X sensors. (a) firmware revision to read data from two identical sensors, (b) user interaction with the hardware: Gesture 1, (c) user interaction with the hardware: Gesture 2, (d) plots obtained from the two different gestures, (e) measurements obtained from the two different gestures. *Source:* Arduino Software.

palm about 10 cm above the VL53L0X found on Port-0, and by moving our hand up and down in a relative small distance.

Figure 4.44 illustrates the 1D gesture recognition by reading two identical VL53L0X sensors. To obtain the results depicted by the figure, the firmware should be updated by changing *const int sensors* value to 2 (Figure 4.44a). Passing our hand on top of VL53L0X sensor found in Port-3, and then on top of VL53L0X sensor in Port-0 (Figure 4.44b), we obtain the plots of the left part of *Serial Plotter* (Figure 4.44d) and *Serial Monitor* (Figure 4.44e). Passing our hand first on top of VL53L0X in Port-0, and then on top of VL53L0X sensor in Port-3 (Figure 4.44c), we obtain the plots of the right part of *Serial Plotter* and *Serial Monitor*, highlighted by the dotted-line frames and arrows.

DAQ Accompanying Software for Distance Measurements

Figure 4.45 illustrates the DAQ software code for obtaining distance measurements from a single VL53L0X sensor (while the DAQ software applying to the 2-sensor setup, is left as exercise for the reader). The code is based on the same philosophy that has been used by the previous examples, that is, the acquired measurements are stored to *measurements.txt* file and plotted via *gnuplot*. However, the code applies to a different technique for reading data from the serial port (i.e. code lines within the green frame). That is, the

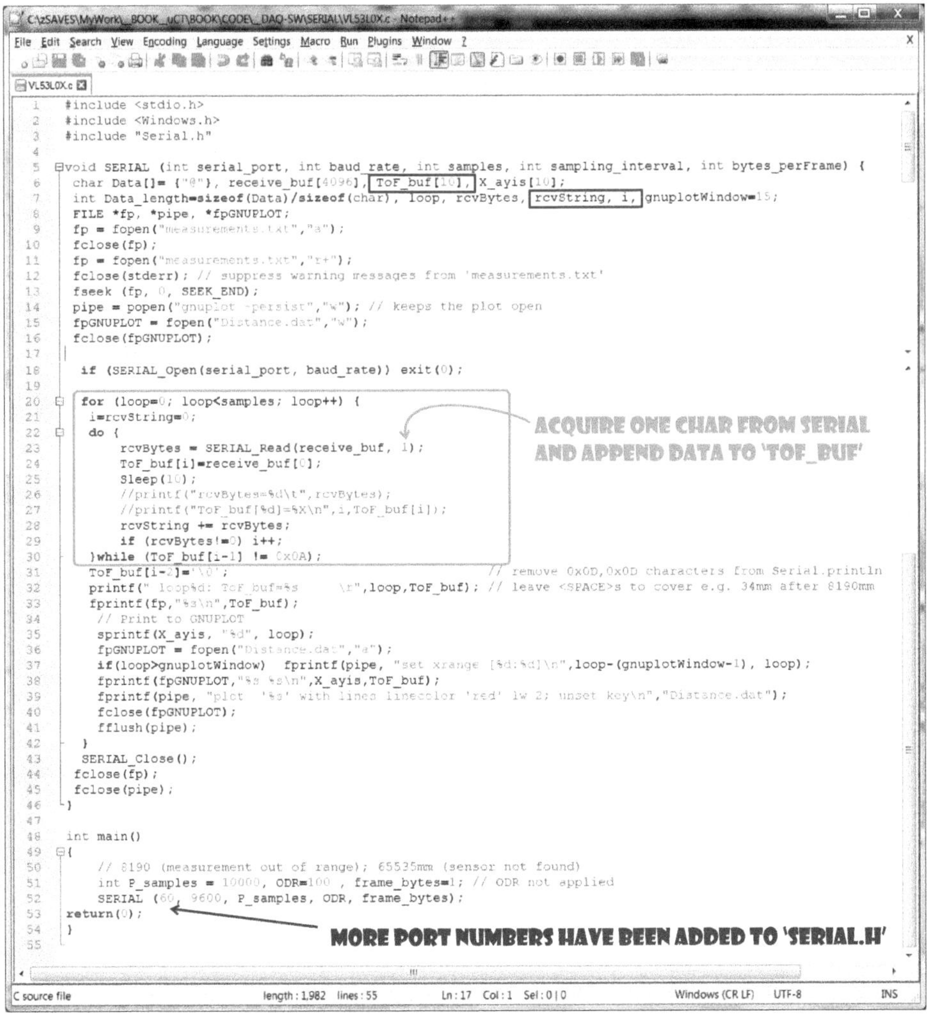

C:\zSAVES\MyWork_BOOK_uC\BOOK\CODE_DAQ-SW\SERIAL\VL53L0X.c - Notepad++

```
1   #include <stdio.h>
2   #include <Windows.h>
3   #include "Serial.h"
4
5   void SERIAL (int serial_port, int baud_rate, int samples, int sampling_interval, int bytes_perFrame) {
6    char Data[]= {"@"}, receive_buf[4096], ToF_buf[10], X_ayis[10];
7    int Data_length=sizeof(Data)/sizeof(char), loop, rcvBytes, rcvString, i, gnuplotWindow=15;
8    FILE *fp, *pipe, *fpGNUPLOT;
9    fp = fopen("measurements.txt","a");
10   fclose(fp);
11   fp = fopen("measurements.txt","r+");
12   fclose(stderr); // suppress warning messages from 'measurements.txt'
13   fseek (fp, 0, SEEK_END);
14   pipe = popen("gnuplot -persist","w"); // keeps the plot open
15   fpGNUPLOT = fopen("Distance.dat","w");
16   fclose(fpGNUPLOT);
17
18    if (SERIAL_Open(serial_port, baud_rate)) exit(0);
19
20    for (loop=0; loop<samples; loop++) {
21     i=rcvString=0;
22     do {
23         rcvBytes = SERIAL_Read(receive_buf, 1);
24         ToF_buf[i]=receive_buf[0];
25         Sleep(10);
26         //printf("rcvBytes=%d\t",rcvBytes);
27         //printf("ToF_buf[%d]=%X\n",i,ToF_buf[i]);
28         rcvString += rcvBytes;
29         if (rcvBytes!=0) i++;
30     }while (ToF_buf[i-1] != 0x0A);
31     ToF_buf[i-2]='\0';                              // remove 0x0D,0x0D characters from Serial.println
32     printf(" loop%d: ToF_buf=%s      \r",loop,ToF_buf); // leave <SPACE>s to cover e.g. 34mm after 8190mm
33     fprintf(fp,"%s\n",ToF_buf);
34      // Print to GNUPLOT
35      sprintf(X_ayis, "%d", loop);
36      fpGNUPLOT = fopen("Distance.dat","a");
37      if(loop>gnuplotWindow)  fprintf(pipe, "set xrange [%d:%d]\n",loop-(gnuplotWindow-1), loop);
38      fprintf(fpGNUPLOT,"%s %s\n",X_ayis,ToF_buf);
39      fprintf(pipe, "plot  '%s' with lines linecolor 'red' lw 2; unset key\n","Distance.dat");
40      fclose(fpGNUPLOT);
41      fflush(pipe);
42    }
43    SERIAL_Close();
44   fclose(fp);
45   fclose(pipe);
46  }
47
48  int main()
49  {
50      // 8190 (measurement out of range); 65535mm (sensor not found)
51      int P_samples = 10000, ODR=100 , frame_bytes=1; // ODR not applied
52      SERIAL (60, 9600, P_samples, ODR, frame_bytes);
53  return(0);
54  }
55
```

ACQUIRE ONE CHAR FROM SERIAL AND APPEND DATA TO 'TOF_BUF'

MORE PORT NUMBERS HAVE BEEN ADDED TO 'SERIAL.H'

C source file   length : 1,982   lines : 55   Ln : 17   Col : 1   Sel : 0 | 0   Windows (CR LF)   UTF-8   INS

Figure 4.45 1D gesture recognition by reading two identical VL53L0X sensors. *Source:* Arduino Software.

code does not send synchronization character to the μC device, instead it obtains one character on each serial read task. The latter task is repeated within the *do..while* statement (lines 22–30) until the 0 × 0A character arrives on the serial port. This way we are able to identify the end of the string, which has been prepared by the firmware code through the *Serial.println()* function.

The aforementioned technique might be somewhat slower because of reading characters one by one, but it is safer in terms of being synchronized with the asynchronous type of communication of the UART interface. It is worth mentioning that we have added more port numbers in the *Serial.h* file, as well, because TinyZero has been attached to COM60 port.

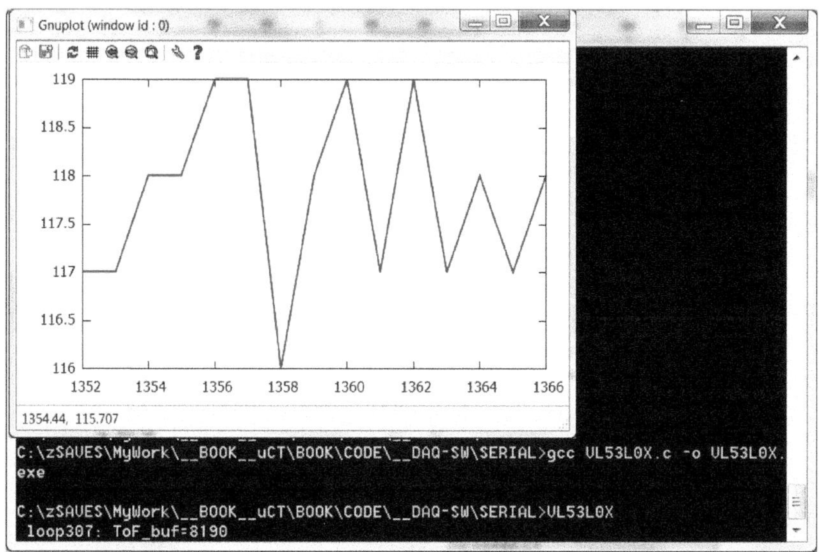

Figure 4.46 DAQ software applying to distance detection (illustrating VL53L0X accuracy). *Source:* Arduino Software.

Figure 4.46 illustrates the compilation command of the DAQ software (applying to distance detection) as well as the printed results arisen from its execution. It is here noted that the plot represents the detection of a fixed distance, which has been selected in order to illustrate the high accuracy of VL53L0X sensor (i.e. max error is 3 mm).

Color Sensing and Wireless Monitoring with Micro:bit

A very popular 32-bit μC motherboard, nowadays, is the Micro:bit designed by BBC. While the Micro:bit is meant to be programmed by the (ease-of-use) drag-and-drop block-based visual programming language of the *Microsoft MakeCode* web-based environment, it can be configured to operate with the Arduino IDE, as well. Micro:bit was originally addressed to introduce microcomputer technology to school students. To this end, the processor of Micro:bit is actually a *2.4 GHz RF System on Chip* that supports Bluetooth 4.1 communication and a fully qualified *Bluetooth low energy* (*BLE*) stack. Moreover, the board's functionality is enriched with motion sensors, LED array, push-buttons, and along with several available daughterboards by third-party industries (such as, *kitronik*), it provides to the students several experimentation possibilities. In addition to the education of children, all these features render the Micro:bit a powerful motherboard, appropriate for the development of sophisticated μC-based systems. Table 4.6 illustrates the board's specs of Micro:bit next to the boards explored before (i.e. TinyZero and Teensy 3.2). It should be noted that the processor might feature more resources than those provided by the Micro:bit board, which holds the unit.

Table 4.6 Specs of Micro:Bit, TinyZero and Teensy 32-bit μC development boards.

Specs	Micro:bit board	TinyZero board	Teensy 3.2 board
Processor	nRF51822 Bluetooth SoC (32-bit ARM Cortex-M0)	ATSAMD21G18A (32-bit ARM Cortex-M0+)	MK20DX256 (32-bit ARM Cortex-M4)
Clock speed	16 MHz	48 MHz (32.728 KHz crystal on board)	72 MHz (32.728 KHz supported, not soldered)
Flash memory	256 kB	256 kB	256 kB
RAM memory	16 kB	32 kB	64 kB
Operating voltage	3V3	3V3	3V3 (5 V pin tolerance)
Digital IO pins	19	14	34
Analog pins	6	6	21
PWM	3	8	12
UART bus	1	1	3
I2C bus	1	1	2
SPI bus	1	1	1
DAC	0	1	1
RTC	1	1	1
External peripherals	**Micro:bit board**	**TinyZero board**	**Teensy 3.2 board**
Motion sensors	3-axis Accelerometer + Magnetometer (LSM303GR)	3-axis accelerometer (BMA250): *optional*	—
Bluetooth antenna	PCB antenna	—	—
Battery connector	Available	Available	Not available
User LEDs	9 × 3 Matrix	1	1
User switches	2 push-buttons	—	—

Arduino Ex.4-9

The current example is addressed to help the learner in getting started with Micro:bit. As before, we need to install a Windows driver (if working on Windows operating system) as well as a board package for the Arduino IDE. The steps are as follows:

1) To install the driver, connect Microbit (Figure 4.47a) to the hostPC, download and run the *mbed Windows serial port driver* installer from the webpage *https://os.mbed.com/handbook/Windows-serial-configuration*. As soon as the process finishes, verify the installation through *Device Manager* (Figure 4.47b).

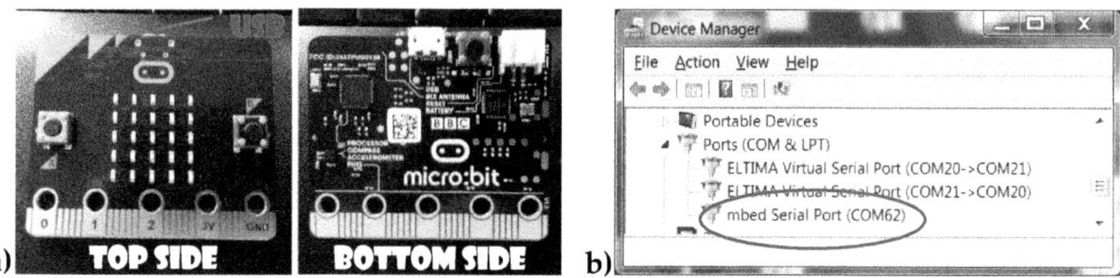

Figure 4.47 *mbed* driver successfully installed. (a) top view of *Micro*:bit board, (b) bottom view of *Micro*:bit board, (c) driver installation for the *Micro*:bit board. *Source:* Arduino Software.

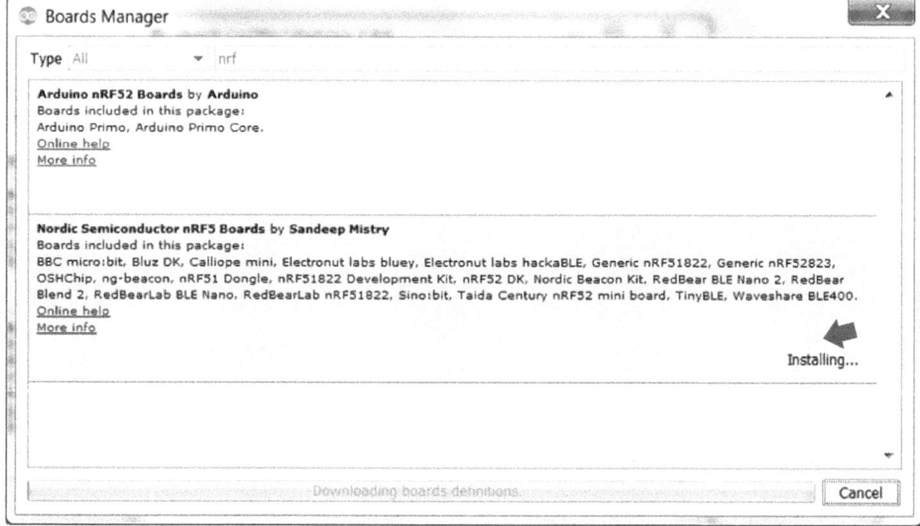

Figure 4.48 Installing *Nordic Semiconductor nRF5 Boards* by *Sandeep Mistry*. *Source:* Arduino Software.

2) To install the board package first delete all **.tmp* files (as they could block the downloading process by the Arduino IDE) from the path:
   C:\Users\<Current-User>\AppData\Local\Arduino15
3) In the Arduino IDE select *File→Preferences* and enter the following URL to the position *Additional Boards Manager URLs*:
   https://sandeepmistry.github.io/arduino-nRF5/package_nRF5_boards_index.json
   If there are other URLs, insert a *coma (,)* before writing the new URL. The installation process is illustrated in Figure 4.48.
4) After the above two installations we are ready to upload code to Micro:bit board, as soon as we proceed to the proper configurations given in Figure 4.49. Consequently, in the Arduino IDE select:
   a) *Tools→Board→"BBC micro:bit"*
   b) *Tools→Softdevice: "S110"*
   c) *Tools→Port→COMxx (BBC micro:bit)*
5) When uploading the Arduino code to the Micro:bit board, you may be asked to approve the warning about *opencd.exe* by selecting *Allow access* as presented in Figure 4.50.

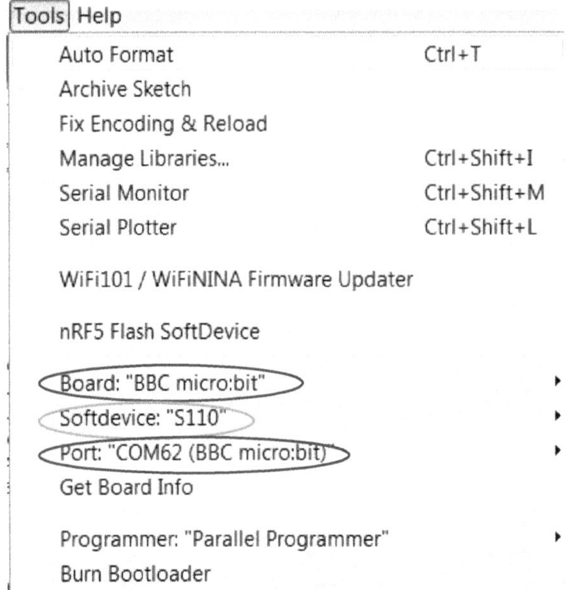

Figure 4.49 Configure the Arduino IDE in order to upload code to Micro:bit. *Source:* Arduino Software.

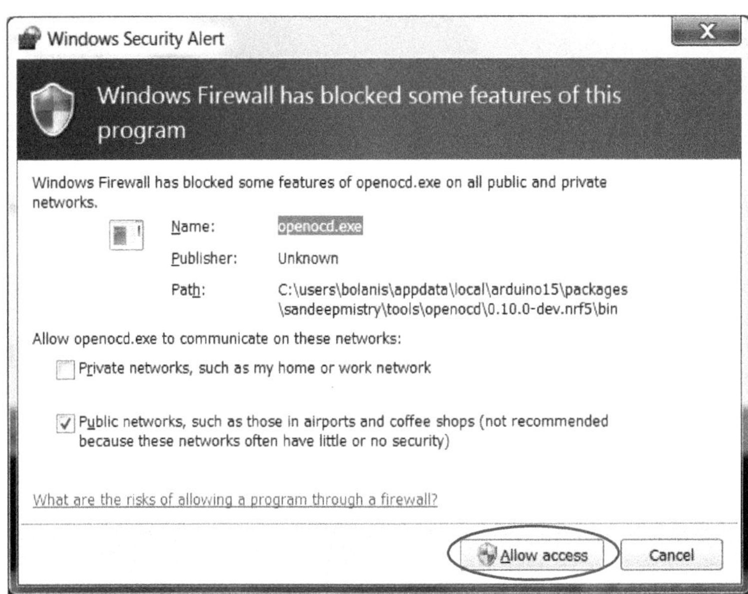

Figure 4.50 Approve the code uploading process to Micro:bit. *Source:* Arduino Software.

Figure 4.51 presents the hardware setup of the example, which blinks a LED on Micro:bit every second. The setup also uses the two push-buttons found on Micro:bit board, which change the time delay between the ON and OFF states of the LED. When pressed by the user, buttons A & B direct PIN5 & PIN11 (respectively) to the ground level. Hence, the μC's input pins admit logical '0', otherwise the pins admit logical '1' (i.e. when buttons are released) as presented by Figure 4.51c. In terms of the LED units, the Micro:bit board

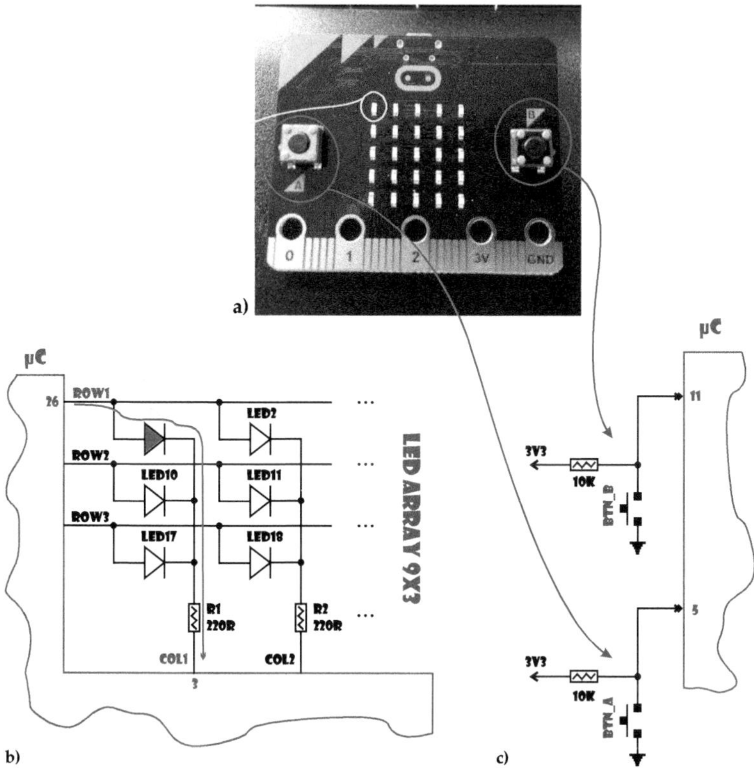

Figure 4.51 Hardware of blinking LED (using push-buttons) in Micro:bit. (a) push-buttons and LEDs on *Micro*:bit board, (b) connection diagram of LED1 to the *Micro*:bit pins, (c) connection diagram of the two push-buttons to the *Micro*:bit pins. *Source:* Arduino Software.

employs a LED array which is entirely driven by the µC pins. For instance, the anode of LED1 unit is driven by ROW1, while its cathode is driven by COLUMN1 (i.e. COL1). Setting COL1 to zero volt (i.e. by clearing PIN3) and applying logical '1' to ROW1 (i.e. by setting PIN26), the LED1 is turned ON as presented by Figure 4.51a. The same LED is turned OFF if we apply logical '0' to ROW1.

Figure 4.52a presents the firmware code of the present example code, which blinks a LED and sends to the serial port the string "Hello World!\n". Figure 4.52b presents the printed results to the serial port. In regard to the firmware code, the red frame presents the pins assigned[7] to the LED and push-buttons. The blue frame illustrates how to change the delay among the ON/OFF states of the LED through an *if()...else if()...else* statement. In detail, when the user presses and holds button A the variable *blinkDelay* changes its value to 100. Thereby, the LED blinks every 100 ms for as long the user keeps pressing the push-button A. On the other hand, when the user presses and holds button B the variable

7 It is here noted that pin assignment for the Micro:bit can be found in the folder where the Arduino package for Micro:bit is installed: *C:\Users\<user-name>\AppData\Local\Arduino15\packages\sandeepmistry\hardware\nRF5\0.6.0\variants\BBCmicrobit.*

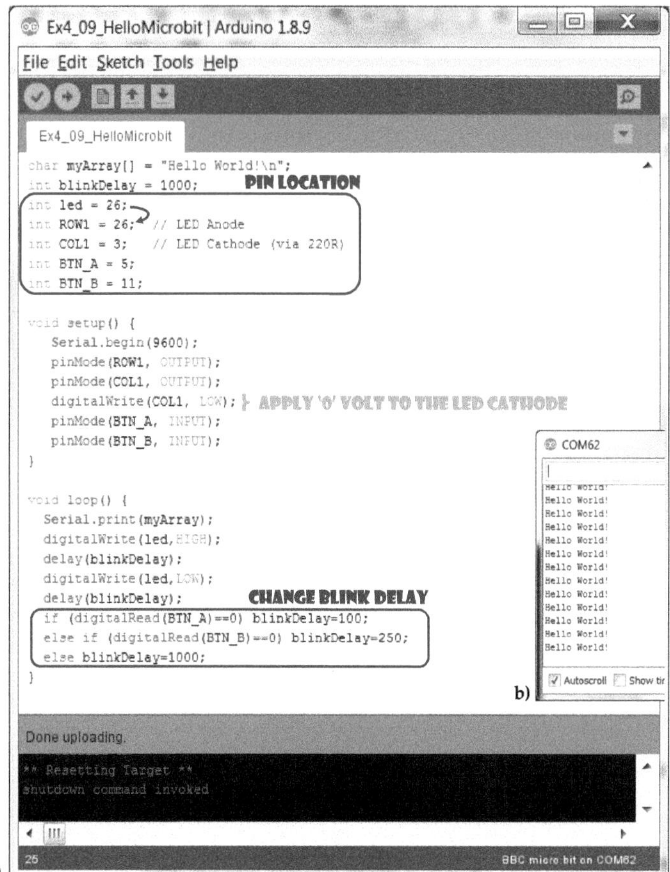

Figure 4.52 Firmware of blinking LED (using push-buttons) in Micro:bit. (a) example firmware, (b) printed results. *Source:* Arduino Software.

blinkDelay changes its value to 250. Consequently, the LED blinks every 250 ms for as long the user keeps pressing push-button B. When both buttons are released, the *blinkDelay* variable is set to 1000 and, hence, the LED blinks every 1000 ms.

Arduino Ex.4-10

The current example applies to TCS3472 *red, green, blue (RGB)* and *clear light* sensor by *Texas Advanced Optoelectronics Solutions* (*TAOS*) Corporation. In the RGB color model (which constitutes one of the most common ways of encoding color in computing), every color is generated by the addition of the three primary colors red, green, and blue. Each of the primary colors admits values from 0 to 255 (i.e. 0×00-FF) and, thereby, the hexadecimal color #FF0000 is composed of 100% red, the color #00FF00 is composed of 100% green, and the color #0000FF is composed of 100% blue. Full intensity of each color (i.e. #FFFFFF) results in white color, while zero intensity (i.e. #000000) results in black color. A colorized

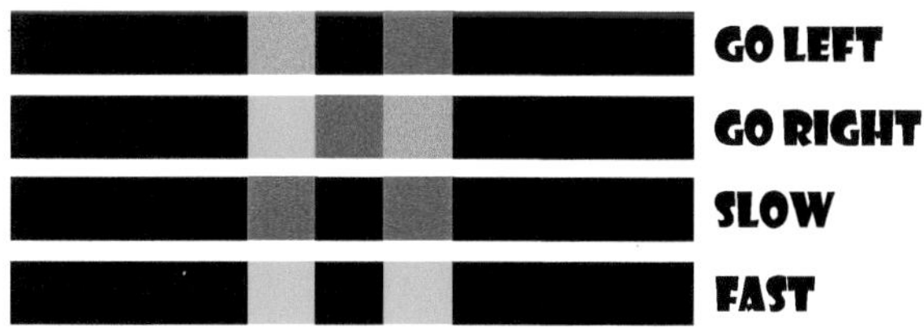

Figure 4.53 Color codes of Ozobot.

hue is generated by applying different intensities to the primary colors. If one of the primary colors has the strongest intensity, the colorized hue is near to that color.

We herein explore the development of a device toward sensing pure red, pure green, and pure blue colors. One popular robot that applies to that kind of identification is the Ozobot (https://files.ozobot.com/). The Ozobot is a line follower educational robot that responds to commands that affect its speed, direction, internal timers, etc. The commands are planed through different sequences of the primary red, green, blue colors, as well as the black color, which additionally forms the line the robot follows. For instance, if the robot identifies the sequence *blue, red, green* it changes its direction and turns *right*, while for the sequence *blue, black, blue* the robot changes its speed and goes faster (Figure 4.53). This way, the kids can learn programming through painting.

Figure 4.54 presents the RGB sensing and testing setup of the present example project. For this example we make use the *RGB sensor module* of Figure 4.54c manufactured by Adafruit Industries. It is worth mentioning that Adafruit delivers, through its website (i.e. www.adafruit.com/), an influx of μC motherboards and daughterboards, as well as an influx of documents and libraries (for Arduino and other development tools embedded computers).

To connect *RGB sensor module* to the edge connector of Micro:bit, we make use of the *Edge Connector Breakout Board* by Kitronik, presented in Figure 4.54a. To implement the hardware setup, we solder a 2.54 female header and make use of the 3V, GND, SCL, and SDA signals of the Kitronik connector (which should be connected to the corresponding signals of the *RGB sensor module*). Figure 4.54b presents the corresponding signals found on the Micro:bit edge connector. After the necessary soldering tasks, the hardware setup of the current example is given in Figure 4.54d.

Figure 4.54e presents three cards of red, green, and blue color, which are scanned and identified by our hardware setup. It is worth mentioning that the *RGB sensor module* incorporates a bright (white) LED so as to illuminate directly to the (colored) surface under identification. Figure 4.54f–h presents the process for scanning the red, blue, and green card, respectively. The corresponding testing results (as arisen by the execution of the firmware code given in Figure 4.55a) are presented (by the *Serial Plotter* tool) in Figure 4.55b. It is worth noting that, when the sensor scans the three cards it does not provide a pure (or almost pure) red, green, or blue color. However, the intensity of the corresponding color allows us to identify which card is being scanned. When no card is placed above the sensor (see setup of Figure 4.54e) we obtain almost zero RGB values as presented (by the *Serial Monitor* tool) in Figure 4.55c.

In consideration of the firmware code of Figure 4.55a, we make use of the *Adafruit_TCS34725.h* library for the I2C control of the RGB sensor device (which can be downloaded

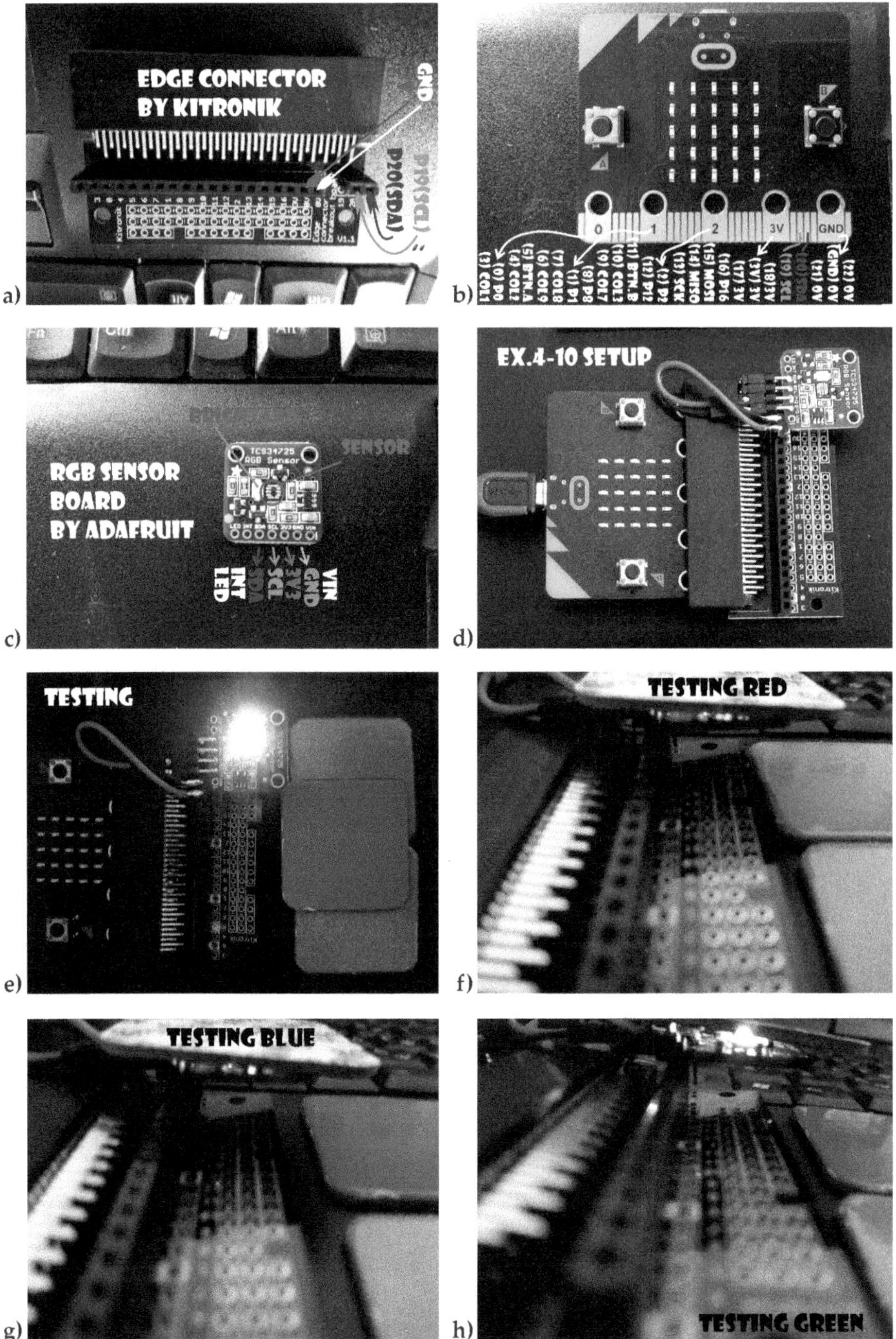

Figure 4.54 Color sensing setup and testing. (a) *edge connector breakout board* for *Micro:bit* by Kitronik, (b) signals on the Micro:bit edge connector, (c) *RGB sensor module* by Adafruit Industries, (d) hardware setup of the current application example, (e) red, green and blue cards for testing the application, (f) demonstration example: scanning the red card, (g) demonstration example: scanning the blue card, (h) demonstration example: scanning the green card.

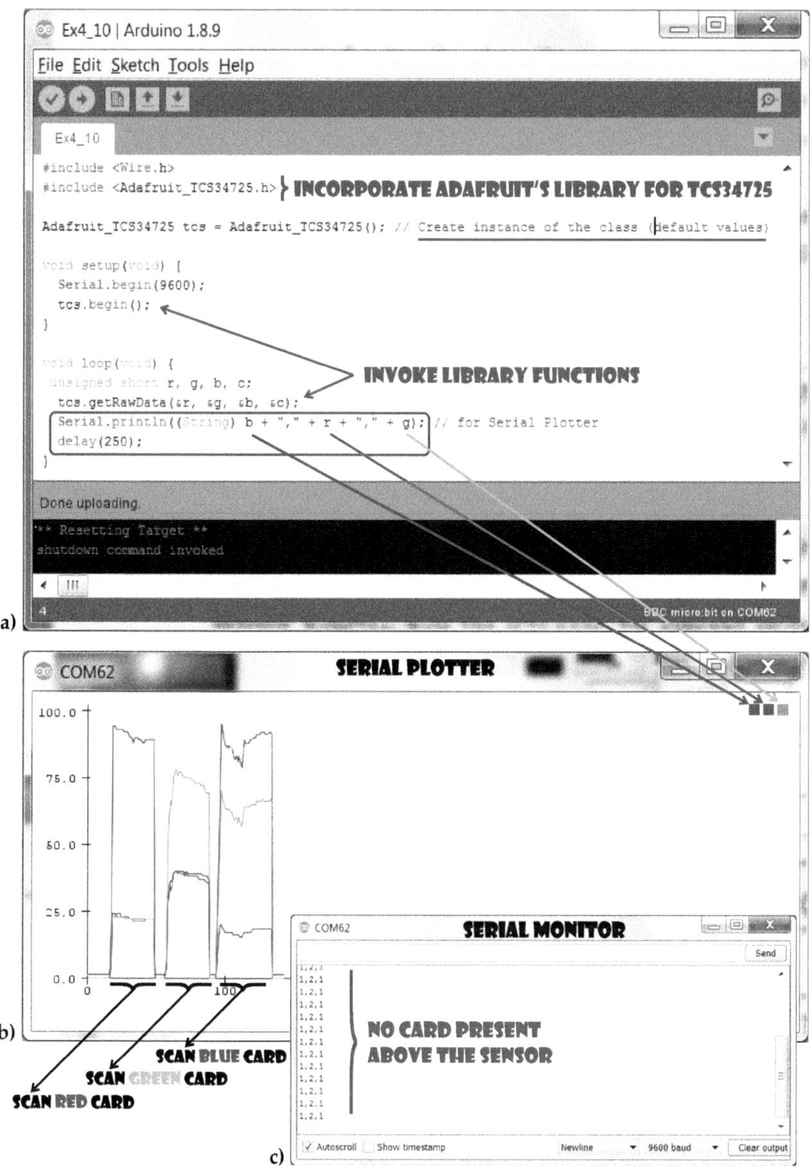

Figure 4.55 Firmware and color sensing results. (a) application firmware, (b) plots of measurements when scanning red, green and blue cards, (c) measurements when no card is placed above the sensor.

from the link: https://github.com/adafruit/Adafruit_TCS34725). The red arrows depict the code lines that access the library functions, while the blue frame illustrates the acquisition of the raw RGB data and their transmission to the serial port every 250 ms. To ease the testing of the color identification, we transmit to the serial port the color values in the exact order: blue, red, and green. We have selected that particular order because the Arduino *Serial Plotter* tool prints the foremost data in blue graph, the subsequent data in red graph, and the final data in green graph. As a result, we have achieved consistency in the color of the each plot (applied by the *Serial Plotter*), to the color of each card scanned by our hardware setup.

Open GL Example Applying to RGB Sensing

Hereafter, we present an Open GL software code, which applies to the RGB color sensing firmware (presented before). The program prints a white rectangle on a new window named *Color sensing*, as well as the RGB values (obtained from sensor module) in the *command prompt* console. When one of the primary colors red, green, or blue is identified by the software code, the color of the rectangle is changed (to red, green, or blue, respectively).

The source code is given in Figures 4.56 and 4.57. We observe that the code is similar to the Open GL example used in *Euler* angles representation. The main difference in terms of the DAQ process is that the current example does not transmit to the serial port any

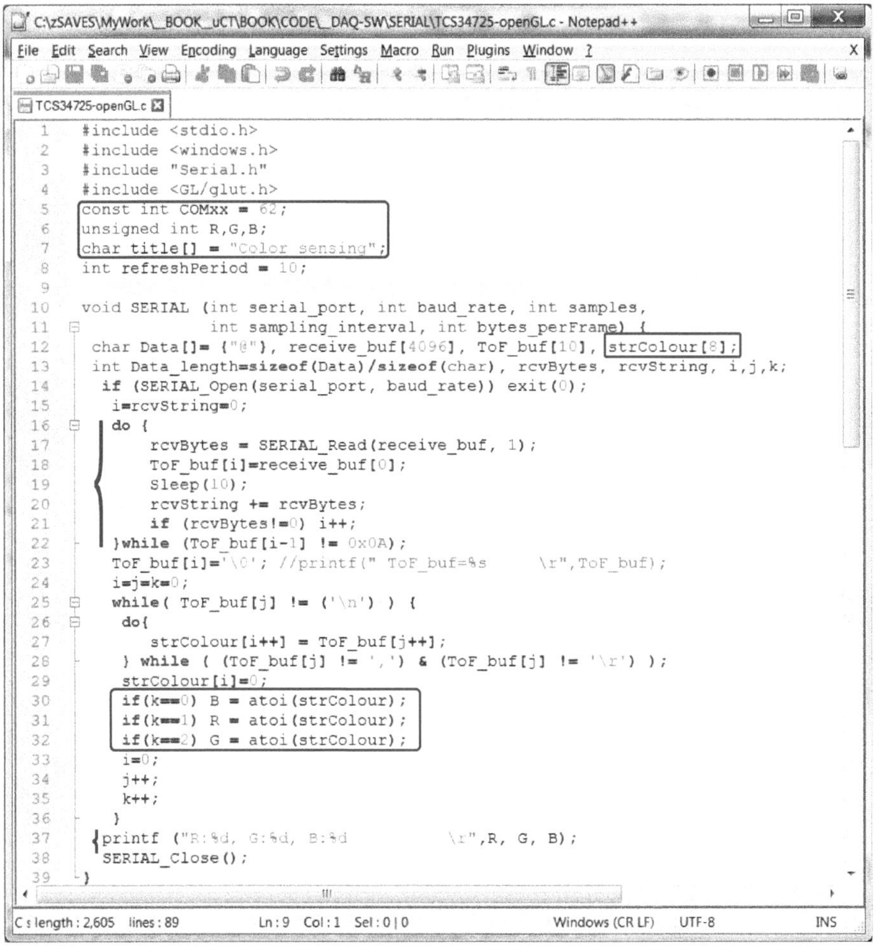

C:\zSAVES\MyWork_BOOK_uCT\BOOK\CODE_DAQ-SW\SERIAL\TCS34725-openGL.c - Notepad++

File Edit Search View Encoding Language Settings Macro Run Plugins Window ?

TCS34725-openGL.c

```c
#include <stdio.h>
#include <windows.h>
#include "Serial.h"
#include <GL/glut.h>
const int COMxx = 62;
unsigned int R,G,B;
char title[] = "Color sensing";
int refreshPeriod = 10;

void SERIAL (int serial_port, int baud_rate, int samples,
                 int sampling_interval, int bytes_perFrame) {
 char Data[]= {"@"}, receive_buf[4096], ToF_buf[10], strColour[8];
 int Data_length=sizeof(Data)/sizeof(char), rcvBytes, rcvString, i,j,k;
  if (SERIAL_Open(serial_port, baud_rate)) exit(0);
   i=rcvString=0;
   do {
       rcvBytes = SERIAL_Read(receive_buf, 1);
       ToF_buf[i]=receive_buf[0];
       Sleep(10);
       rcvString += rcvBytes;
       if (rcvBytes!=0) i++;
   }while (ToF_buf[i-1] != 0x0A);
   ToF_buf[i]='\0'; //printf(" ToF_buf=%s      \r",ToF_buf);
   i=j=k=0;
   while( ToF_buf[j] != ('\n') ) {
    do{
       strColour[i++] = ToF_buf[j++];
    } while ( (ToF_buf[j] != ',') & (ToF_buf[j] != '\r') );
    strColour[i]=0;
    if(k==0) B = atoi(strColour);
    if(k==1) R = atoi(strColour);
    if(k==2) G = atoi(strColour);
    i=0;
    j++;
    k++;
   }
  printf ("R:%d, G:%d, B:%d                \r",R, G, B);
  SERIAL_Close();
}
```

C s length : 2,605 lines : 89 Ln : 9 Col : 1 Sel : 0 | 0 Windows (CR LF) UTF-8 INS

Figure 4.56 RGB sensing example with Open GL (1 of 2).

characters; instead, it only obtains packets by acquiring characters one by one (i.e. as applied to the 1D gesture recognition example with the VL53L0X sensors). The DAQ is performed in code lines 16–22 (Figure 4.56). Contrary to the *Euler* angles example, the RGB data are assigned to an *integer* datatype, and hence, code lines 30–32 address the built-in C function *atoi()* in order to interpret the acquired RGB *numeric strings* into *integers*.

The color of the rectangle is modified within the *display()* function in code lines 41–59 (Figure 4.57). In detail, this code part invokes *SERIAL()* function in code line 43 so as to obtain the new RGB data by the sensor device. Then the *if(). . .else if(). . .else* statement in code lines 48–51 determine the intensity of red, green and blue colors. Thereby, if the intensity of the red (sensed) color is greater than the intensity of the green color plus 10, and greater than the intensity of the blue color plus 10 (i.e. code line 48), the color of the box changes to red via the *glColor3f(1.0, 0.0, 0.0)*. An analogous

C:\zSAVES\MyWork_BOOK_uCT\BOOK\CODE_DAQ-SW\SERIAL\TCS34725-openGL.c - Notepad++

File Edit Search View Encoding Language Settings Macro Run Plugins Window ?

TCS34725-openGL.c

```
41 void display() {
42     int P_samples = 100, ODR=250 , frame_bytes=32;
43       SERIAL (COMxx, 9600, P_samples, ODR, frame_bytes);
44       glClear(GL_COLOR_BUFFER_BIT | GL_DEPTH_BUFFER_BIT);
45       glMatrixMode(GL_MODELVIEW);
46       glLoadIdentity();
47       glTranslatef(0.0, 0.0, -20.0);
48       if ( (R>10+G) && (R>10+B) ) glColor3f(1.0, 0.0, 0.0);
49       else if ( (G>10+R) && (G>10+B) ) glColor3f(0.0, 1.0, 0.0);
50       else if ( (B>10+R) && (B>10+G) ) glColor3f(0.0, 0.0, 1.0);
51       else glColor3f(1.0, 1.0, 1.0);
52       glBegin(GL_POLYGON);
53        glVertex3f(2.0, 4.0, 0.0);
54        glVertex3f(8.0, 4.0, 0.0);
55        glVertex3f(8.0, 6.0, 0.0);
56        glVertex3f(2.0, 6.0, 0.0);
57       glEnd();
58       glutSwapBuffers();  //double buffering
59 }
60
61 void timer(int value) {
62       glutPostRedisplay();
63       glutTimerFunc(refreshPeriod, timer, 0);
64 }
65
66 void reshape(GLsizei width, GLsizei height) {
67       GLfloat aspect = (GLfloat)width / (GLfloat)height;
68       glViewport(0, 0, width, height);
69       glMatrixMode(GL_PROJECTION);
70       glLoadIdentity();
71       gluPerspective(45.0f, aspect, 0.1f, 100.0f);
72 }
73
74 int main(int argc, char** argv) {
75       glutInit(&argc, argv);
76       glutInitDisplayMode(GLUT_DOUBLE);
77       glutInitWindowSize(640, 480);
78       glutInitWindowPosition(50, 50);
79       glutCreateWindow(title);
80       glutDisplayFunc(display);
81       glutReshapeFunc(reshape);
82       glutTimerFunc(0, timer, 0);
83       glutMainLoop();
84     return(0);
85 }
```

CHANGE COLOR ACCORDING TO INTENSITY

C source file length : 2,592 lines : 85 Ln : 73 Col : 1 Sel : 0 | 0 Windows (CR LF) UTF-8 INS

Figure 4.57 RGB sensing example with Open GL (2 of 2).

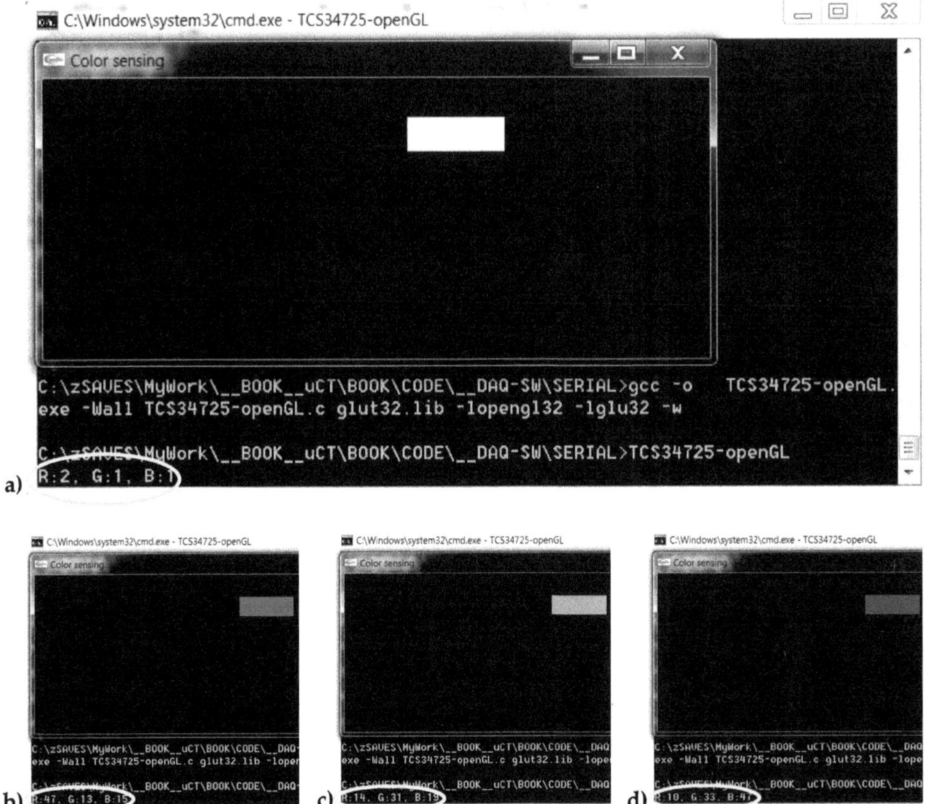

Figure 4.58 Execution of the RGB sensing example with Open GL. (a) compilation and execution of the application software (no colour detected), (b) red colour is identified by the application software, (c) green colour is identified by the application software, (d) Blue colour is identified by the application software.

evaluation is performed for the green (sensed) color in code line 49, and for the blue (sensed) color in code line 50. If none of the above assessments is verified, the color of the box is set to white in code line 51.

Figure 4.58a depicts the compilation as well as execution of the application software code when none of the three primary colors (i.e. RGB) is sensed, and hence, the color of the box is set to white. Figure 4.58b–d depicts the results of the executed code when the color red, green, and blue, respectively, is identified by the application.

Arduino Ex.4-11

As mentioned earlier in the chapter, the processor of Micro:bit is actually a *2.4 GHz RF System on Chip,* which supports Bluetooth 4.1 communication and a fully qualified *BLE* stack. The current example illustrates how to upgrade the previous RGB sensing application in order to forward the acquired data to a mobile phone via BLE feature. Such an application would be useful in case the user would like to install a (battery-operated) data logger at some place in nature. For instance, a data logger could be used to read temperature of an active volcano, or the oxygen of a river. The BLE connectivity of a stand-alone device allows us to perform a quick test in order to confirm that the installed device is working properly, without interrupting the data logging procedure.

To implement BLE connectivity to Micro:bit we make use of an Adafruit library, named *Adafruit_Microbit*, which invokes the corresponding Arduino driver,[8] but makes things simpler as it encrypts too many technical details. The Adafruit library can be downloaded from the link https://github.com/adafruit/Adafruit_Microbit (and should be extracted at the Arduino folder: *C:\arduino-1.8.9\libraries*). It is here noted that, if the Micro:bit is not programmed correctly, the user might need to install a Bluetooth advertising example. A corresponding *.hex* code can be found at Adafruit's website: https://cdn-learn.adafruit.com/assets/assets/000/046/777/original/microbit-adv.hex?1506701272 and could easily installed in Micro:bit board with a *drag-and-drop* task into the MICROBIT drive (which appears on our computer as soon as we connect the board to the host PC).

The firmware of the example is presented in Figure 4.59. The code lines within the red frames first invoke the Adafruit library and then create an instance of the class. The code

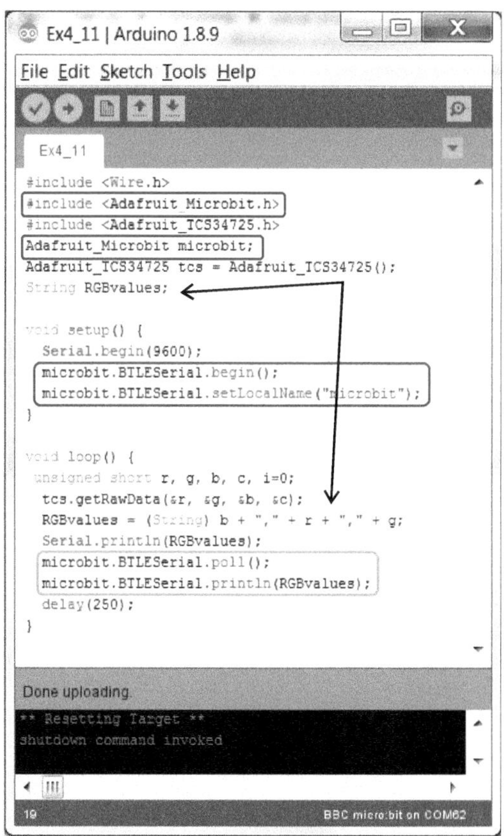

Figure 4.59 RGB sensing firmware with BLE connectivity.

8 Explore the folder *C:\Users\<User-name>\Documents\Arduino\libraries\BLEPeripheral* for more information.

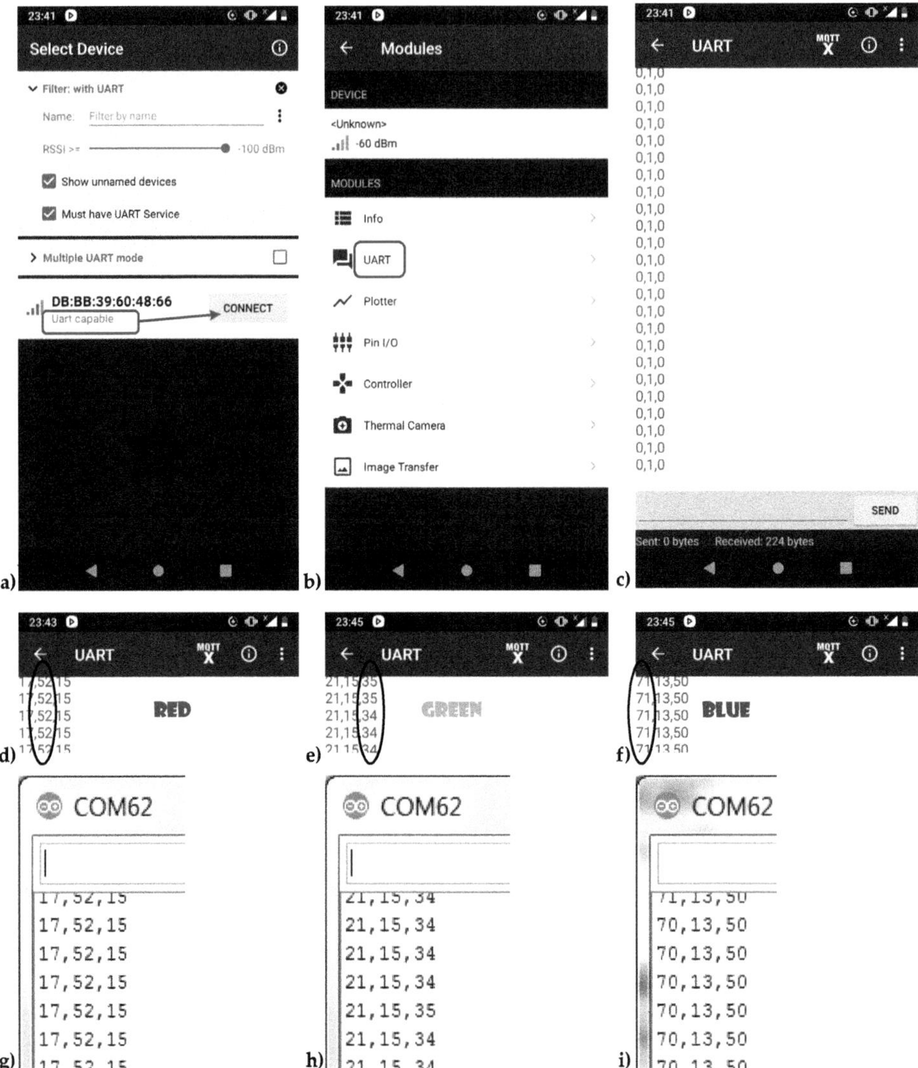

Figure 4.60 Interfacing with a mobile phone through BLE (*Bluefruit Connect* app). (a) the initial screen on the smartphone when starting the *Bluefruit Connect* app, (b) the subsequent screen where the *UART* module is selected, (c) RGB measurement obtained from the Micro:bit board, (d) measurements on the smartphone when scanning red card, (e) measurements on the smartphone when scanning green card, (f) measurements on the smartphone when scanning blue card, (g) measurements on *Arduino serial monitor* when scanning red card, (h) measurements on *Arduino serial monitor* when scanning green card, (i) measurements on *Arduino serial monitor* when scanning blue card.

lines within the blue frame invoke the necessary functions to initialize the BLE peripheral device. Finally, the code lines within the green frame first poll the radio for events and then transmit to the mobile phone the acquired RGB values. Due to simplicity reasons, the transmitted string is arranged via the *String constructor* named *RGBvalues* (pointed by the black arrows).

For the purpose of this example, we make use of the mobile device *Nokia 3.1* (which supports Bluetooth 4.2), where we have installed the Adafruit app for Android device, named *Bluefruit Connect*. Figure 4.60a depicts the foremost screen appearing on our mobile phone, when we first run *Bluefruit Connect* app. We track down the peripheral device running on the Micro:bit board (which should be referred to as *Uart capable*) and press *CONNECT*. Figure 4.60b depicts the next screen on *Bluefruit Connect* app where we select the *UART* option. The next screen should print the RGB measurements, as presented by Figure 4.60c. If we place the red, green, or blue card above the sensor device found on our Micro:bit setup, we respectively obtain the results given in Figure 4.60d–f (as well as the identical results in Arduino *Serial Monitor* tool given in Figure 4.60g–i).

Figure 4.61a depicts the revisions that should be made in the firmware code so as to read, but also write data from the mobile phone to the peripheral device. The *while()* statement within the blue frame waits to receive any character from the mobile phone and if it does, it forwards the RGB values to the smartphone. Figure 4.61a depicts the printed results for sensing no card, red card, green card, and blue card.

It is here noted that the reader may also work with *apps* development for the mobile phone. One possible solution is the *MIT App Inventor* web application IDE, which is

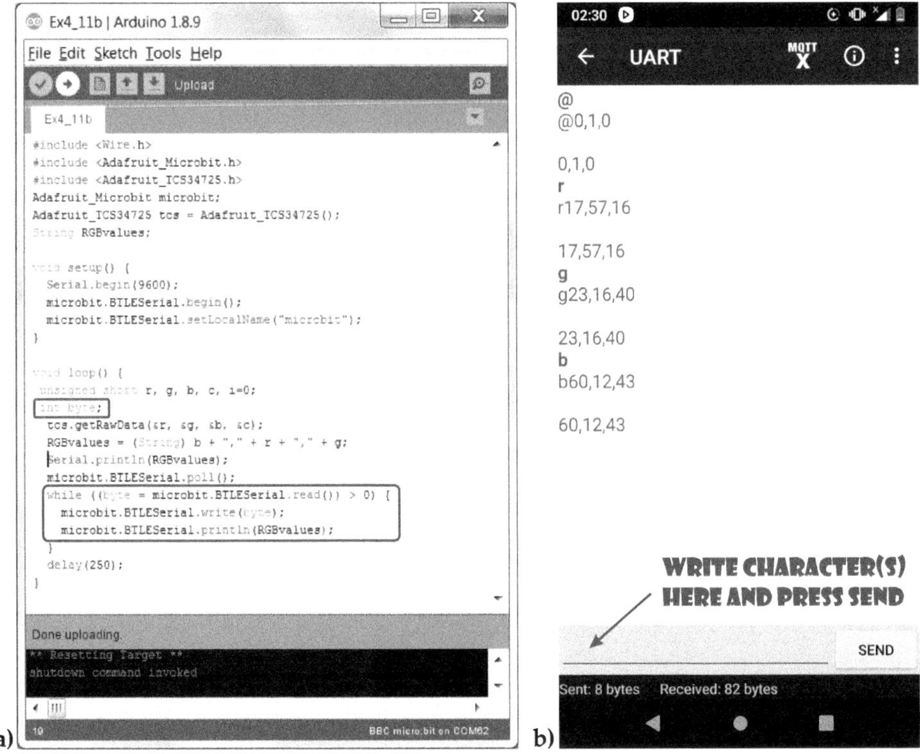

Figure 4.61 RGB sensing firmware with BLE connectivity (central device read & write). (a) revised firmware to transmit as well as receive data to/from the smartphone, (b) the *Bluefruit Connect* app exchanges information with the updated firmware.

addressed for newcomers to computer programming and supports apps development for Android (for iOS operating system not currently available). However, this area of study is out of the scope of the current book.

Conclusion

Modern sensor devices constantly pave the wave for new solutions in embedded design and give rise to an opportunity for creativity and innovation. This chapter has been focused on the study of some of the most interesting sensor devices, as well as the process of collecting and (real-time) monitoring data to a PC, using freeware development tools. Detection of orientation, motion, gesture, distance, as well as color sensing, are some of the issues discussed by the chapter. The process of interfacing with a mobile phone, through Bluetooth protocol, has also been explored. Hereafter are some unsolved problems for the reader.

Problem 4–1 (Data Acquisition of Atmospheric Pressure)

Upgrade the code example **Ex.4–2** so that the firmware code calculates the absolute height in-between the sensor devices (a useful model for *indoor navigation* systems).

Tips: Place the two sensors in different *heights (h)* and use the calculations given in Formula 4.2 to estimate the in-between h distance.

Problem 4–2 (Fusion of Linear Acceleration and Barometric Altitude)

Write a firmware code for Teensy3.2 board and Pesky's BNO055+BMP280 daughterboard, which estimates the absolute height between two positions when identifying motion of the sensor in the z axis.

Tips: This particular example could be utilized in a drone system, which identifies and determines elevation.

Problem 4–3 (1D Gesture Recognition)

Write a firmware code for TinyZero board when the latter acquires data from 2×VL53L0X distance sensors and emulates the opening of a light lamp with 1D gesture recognition. In detail, when the system identifies that the user's palm has passed over Sensor 1, and then, over Sensor 2 (with that particular order), it turns *ON* the LED found on TinyZero board. Otherwise, when the system identifies that the user's palm has passed over Sensor 2 first, and then, over Sensor 1, it turns the LED *OFF*.

Problem 4–4 (Color Sensing)

Update the firmware code example **Ex.4–11** so that it emulates some of the functions of Ozobot. For instance, if the TCS34725 identifies the sequence of colors *red*, *black*, and *blue*, it sends to the host PC and mobile phone the *string* "FAST", or for the identified colors *blue*, *red*, *green*, it sends to the PC/mobile the string "GO RIGHT", and so forth.

5

Tinkering and Prototyping with 3D Printing Technology

3D printing processes become more and more popular lately. With merely a few hundred Euros, makers can purchase a reliable 3D printer and manufacture the enclosure of a custom-designed *microcontroller* (*μC*) product at home. This chapter explores the creation of a prototype microcontroller-based electronic system, as well as the process of producing and printing 3D models with the utilization of freeware development tools. Before the prototyping of an Arduino-based electronic system, we explore how to *tinker* (i.e. repair and/or improve) electronic products with Arduino.

Tinkering with a Low-cost RC Car

The present example explores how to turn a low-cost *remote control* (*RC*) car into smartphone-controlled car (via Bluetooth). For this example we exploit a *360 Cross* RC car, delivered by Exost. The RC car, which is presented in Figure 5.1a, can be found at www.exostrc.com/product/360-cross-2/. If we remove the chassis of the RC car we observe that it incorporates two direct current (DC) motors, where DC motor 1 controls the two wheels found on the top, and DC motor 2 controls the two wheels found on the bottom of Figure 5.1b.

In Figure 5.1c we have traced and unsoldered the power cables that are connected to 4×AA batteries in series (placed at the bottom of the RC car). Accordingly, the power from the batteries is equal to $4 \times 1.5\,\text{V} = 6\,\text{V}$ (when fully charged). In the positive (i.e. red) cable we have attached the anode of a general-purpose 1N4007 rectifier, so as to prevent reverse connection of the power to the boards. The diode's primary characteristics specify approximately 1 V forward voltage drop for the maximum average forward rectified current (that is, 1 A). Accordingly, obtaining the system's power from the cathode of the rectifier we achieve a power of approximately $(4 \times 1.5\,\text{V}) - 1\,\text{V} \cong 5\,\text{V}$.

In Figure 5.1d, we have attached two JST PH2.0 (two-pin) male connectors to power the boards of our system, where the 1N4007 rectifier is placed within a heat-shrink plastic tube highlighted by green color. In addition, we have attached one JST PH2.0 (four-pin) male connector to drive the two motors of the RC car.

The first thing we need to consider is what kind of peripheral modules are needed in order to implement the smartphone (Bluetooth)-controlled RC car. Despite the

Microcontroller Prototypes with Arduino and a 3D Printer: Learn, Program, Manufacture, First Edition.
Dimosthenis E. Bolanakis.

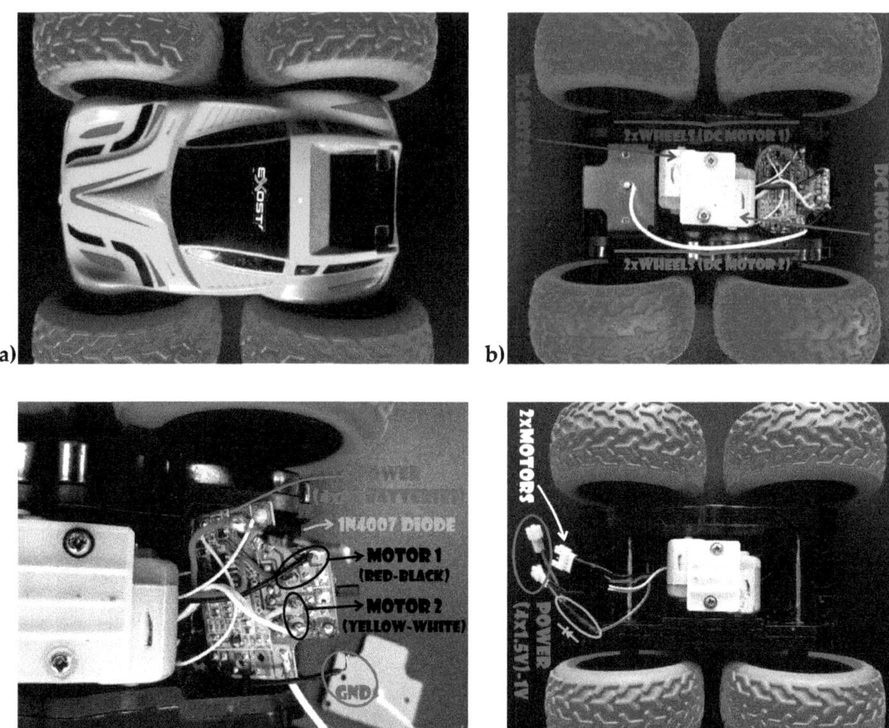

Figure 5.1 Tinkering a *360 Cross* RC car by Exost. (a) the 360 Cross RC car by Exost, (b) reverse engineering of the RC car (step 1), (c) reverse engineering of the RC car (step 2), (d) reverse engineering of the RC car (step 3).

microcontroller motherboard, we are going to need a Bluetooth device in order to collect the commands sent by the mobile device, as well as a motor driver. The latter module is needed because the microcontroller integrated circuit (IC) is not able to supply the high current needed to drive a DC motor. The next issue that should be taken into account (before proceeding to the selection of the aforementioned boards) is the available space, in the interior of the RC car, when removing the pre-installed boards. For instance, while the Micro:bit board holds a 2.4 GHz radio frequency (RF) system on chip (SoC), which supports Bluetooth 4.1 communication (and, hence, it is particular suitable for such a project), it does not fit inside the current RC car.

A possible solution is the Teensy 3.2 board (explored earlier by the book), which fits precisely within the available inner space of the RC car. Because the Teensy 3.2 board does not embed a Bluetooth device, we select the *nRF52 add-on for Teensy* by Pesky (www.tindie.com/products/onehorse/nrf52-add-on-for-butterfly-and-teensy/). In consideration of the motor driver chip, the original board found on the RC car is the MX1919. The latter IC constitutes a low-cost solution that can be purchased from sites such as the *aliexpress*, *banggood*, and so forth. According to the information found on those sites, the output average/peak current that can be delivered by the MX1919 chip is 1.5 A/2.5 A, respectively. We herein choose a module that employs a well-known IC of comparable specs. That is,

Table 5.1 BOM of the tinkered RC car project.

Part (manufacturer)	Description	Unit
Teensy 3.2 (by Electronic Project Components)	μC motherboard	1
2×1.2A DC motor driver (by DFRobot)	DC motor driver	1
nRF52 add-on for Teensy (by Pesky)	Bluetooth 5.2 SoC	1
1N4007 (by ON Semiconductors)	General-purpose rectifier	1
PH2.0/four-pin male connector (OEM)	Connector with wires (2 mm pitch)	2
PH2.0/four-pin female connector (OEM)	Connector with wires (2 mm pitch)	2
PH2.0/two-pin male connector (OEM)	Connector with wires (2 mm pitch)	2
PH2.0/two-pin female connector (OEM)	Connector with wires (2 mm pitch)	2
Grove wrapper: 1×big and 1×small (by SEEED Studio)	Lego-compatible case for Grove modules	1+1

the TB6612FNG chip by Toshiba, which is able to deliver output average/peak of 1.2 A/3.2 A, respectively, while the supply voltage can vary from 2.5 V to 13.5 V. The selected module for our project (which incorporates that particular chip) is the *2×1.2A DC Motor Driver* by DFRobot. The *bill of material* (*BOM*) of the tinkered RC car is given in Table 5.1.

Figure 5.2 presents the connection diagram of the proposed tinkered RC car. The 4×AA batteries (placed at the bottom of the RC car) provide power to the Teensy 3.2 μC motherboard, through a general purpose rectifier. Due to the printed-circuit board (PCB) thermal dissipation limits, the Teensy 3.2 board recommends connecting 3V6 to 6 V power to the corresponding power pins. However, the on-board regulator chip supports the connection of up to 10 V input with non-universal serial bus (USB) power sources. Accordingly, there is no danger of damaging Teensy 3.2 because of an overvoltage power connection. The same power through the rectifier is directed to the VM input power of the DC motor module. In consideration of the DFRobot's specs, the module works for voltage range: 2V5–12 V.

The two DC motors of the RC car are driven by the *M1+/M1−*, *M2+/M2−* outputs of the motor driver module, where the motor speed is controlled by *PWM1*, *PWM2* inputs, and the motor direction (i.e. clockwise or counterclockwise) is controlled by *DIR1*, *DIR2* inputs. Thereby, the PIN23 and PIN20 of Teensy board are configured output pins, while PIN22 and PIN21 are configured pulse width modulation (PWM) pins. On the other hand, the *nRF52 Bluetooth* module is controlled over universal asynchronous receiver/transmitter (UART) serial interface, and in particular, over the *Serial1* port of Teensy 3.2 board. The *nRF52 Bluetooth* module admits 3V3 power input, which is obtained from Teensy 3.2 board. To control the proposed setup system we need an Android device with Bluetooth connectivity, which could run the *Bluefruit Connect* App by Adafruit (explored earlier in the book).

Figure 5.3 presents the boards used by the proposed tinkered project. In detail, the top and bottom view of the DC motor driver is presented in Figure 5.3a, the top and bottom view of the nRF52 Bluetooth module is illustrated in Figure 5.3b, and the Teensy 3.2 μC motherboard is depicted in Figure 5.3c. The latter board is attached to a Grove Wrapper of $X{\cdot}Y$ dimensions equal to 40 mm×25 mm. The latter box is a Lego-compatible case

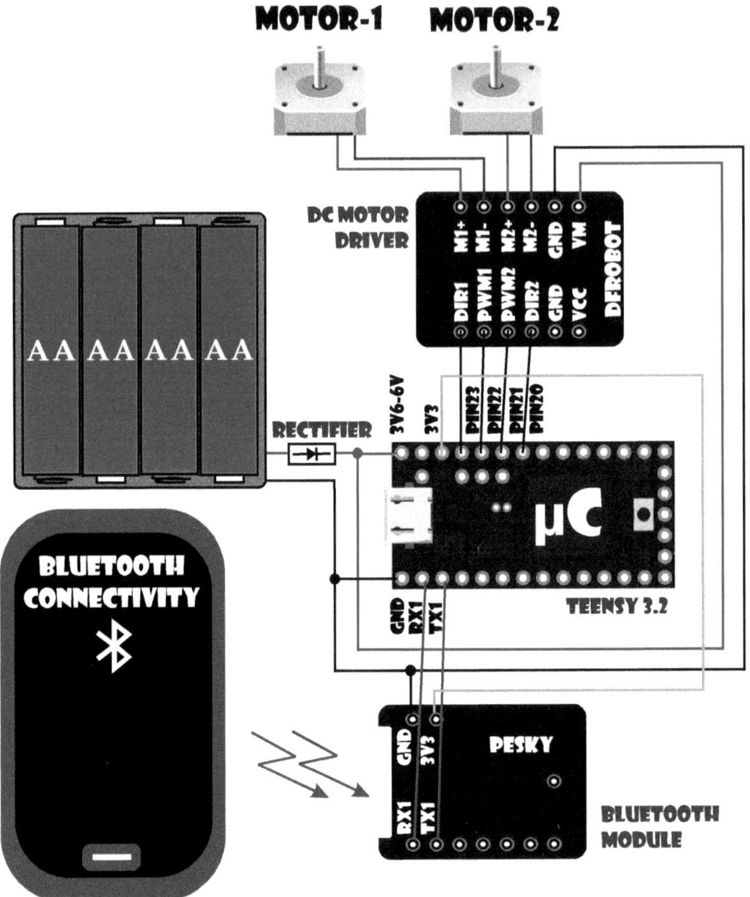

Figure 5.2 Connection diagram of the tinkered RC.

appropriate for the so-called *Grove Modules* delivered by Seeed Studio; however, it constitutes a suitable case for the Teensy 3.2 board.

Figure 5.3d presents the PH2.0/four-pin as well as /two-pin cables attached to the DC motor driver and Teensy 3.2 boards. The nRF52 Bluetooth module is stacked on the top of Teensy 3.2 board. The PH2.0/four-pin female connector highlighted by blue color should be connected to the corresponding male connector of the two DC motors. The two PH2.0/two-pin female connectors highlighted by green color should be connected to the corresponding male connectors of the battery power (after the rectifier). Finally, the PH2.0/four-pin male and female connectors highlighted by red color should be connected to each other in order to pass control of the DC motors to the microcontroller pins. The final result of the tinkered setup, installed within the RC car, is given in Figure 5.4. It should be noted that the PH2.0/four-pin male–female connection of the two DC motors with the DC driver is placed below the latter module, and hence, it is not visible in the current figure. Moreover, it is worth mentioning that the *BNO055+BMP280 Pesky* module soldered on the top of Teensy 3.2 board is not used by the current project, but it could be used for a future upgrading toward a more creative control of the RC car.

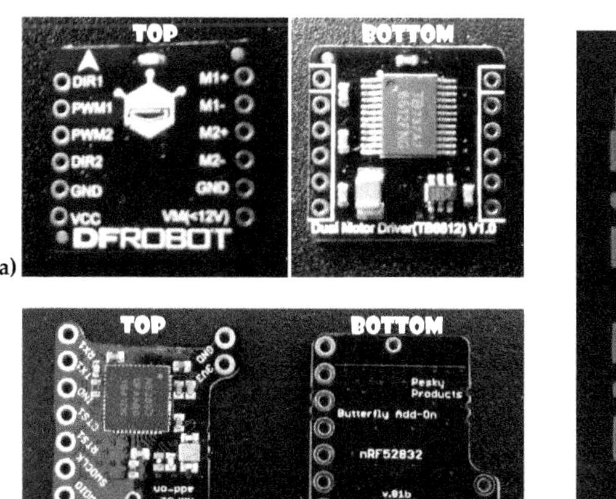

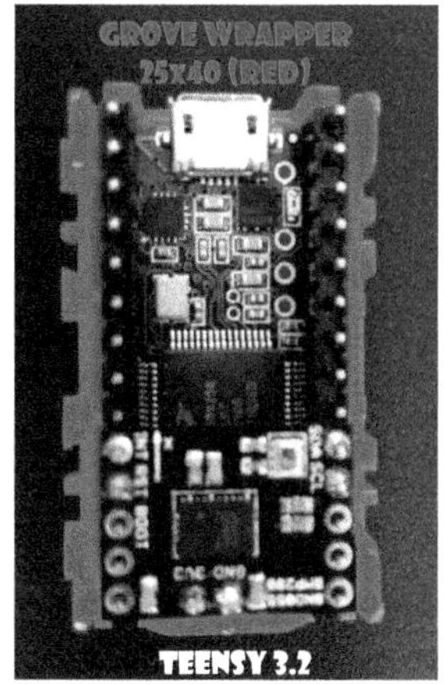

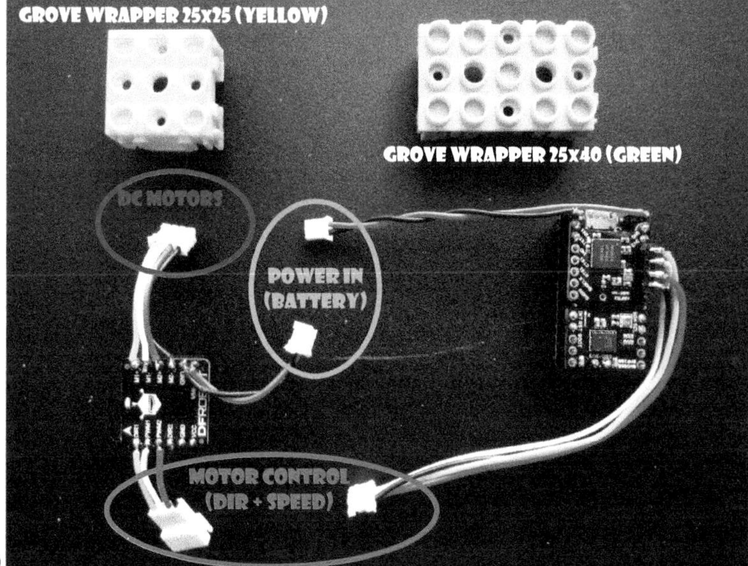

Figure 5.3 Connection diagram of the tinkered RC. (a) top and bottom side of the *Dual Motor Driver* module by DFRobot, (b) top and bottom side of the *nRF52 Bluetooth* module by Pesky Products, (c) teensy 3.2 μC motherboard within the *Grove Wrapper* case by Seeed Studio, (d) PH2.0 connectors attached to the DC motor driver and Teensy 3.2 motherboard.

Arduino Ex.5.1

Before building the firmware that controls the RC car we start with the development of a trial code, in order to test the Bluetooth connectivity between the Teensy 3.2 board and a smartphone (running the Adafruit app for Android devices, that is, the *Bluefruit Connect* app that was tested earlier by this book). The trial code is given in Figure 5.5a. The *setup()* function initializes two serial ports, that is, the *Serial* and *Serial1* UART ports. The former is used for transferring data to the personal computer (PC), while the latter is used for accessing the *nRF52 Bluetooth* module of the current setup. The module is programmed as

Figure 5.4 Tinkering a *360 Cross* RC car by Exost.

a Bluetooth low-energy (BLE)/UART bridge and herein holds a *Peripheral* role. Accordingly, the Teensy 3.2 board streams data directly to the smartphone device over *Serial1* port. In the example presented herein, the *loop()* function obtains incoming data (if available) over *Serial1* port (i.e. data sent by the smartphone device using *Bluefruit Connect* app), and forwards those data to the PC through the *Serial* port. The overall process is repeated every 100 ms. Figure 5.5b presents the printed results on the PC port when pressing the four different arrows available at the *Control Pad* function of the *Bluefruit Connect* app.

The initiation of the *Control Pad* function (embedded in the *Bluefruit* app) is given in Figure 5.6. Having found an opened BLE/UART bridge, the user presses the *CONNECT* button presented in Figure 5.6a. Then, the user selects the *Controller* module of *Bluefruit* app, presented in Figure 5.6b, and then the *Control Pad* function of Figure 5.6c. The interface of the latter function is illustrated in Figure 5.6d. As soon as the user presses the "Up" (↑) arrow on the *Bluefruit* app, the smartphone transmits to the BLE hardware the ASCII text "!B516". As soon as the user releases the "Up" (↑) arrow, the smartphone transmits to the BLE hardware the string "!B507". Likewise, when the user presses and releases the "Right" (→) arrow, the *Bluefruit* app transmits the strings "!B813" and "!B804", respectively. In the same way, when the user presses and releases the "Down" (↓) arrow, the *Bluefruit* app transmits the strings "!B615" and "!B606", respectively. Finally, when the user presses and releases the "Left" (←) arrow, the *Bluefruit* app transmits the strings "!B714" and "!B705", respectively. It is here noted that every string is terminated by the *new line (0×0A)* ASCII character.

The trial code of Figure 5.7a, decodes the aforementioned ASCII commands send by the *Bluefruit* app. In detail, the received string is assigned to the *unsigned char* array named *serialData[]*, which reserves seven memory addresses, within the *while()* statement. The foremost *if()* statement exploits the fact that the fourth element of each transmitted string is equal to the ASCII char '1' or '0', in case the user either presses or releases an arrow. Those '1' and '0' values appear only on that particular array element and, hence, the

Ex5_01 | Arduino 1.8.9

File Edit Sketch Tools Help

Ex5_01

```
void setup() {
  Serial1.begin(115200);
  delay(1000);
  Serial.begin(115200);
  delay(100);
}

void loop() {
 unsigned char serial_byte;
  while(Serial1.available())
  {
    serial_byte = Serial1.read();
    Serial.write(serial_byte);
  }
  delay(100);
}
```

READ BLUETOOTH DEVICE

SEND TO PC

Done uploading.

```
Sketch uses 9564 bytes (3%) of program storage space. Maximu
Global variables use 3444 bytes (5%) of dynamic memory, leav
```

Teensy 3.2 / 3.1 on usb:0/1A0000/0/1

a)

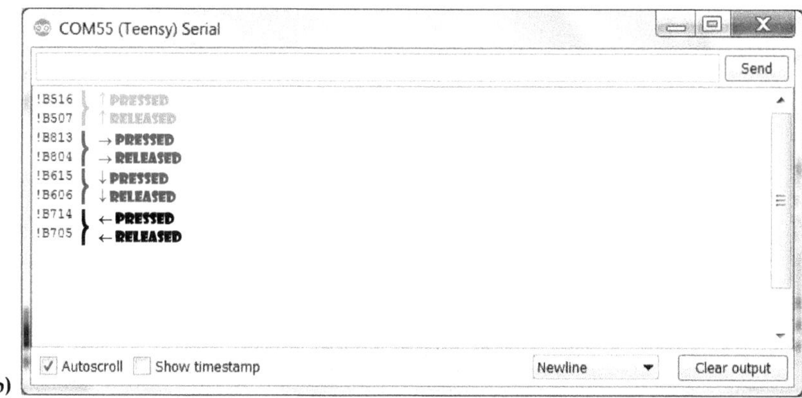

b)

Figure 5.5 Tinkering an RC car: testing *Control Pad* commands (Arduino hardware). (a) application firmware, (b) printed results. *Source:* Arduino Software.

current *if()* statement transmits to the PC the string “STOP” if the value of the evaluated element is equal to ‘0’.

The subsequent *if()* statements explore if the fourth array element is equal to the ASCII char ‘1’, which means the user has pressed an arrow. With the utilization of the *logical AND (&&)* operator, the *if()* statements explore the value of the third array element in order to decode which arrow is pressed. Accordingly, if the user presses the “Up” arrow the third array element holds the ASCII char ‘5’, if the user presses the “Down” arrow the element

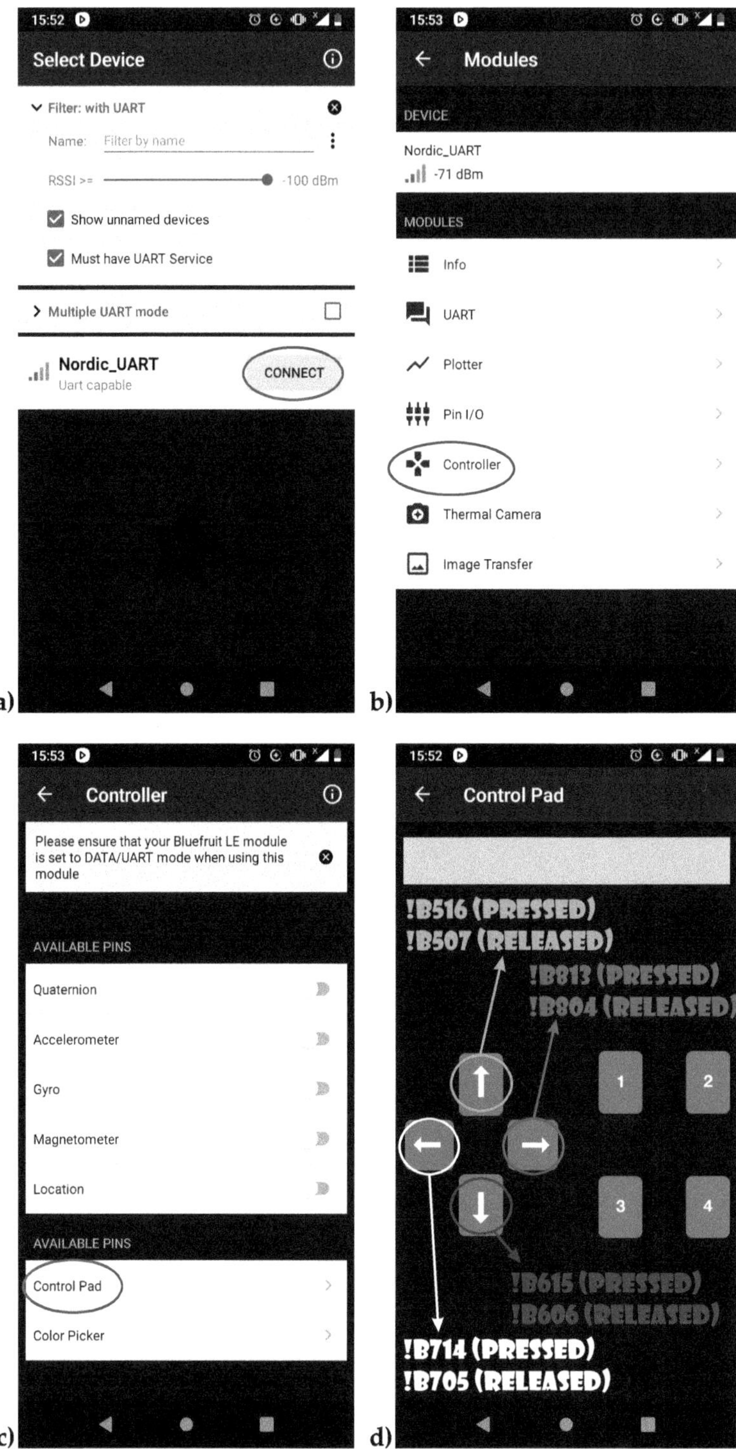

Figure 5.6 Tinkering an RC car: testing *Control Pad* commands (*Bluefruit* app). (a) select *UART* module from the *Bluefruit Connect* app, (b) Select *Controller* module from the *Bluefruit Connect* app, (c) select *Control Pad* panel from the *Bluefruit Connect* app, (d) command description of the controls available at the *Control Pad* panel. *Source:* Arduino Software.

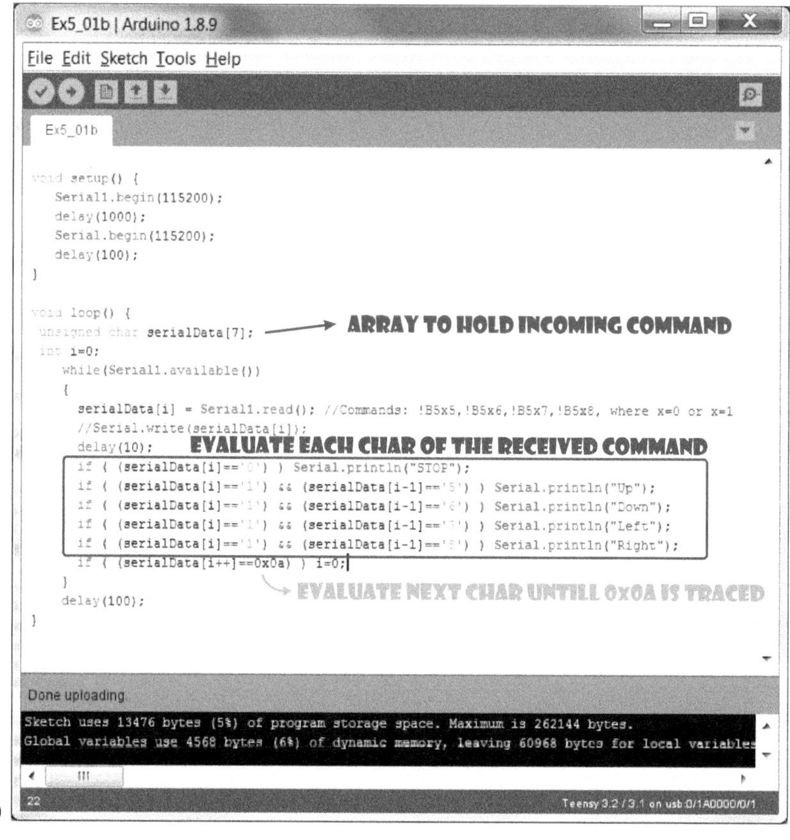

a)

COM55 (Teensy) Serial

Send

Up ↑ PRESSED
STOP ↑ RELEASED
Right → PRESSED
STOP → RELEASED
Down ↓ PRESSED
STOP ↓ RELEASED
Left ← PRESSED
STOP ← RELEASED

Autoscroll Show timestamp Newline Clear output

b)

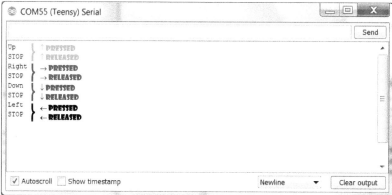

Figure 5.7 Tinkering an RC car: decoding *Control Pad* commands (Arduino hardware). (a) application firmware, (b) printed results. *Source:* Arduino Software.

holds the value '6', if the user presses the "Left" arrow the element holds the value '7', and if the user presses the "Right" arrow the element holds the value '8'. The ultimate *if()* statement explores if we have reached to the final element of the array, and if so, it resets the *i* counter so as to prepare the evaluation of the subsequent received command. Otherwise, the *i* counter is unary increased and the content of the next array element is evaluated. The printed results of the current firmware code are given in Figure 5.7b, where the user successively presses and releases the arrows "Up" (↑), "Right" (→), "Down" (↓), and "Left" (←).

Arduino Ex.5.2

Having decoded the Bluetooth commands of the *Control Pad* function (found in the *Bluefruit* app) we may now proceed to the development of the firmware code that controls the RC car through a smartphone device. As before, we address *Nokia 3.1* mobile device for the evaluation of the book examples.

The firmware code of the current example is given in Figure 5.8a. It is here reminded that the motors' speed is controlled by PWM1 (PIN22) and PWM2 (PIN21) inputs, while the motors' direction (i.e. clockwise or counterclockwise) is controlled by DIR1 (PIN23) and DIR2 (PIN20) inputs. Supplementary to the previous code that was used to decode the commands derived from the *Bluefruit* app, the main loop of the current example acts respectively to those commands, i.e. by controlling the speed and direction of the two motors within the corresponding *if* statement.

In detail, the foremost of the six total *if()* statements is used to STOP the RC car. To do so, the PWM signals that control the two motors are set to zero value, and in addition, the direction of the motors is set to counterclockwise motion. The latter task causes an *instant stop* of the RC car. Instead, to cause a *short brake,*[1] the corresponding code lines should be revised as follows: *digitalWrite(DIR1,LOW)* and *digitalWrite(DIR2,LOW).*

The *if()* statement within the blue frame is used to drive both motors in clockwise motion using DC=75%, and hence, it forces the RC car to move *forward*. It is here noted that the DC is set to 75% only when it is attempted a movement of the RC car, otherwise it is set to 0%. The designer might change the DC to a higher or lower value through the corresponding global variable at the very beginning of the source code, hence, causing a higher or lower motor speed (i.e. when DC = 192 the *duty cycle* is set to 192/256 = 75%). Accordingly, the *if()* statement within the black frame is used to drive both motors in counterclockwise motion (using DC = 75%), and hence, it forces the RC car to move *backward*.

The *if()* statement within the green frame causes one of the two motors to move *forward*; that is, the motor that controls the two wheels highlighted by green color in Figure 5.8b. The latter task forces the RC car to a *forward left-turn*. Likewise, the *if()* statement within the red frame causes the other motor to move *forward*; that is, the motor that controls the two wheels highlighted by red color in Figure 5.8b. Hence, the RC car is forced to a *forward right-turn*.

A good practicing example for the reader would be to decode the commands of the *Control Pad* function (found in the *Bluefruit* app), which corresponds to the keys "1", "2", "3", "4" (Figure 5.9) and implement a *backward left-turn* and *backward right-turn*, currently missing by the present example. In addition, the designer may utilize the remaining two function keys of *Bluefruit* app to force the car in 360° rotation (either clockwise or counterclockwise). Such a task is performed by forcing the two motors in diverse motion (i.e. the first in clockwise motion and the second in counterclockwise motion, and vice versa).

---

1 For more information the reader may refer to TB6612 chip, which drives the motors as well as to the modes of breaking motors.

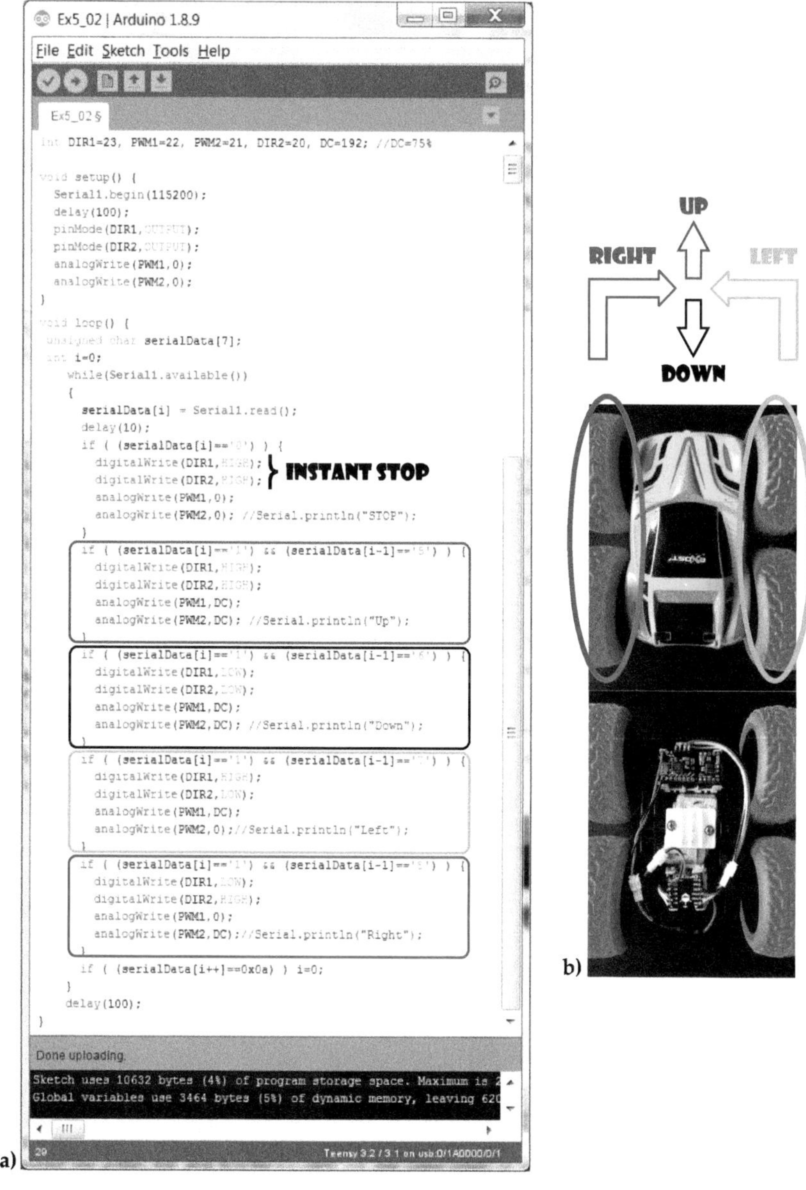

Figure 5.8 Tinkering an RC car: firmware to control the RC car though a smartphone. (a) application firmware, (b) anticipated reaction the RC car from the commands decoded by the firmware. *Source:* Arduino Software.

A Prototype Interactive Game for Sensory Play

Having dealt with the theoretical and practical topics covered by the previous chapters (mainly by Chapters 2–4), we are now ready to proceed to the development of prototype microcontroller-based electronic systems. We herein explore the development of a prototype interactive game for sensory play, appropriate mainly for toddlers as well as for elder

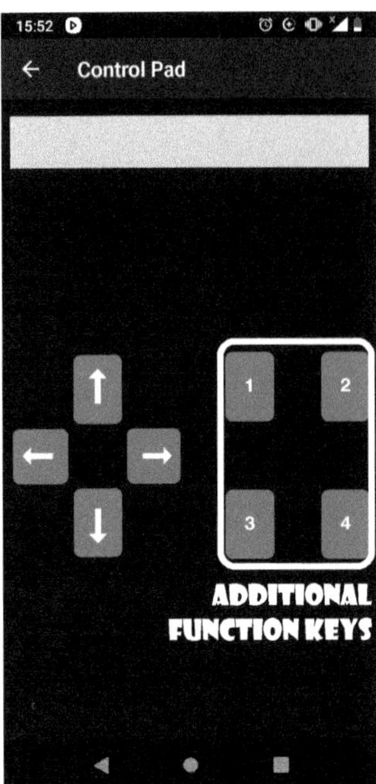

Figure 5.9 Upgrading the control of the RC car via the additional *Bluefruit* app commands. *Source:* Arduino Software.

kids within the *early childhood education*.[2] The term *sensory play* refers to any play activity, which engages one or more of the children's senses. In general, the human's body obtains sensory information from the environment, which is subsequently processed by our brain. Educational researchers sustain that providing to the children opportunities to use their senses can pique their curiosity and support their brain development [58].

The proposed educational game presented hereafter, engages three of children's senses, that is, the *sight*, *hearing*, and *proprioception*. The first two are quite familiar to humans, yet, the last one is particularly important as well. *Proprioception* is related to the sensory information arrived from muscles, tendons, and ligaments and represents the body's ability to know where it is in space. For instance, if we close our eyes we are still able to touch our nose with our forefinger because of the sense of proprioception.

Before exploring the partial steps toward the design and implementation of the proposed system, it would be wise to get an idea of what we expect to build. Accordingly, an initial presentation of the proposed educational game is illustrated in Figure 5.10. Starting with the sense of *sight* we need an output unit able to change colors. Because what we see is regularly what we place in the front side of a system, we initiate the specs of our system with the selection of the proper output unit. In consideration of the sense of *hearing*, we select a speaker that could be embodied inside our apparatus (as it does not necessarily need to be visible to the end user). During the play, the children will be asked to react on particular colors (outputted by our game), using their sense of *proprioception*. Our interactive game detects children's response via a barometric sensor that emulates the operation of a push-button, as well as two distance sensors with apply to 1D gesture recognition.

Hardware Boards of the Prototype System

For the sense of *sight* we select the *ZIP Halo* board by Kitronik. This board is designed to fit in the Micro:bit board, as presented in Figure 5.11. The *ZIP Halo* output unit incorporates 24 color addressable light-emitter diodes (LEDs) connected to Micro:bit pin P0. The *ZIP Halo* attached to the Micro:bit board constitutes a good-looking setup to start building our project on it. However, the first thing we need to consider when selecting *ready-to-use* boards for a prototype system, is the possibility to attach the requisite peripheral units of

2 The *early childhood education* branch refers to the teaching of children from birth up to the age of 8.

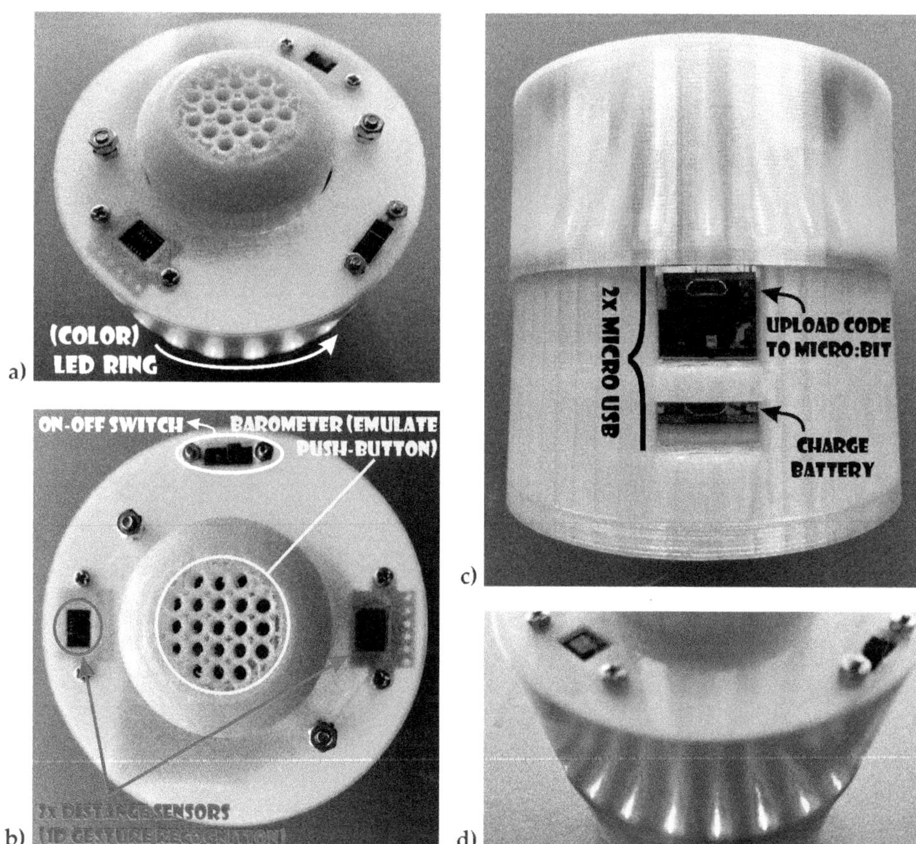

Figure 5.10 The proposed interactive game for sensory play. (a) LED ring displaying green colour, (b) ON/OFF switch and sensors, (c) USB connectors for uploading code and charging the system battery, (d) different colour displayed by the LED ring (red).

the overall system to the selected setup (i.e. herein the possibility to attach the system's sensors and speaker to the *ZIP Halo* + *Micro:bit* setup).

As presented by Figure 5.11a, the *ZIP Halo* board incorporates breakout pads connected to Micro:bit P1 and P2 pins, as well as to the 3V3 power and GND. Those four pads are enough to connect a sensor device using a software-implemented I2C (as P1 and P2 pins are not the dedicated I2C pins of Micro:bit board). However, the speaker requires one additional signal and it is not wise to share that signal with one of the software-implemented I2C pins, as it is possible to hear noise through the speaker while accessing the I2C sensors. Moreover, the two identical distance sensor devices that will be used to implement the 1D gesture recognition (as explored by the previous chapter), require another extra pin in order to deactivate one of the two sensors (of same I2C address) when talking to the other one.

Looking at the front and rear side of *ZIP Halo* + *Micro:bit* setup, there are two possible positions where we could "steal" the additional signals required to build the prototype system, that is, either from the Micro:bit bus found at bottom of the rear side view of

Figure 5.11 *ZIP Halo* board with *Micro:bit* (for the sense of *sight*). (a) top side, (b) bottom side.

Figure 5.11b, or from the available push-buttons at the front side view of Figure 5.11a (which are not used by the current project). We herein make the latter choice, as it is more difficult and risky to solder cables to the Micro:bit bus (because of the small space between the adjacent pads).

Figure 5.12 presents the rest of the peripheral units (apart from the *ZIP Halo + Micro:bit*) used by the prototype interactive game. In detail, Figure 5.12a presents a module by DFROBOT, which incorporates BNO055 + BMP280 sensors. In the current project we take advantage of the barometric sensor only (i.e. BMP280), however, the incorporation of BNO055 device supports a future enrichment of the system with additional interactive options. Figure 5.12b presents two identical distance sensor (original equipment manufacturer [OEM]) modules,[3] which will be used to implement the 1D gesture recognition. It is worth mentioning that the mounting of two identical VL53L0X distance sensors directly on the microcontroller's I2C pins (i.e. without an intermediate multiplexer device as performed in the previous chapter), requires using the extra XSUT pin that deactivates the sensor device. The latter pin is an active-low shutdown input, which is (by default) pulled up in the module. This pin can be used during power up in order to deactivate one of the two sensors, and then, change the I2C address of the active sensor. This way, we configure two identical VL53L0X sensors of different I2C address.

Figure 5.12c presents the PowerBoost 1000 Charger by Adafruit Industries. Since our system is battery-operated, we need a module that boosts and regulates up to 5 V the *lithium polymer (LiPo)* nominal voltage of 3.7 V. The battery is connected to the module via JST PH2.0 (two-pin) connector. The module is also used to charge the battery through a regular USB charger, the same that is used by mobile phones. It is worth mentioning that the module features the so-called *enable (EN)* signal, which powers *OFF* the device. Figure 5.12d presents a speaker module by Waveshare Electronics, with a volume adjust trimmer. The speaker module is interfaced via JST PH2.0 (three-pin) connector, which carries the power (i.e. 3V3 and GND) as well as the signal that controls the speaker. The 5 V DC fan of Figure 5.12e is used along with the BMP280 sensor to emulate a push-button, while the

3 The term *OEM* refers to a company that produces parts marketed by another manufacturer. Such parts can be purchased, at low cost, through online stores, such as www.aliexpress.com.

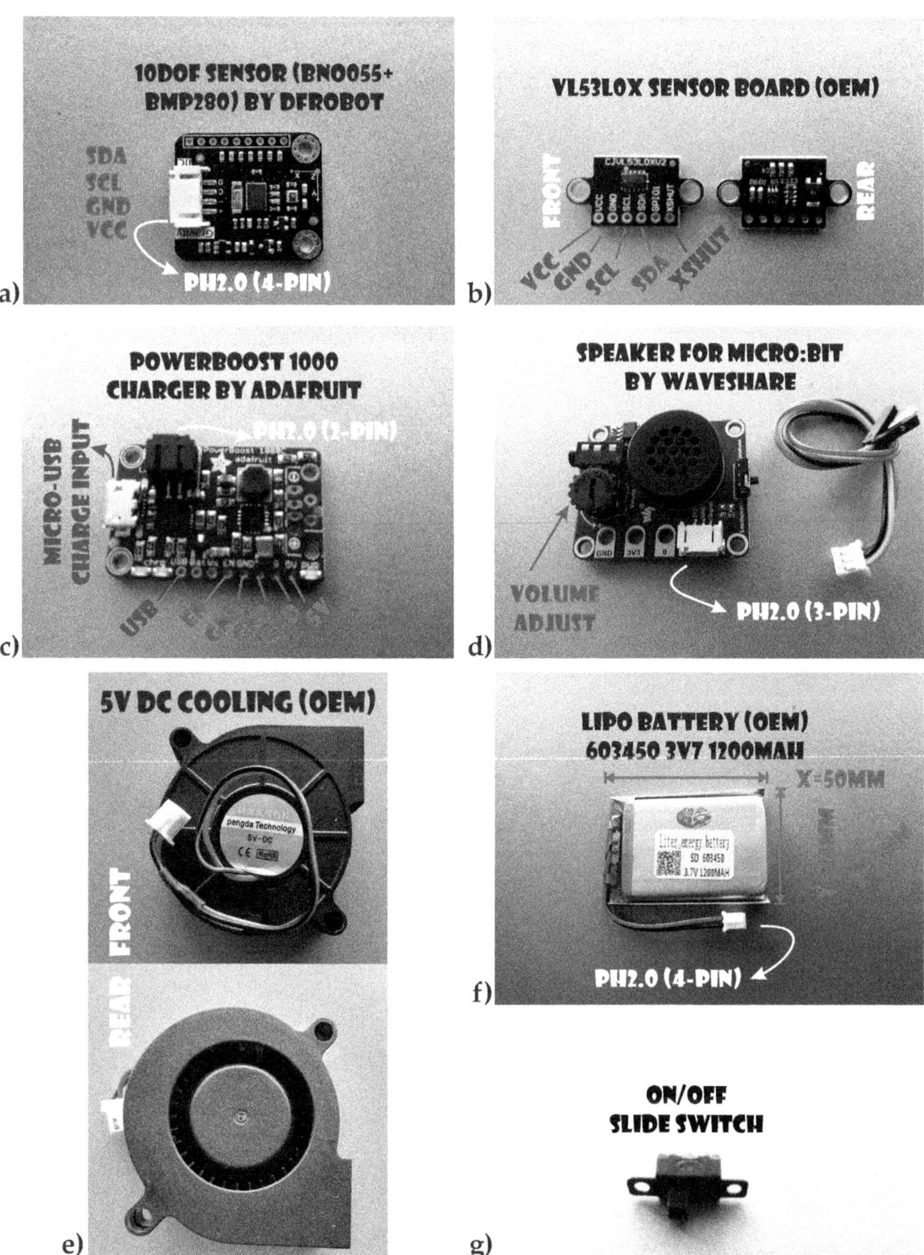

Figure 5.12 Peripheral units used by the prototype game (except *ZIP Halo+Micro:bit*). (a) *10DOF sensor* module by DFRobot, (b) *Distance sensor* board (OEM product), (c) *powerboost 1000 charger* by Adafruit, (d) *speaker module* by Waveshare, (e) *5V fan* (OEM product), (f) LiPo battery / 1200mAh (OEM product), (g) ON/OFF slide switch (OEM product).

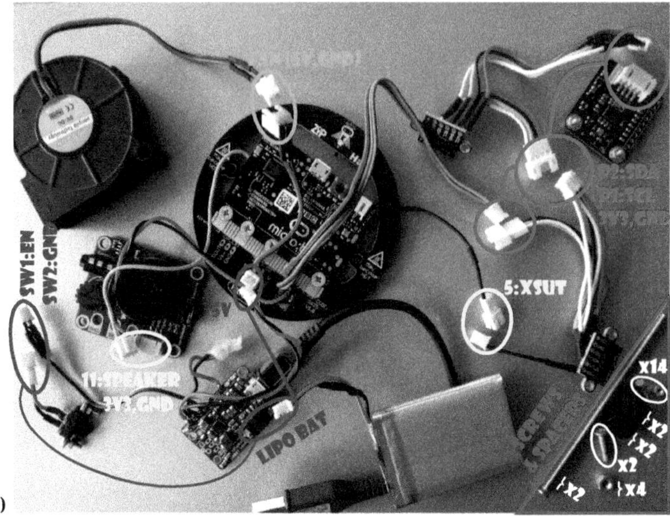

a)

b)

Figure 5.13 Interconnection between the peripheral units of the prototype interactive game. (a) how to connect the available PH2.0 connectors of the system, (b) PH2.0 connectors properly connected.

LiPo battery as well as the ON/OFF power switch (also necessary for our system) are respectively presented in Figure 5.12f, g.

Figure 5.13 presents the various parts and modules used by the prototype system along with the soldered cables and connectors. Figure 5.13a circles the female–male connectors that should be attached together. In detail, the red circles depict the software-implemented I2C bus driven by Micro:bit pins P2 (serial data [SDA]) and P1 (serial clock [SCL]). The yellow circles depict: (i) the XSHUT pin of the distance sensor module driven by Micro:bit P5 pin ("stolen" by button A) and (ii) the speaker's input signal driven by Micro:bit P11 pin ("stolen" by button B on the front side of Micro:bit). It is worth

Table 5.2 BOM of the prototype interactive game.

Part (manufacturer)	Description	Unit
Micro:bit (by BBC)	μC motherboard	1
ZIP Halo (by Kitronik)	Color LED Ring board for Micro:bit	1
10 DOF (by DFRobot)	BMP280 + BNO055 sensor board	1
CJVL53L0XV2 (OEM)	VL53L0X distance sensor board	2
PowerBoost 1000 charger (by Adafruit)	LiPo charger and DC/DC boost converter	1
Speaker for Micro:bit (by Waveshare)	Volume adjustable speaker module	1
5V DC FAN 5015 (OEM)	DC cooling blower fan	1
Power switch (OEM)	ON/OFF slide switch	1
PH2.0/four-pin male connector (OEM)	Connector with wires (2 mm pitch)	3
PH2.0/four-pin female connector (OEM)	Connector with wires (2 mm pitch)	2
PH2.0/two-pin male connector (OEM)	Connector with wires (2 mm pitch)	5
PH2.0/two-pin female connector (OEM)	Connector with wires (2 mm pitch)	3
PH2.0/three-pin male connector (OEM)	Connector with wires (2 mm pitch)	1
Screw M3 × 80 mm (OEM)	Flat head countersunk screws **+ 2 Nuts**	2
Screw M3 × 12 mm (OEM)	Flat head countersunk screws **+ 2 Nuts**	2
Screw M2 × 8 mm (OEM)	Low profile screws **+ 14 Nuts**	10
Spacers M3 × 10 mm (OEM)	Spacers for the boards' separation	2
Spacers M3 × 11 mm (OEM)	Spacers for the boards' separation	2

mentioning that all interconnections can be carried out through JST PH2.0 connectors of two pins, three pins, and four pins. Figure 5.13b illustrates the various parts and modules of the prototype system linked together via the connectors. It is worth noting the additional screws and spacers found at the right bottom area of Figure 5.13a, which are used for the assembly of prototype game, using the (custom-designed) 3D printed parts of the enclosure (presented next in the chapter). The *BOM* of the prototype interactive game is given in Table 5.2.

Assembly Process of the 3D Printed Parts of the System's Enclosure

Having reached to the agreement that the *ZIP Halo* diameter determines the limits of our 3D printed enclosure, all the peripherals are selected to fit those limits. Before exploring the enclosure design and its 3D printing processes (as well as the development of the firmware examples that will run on the proposed system), we overview the assembly process of our system with the 3D printed parts that have been developed. The overall assembly is illustrated by the enumerates steps of Figures 5.14–5.16, while screws and spacers that are used for mounting the parts are illustrated in Figure 5.17.

Starting the assembly process from ***step 1*** of Figure 5.14, the 3D printed part on the left of the figure constitutes the floor of the enclosure. The two holes are indented for the

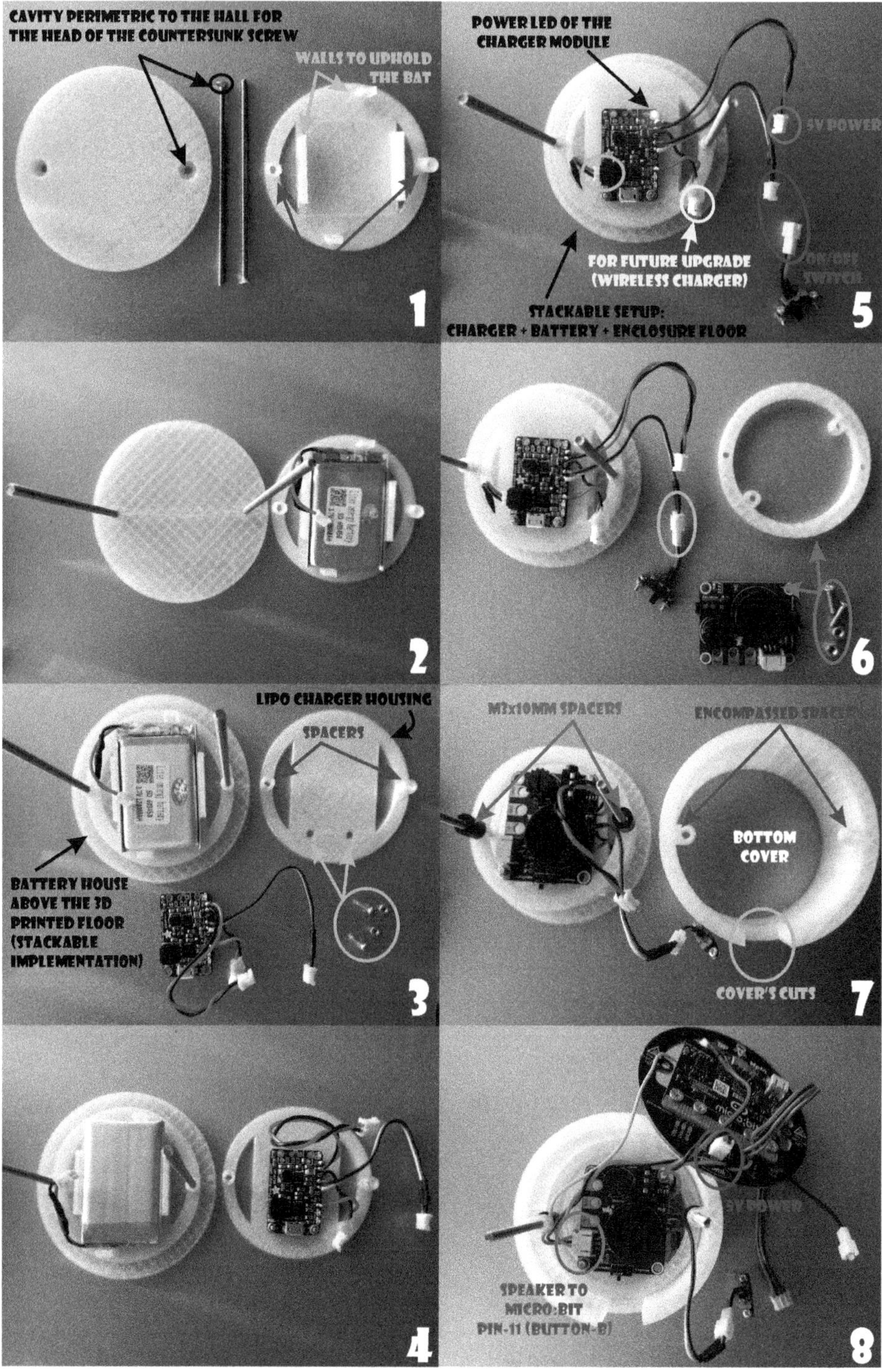

Figure 5.14 Assembly of the system's enclosure (1 of 3).

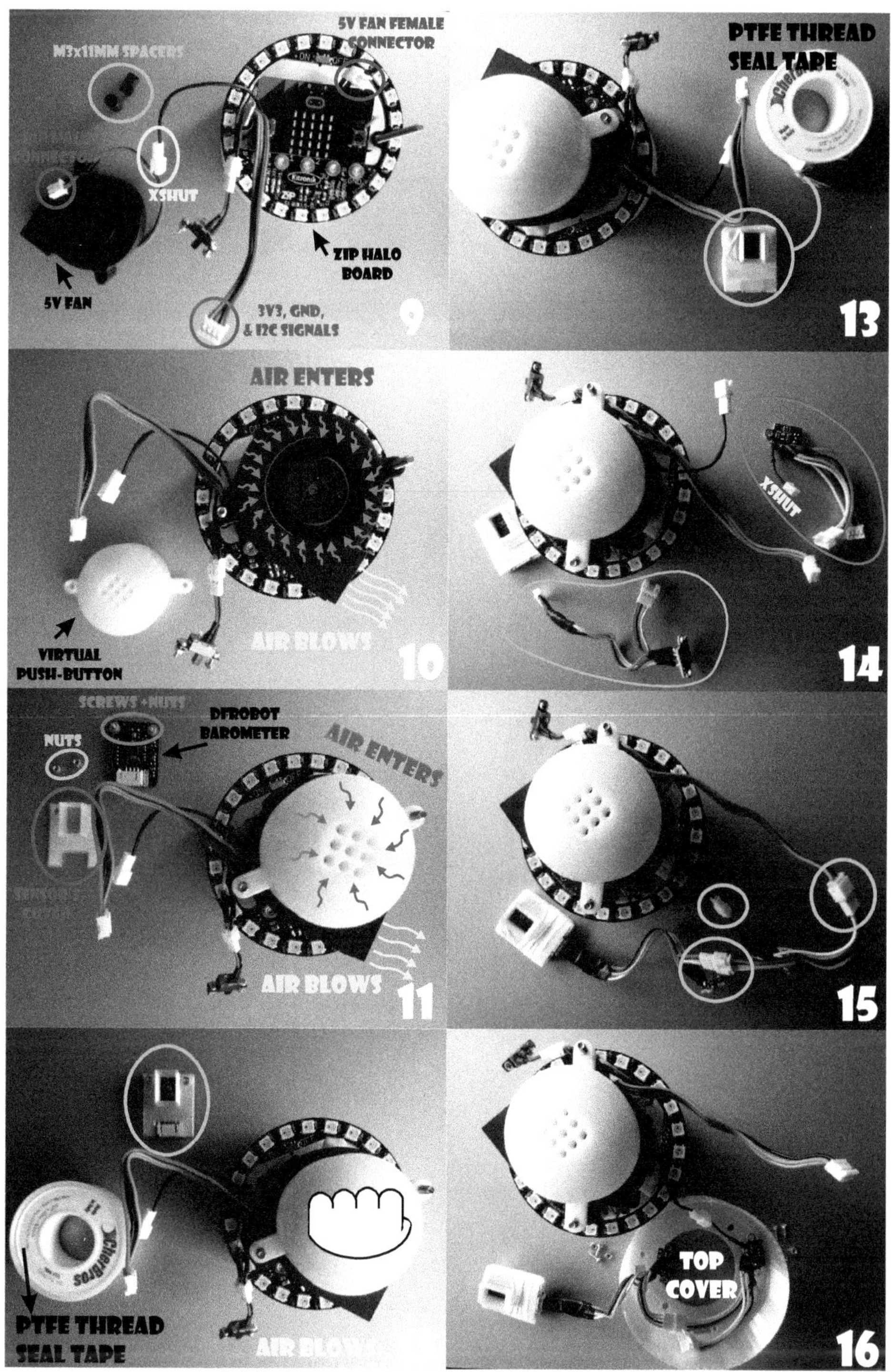

Figure 5.15 Assembly of the system's enclosure (2 of 3).

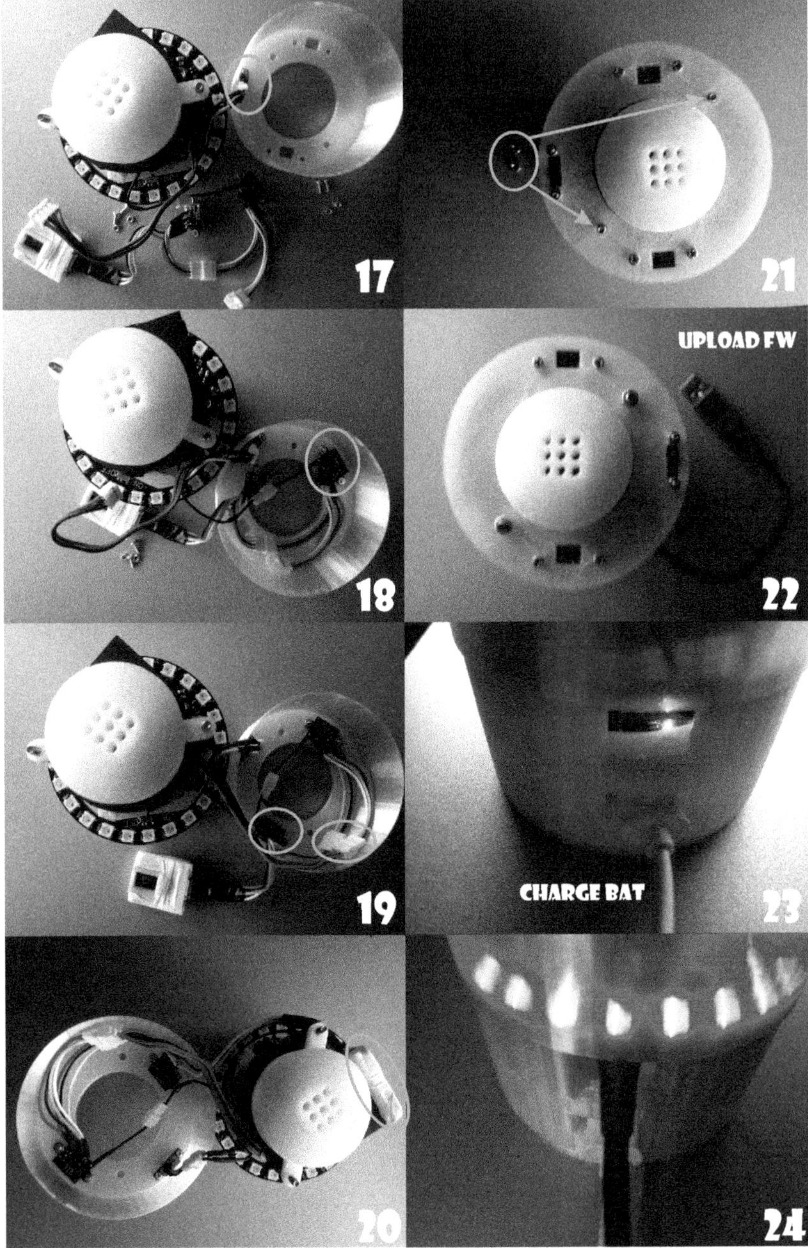

Figure 5.16 Assembly of the system's enclosure (3 of 3).

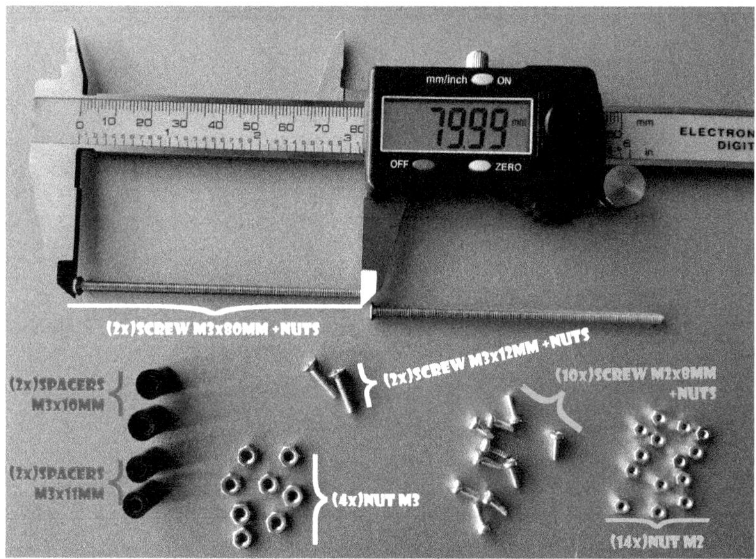

Figure 5.17 Screws, nuts, and spacers required for the system's assembly.

M3×80 mm screws that hold together all the internal 3D printed parts of the enclosure (as we will see next). The little cavity of the two holes allows the head of the countersunk screws to be accommodated within the cavity. The 3D printed part on the right of the figure constitutes the battery holder, where the four barriers keep steady the battery. It is worth noting that two spacers have been encompassed in the latter 3D design.

In ***step 2***, we have placed the screws to the 3D printed part on the left, as well as the battery to the 3D printed part on the right of the figure. In ***step 3***, we have placed the battery (attached to the 3D printed battery holder) above the 3D printed floor (as presented on the left of the figure). It is here noted that the 3D printed parts of the overall enclosure are designed to implement a stackable assembly. The 3D printed housing of the *LiPo charger and DC/DC booster* on the right of the figure, encompasses two spacers, which support the stackable assembly, but also two holes where the module will be attached (via M2×8 mm screws and corresponding nuts highlighted by the green circle). In ***step 4***, we present the attachment of the *LiPo charger and DC/DC booster* to its housing.

In ***step 5*** (and following the stackable assembly of the enclosure), we have placed the charger housing above the LiPo battery. In addition, we have attached PH2.0 connector of LiPo bat to the corresponding socket of *LiPo charger and DC/DC booster* module, as highlighted by the green circle. As soon as finishing that particular interconnection the power (blue) LED of the *LiPo charger and DC/DC booster* is turned ON. To turn OFF the module's power we should connect the ON/OFF switch to the PH2.0 junction highlighted by the blue circle. The ON/OFF switch grounds (or leaves floating) the *enable* (and pulled up) input of the module. The PH2.0 male connector highlighted by the red circle constitutes the 5 V power, which supplies our system (i.e. to be connected to the *ZIP Halo* module). Finally, the PH2.0 male connector highlighted by the yellow circle is not currently used by the system. That particular connector powers the *LiPo charger and DC/DC booster* and is supposed to be used with a wireless charger (so as to charge LiPo battery wirelessly and not through the USB connector).

In ***step 6*** we have connected ON/OFF switch to the corresponding PH2.0 junction (highlighted by the green circle) and turned OFF the power of the *LiPo charger and DC/DC booster*. On the right of the figure we observe the speaker housing along with the M3×12 mm screws and corresponding nuts (highlighted by the red circle), which are used to attach the speaker module on the 3D printed housing. In ***step 7*** we have stacked the speaker housing and module over the *LiPo charger and DC/DC booster*. We have also placed the M3×11 mm spacers above the speaker's housing (as pointed by the red arrows). On the right of the figure we observe the *bottom cover* of our system, which encloses all the aforementioned modules and corresponding housings. There are two cuts on that cover (highlighted by the green circle), which give us access to the micro-USB connectors of the system. The USB connectors are used for: (i) charging the battery (via the *LiPo charger and DC/DC booster*), and (ii) uploading code to the microcontroller device (employed by the Micro:bit board). The *bottom cover* encompasses two spacers for the attachment of the Kitronic *ZIP Halo* board.

In ***step 8*** we have attached the speaker's PH2.0 connector (highlighted by the red circle), which is driven by P11 pin (originally meant to read push-button B of Micro:bit) as well as the PH2.0 connector, which supplies 5 V power to the *ZIP Halo* board from the *LiPo charger and DC/DC booster* (highlighted by the blue circle).

Continuing the assembly process from ***step 9*** of Figure 5.15, the Kitronic *ZIP Halo* board (which holds the Micro:bit) is placed above the 3D printed *bottom cover*. The two spacers (M3×11 mm) will be placed above the *ZIP Halo* board in order to leave the requisite room for the 5V fan, which is supposed to be positioned above the *ZIP Halo* board. The latter board holds connectors that are addressed for (i) the I2C interface (that is, the PH2.0 connector highlighted by red color), (ii) the attachment of the 5V fan (highlighted by blue color), and (iii) the shutdown input (i.e. XSHUT pin) of one of two distance sensors (highlighted by yellow color). The XSHUT pin is controlled by P5 pin of Micro:bit (originally meant to read push-button A of Micro:bit).

In ***step 10*** we have placed the spacers and then the 5 V fan above the *ZIP Halo (+Micro:bit)* setup. It is here noted that the cooling fan absorbs the air from above (as illustrated by the red arrows) and blows the air through the exit gate of the fan (as illustrated by the yellow arrows). At the left bottom area of the figure, we have designed a 3D part that will emulate the operation of a push-button. To achieve this goal, the 3D virtual button has been designed to fit the perimeter of the fan cover, without blocking the air entering the fan. This is the reason why this 3D print has holes on the top side (i.e. to allow the air to go into the fan).

In ***step 11*** we have placed the 3D virtual button above the cooling fan and we have also designed a cover to be placed at the front side of the DFRobot's barometric sensor. We have already attached two M2×8 mm screws along with two nuts on the DFRobot module, as highlighted by the red circle. The nuts will provide the necessary space between the module's cap (highlighted with the blue circle) and the electronic components found on the module's PCB. Above those nuts we should place the module's 3D cap, which is then tied up with two extra nuts (highlighted by the green circle).

In ***step 12*** we present the 3D cap tied up to the DFRobot module (highlighted with the green circle). It is worth noting that the 3D cap features a rectangle projection, which serves two goals. The first goal is to accurately fit inside the exit gate of the cooling fan. The second is to allow the blowing air to reach the BMP280 barometric pressure sensor employed by the DFRobot module. Consequently, when the cooling fan is working, the blowing air is identified by the BMP280 sensor as an augmented air pressure. If the user blocks with this palm the holes of the 3D virtual button (as presented by the figure of the current step) the air pressure decreases. Again, when the user removes his/her palm from the 3D virtual button, additional air is inserted through the fan and the atmospheric pressure (identified by the sensor) increases. With a simple data acquisition (DAQ) software we could easily determine the approximate fluctuating level of air pressure, as blocking/unblocking the holes of the 3D virtual button, and then interpret the fluctuation to a closed/opened switch. We will come again to this topic next in the chapter, but it is worth making a reference to how creative we can get with modern sensor devices, as we herein transform a 3D printed part with holes to a push-button.

In ***step 13*** we have wrapped around the *DFRobot module + 3D cap* setup with *polytetrafluoroethylene (PTFE)* thread seal tape (*aka* Teflon) so as to minimize the air leakage, and hence, increase the changing level of air pressure as pressing/releasing our virtual button.

In ***step 14*** we prepare the interconnection of three I2C sensors, that is, the BMP280 and the two VL53L0X distance sensor. The VL53L0X module within the red frame incorporates the extra XSHUT pin, which enables/disables the sensor device. In ***step 15*** we have performed the necessary connections of the three I2C devices and, in ***step 16***, we prepare the attachment of the VL53L0X modules, as well as the ON/OFF switch to the (3D printed) *top cover*.

Continuing the assembly process from ***step 17*** of Figure 5.16, we have there attached the ON/OFF switch at the *top cover*. In ***step 18*** we have attached the VL53L0X sensor with extra XSHUT pin at the *top cover*, while in ***step 19*** we have attached the second VL53L0X senor. Having finished with the preparation of the *top cover*, in ***step 20*** we have placed the BMP280 sensor module at the exit gate of the cooling fan (with the rectangle projection of the 3D part being placed inside the gate). In ***step 21*** we have adjusted the two holes of *top cover* to the long screws that hold the overall setup, and we are ready to place the two nuts that will render the overall setup steady. The assembly process is concluded in ***step 22*** where we have also connected the USB cable in order to download firmware to the Micro:bit. In ***step 23*** we have connected the second USB cable, which charges the LiPo battery, where the orange LED declares that the battery is being charged (and when a green LED turns *ON* instead, then the battery is fully charged). Finally, in ***step 24*** we have tested a firmware, which turns *ON* the LED ring of the *ZIP Halo* board.

Firmware Code Design and User Instructions

Having finished the assembly of the prototype (3D printed) system, we start developing a couple of test codes. Because we have created a composite system, which encompasses different and complex subunits, we apply the so-called *bottom-up* design approach. That is, we decompose (i.e. break down) the firmware development of the overall system into smaller segments of code, and then, the various segments are composed together to form the overall application firmware. Herein, the segments of code are addressed to control each one of the IO units incorporated by the system.

Figure 5.18 depicts the three subunits that will be first tested by three individual firmware codes. That is, (i) the LED ring which triggers the kids' *sight*, (ii) the virtual button,

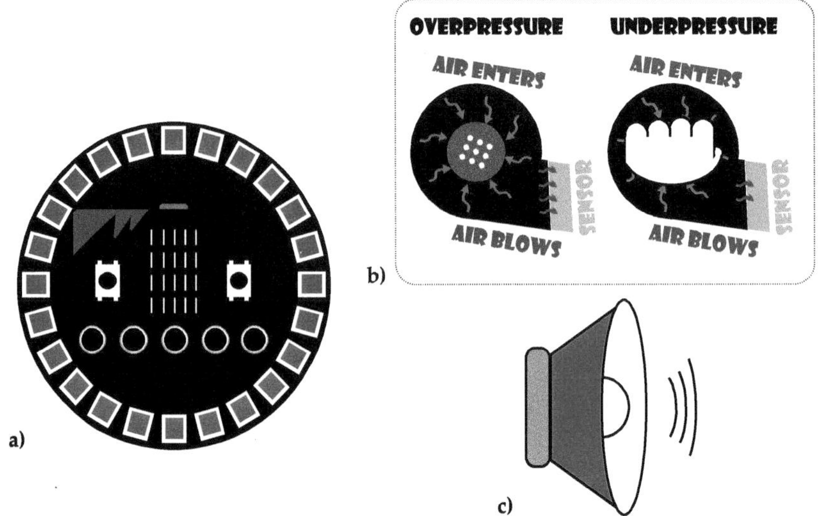

Figure 5.18 Bottom-up design method: decomposing the application firmware into three segments. (a) the LED ring applying to the *sight* sense, (b) the virtual button applying to the *proprioception* sense, (c) the speaker applying to the *hearing* sense.

which obtains information from the kids reaction (i.e. related to the children's *proprioception*), and (iii) the speaker, which triggers the kids' *hearing* sense. The overall connection diagram of the prototype system is depicted in Figure 5.19.

Arduino Ex.5.3

We start the bottom-up design method with a firmware that controls the LED ring of *Zip Halo* board. The test firmware code, presented in Figure 5.20, alternates the red and green colors on the 24 LEDs of the Zip Halo board every 250 ms. To apply a more realistic implementation of this alternation between red and green colors, we address a *pseudo-random number generator (PRNG)*. A PRNG sequence is not entirely random because it is determined by an initial value, aka *PRNG's seed*. Hence, while sequence might be long and random, it is always the same (i.e. it is repeated periodically).

Arduino embeds the *randomSeed()* function to initialize a PRNG. The *randomSeed()* function admits an *unsigned long* number, while that number is regularly acquired by a fairly random input. Accordingly, the values generated by the PRNG, during the execution of the Arduino *random()* function, differ upon subsequent executions of a sketch. A regular technique in Arduino is to obtain that number from the *analogRead*() function driven by an unconnected pin, e.g. *randomSeed(analogRead(0))*. The latter command resets the PRNG to a random initial value, picked up from random noise generated by the unconnected analog pin 0 (PINA0). While the PRNG sequence is predictable, the initial value of PRNG varies because of the random noise of the surrounding environment (obtained from PINA0). The Arduino *random(min,max)* function admits two arguments and generates PRNG sequence from the number *min* to *max*, where *max* is excluded from the sequence. For instance the syntax *random(1,3)* will generate numbers 1 and 2 randomly.

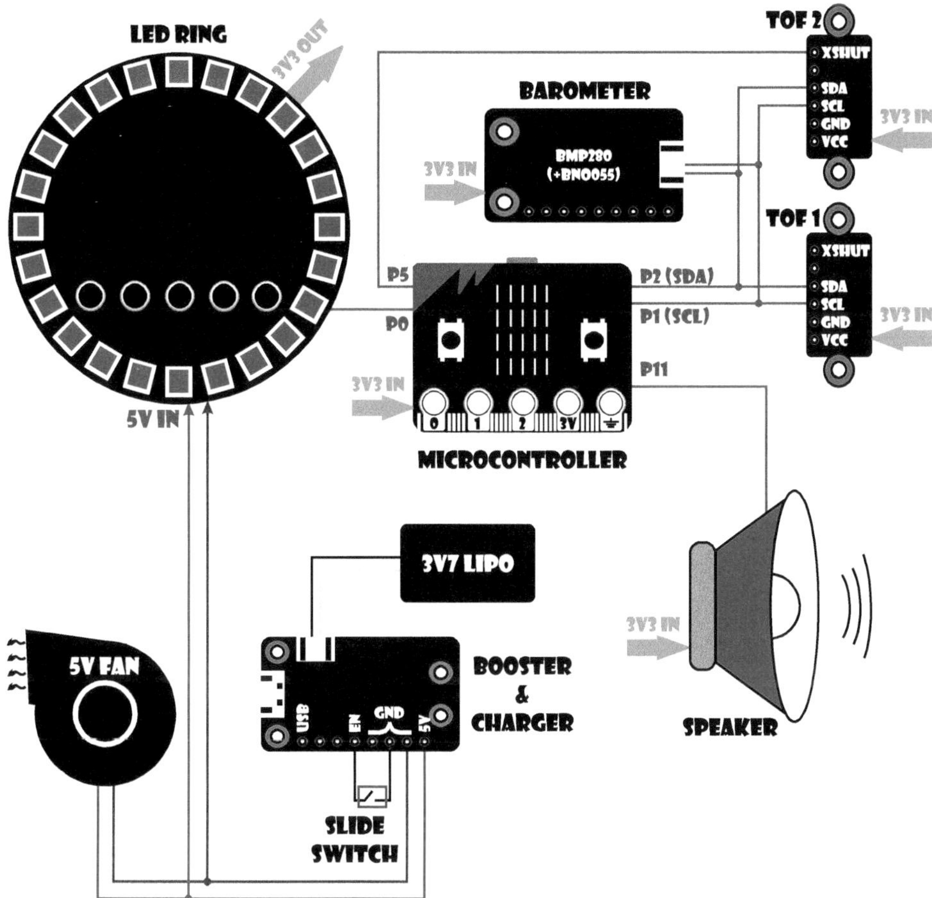

Figure 5.19 Connection diagram of the prototype system.

To control the LED ring of the current example we make use of the *Adafruit_NeoPixel* library, delivered by Adafruit, which can be downloaded from the link: https://learn.adafruit.com/adafruit-neopixel-uberguide/arduino-library-use. Having stored the library to the Arduino root directory, the first code line of the firmware example (in Figure 5.20) invokes that particular library for the control of the LED ring of *Zip Halo* board. Since the latter board incorporates a 24-LED ring, controlled by the pin P0 of Micro:bit, the subsequent two code lines declare two constants to hold the corresponding information used by the library. Next, in the code, we create an instance of *Adafruit_NeoPixel* class, herein named *ring*, which admits as *arguments* the aforementioned *constants*. The code lines within the black frame establish objects of the class associated with fundamental colors that may be used by the LED ring. To ease the code development process, those colors are declared by the identical *variables* named *red*, *green*, *blue*, *cyan*, *magenta*, *yellow*, *white*, and *black*, where the latter variable is used to turn the LED ring *OFF*.

Inside the Arduino *setup()* function, the *ring.begin()* prepares the data pin of *Adafruit_NeoPixel* library. Next, the *ring.setBrightness()* adjusts the brightness of the LED ring (where

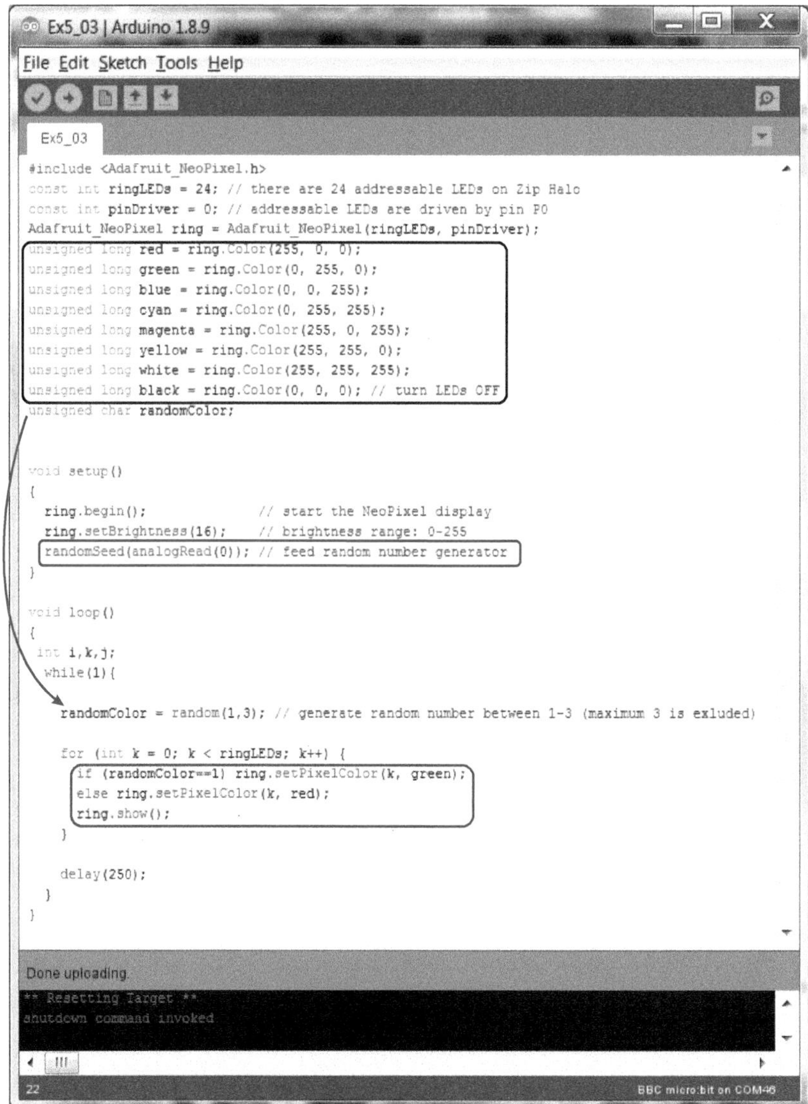

Ex5_03 | Arduino 1.8.9
File Edit Sketch Tools Help
Ex5_03
#include <Adafruit_NeoPixel.h>
const int ringLEDs = 24; // there are 24 addressable LEDs on Zip Halo
const int pinDriver = 0; // addressable LEDs are driven by pin P0
Adafruit_NeoPixel ring = Adafruit_NeoPixel(ringLEDs, pinDriver);
unsigned long red = ring.Color(255, 0, 0);
unsigned long green = ring.Color(0, 255, 0);
unsigned long blue = ring.Color(0, 0, 255);
unsigned long cyan = ring.Color(0, 255, 255);
unsigned long magenta = ring.Color(255, 0, 255);
unsigned long yellow = ring.Color(255, 255, 0);
unsigned long white = ring.Color(255, 255, 255);
unsigned long black = ring.Color(0, 0, 0); // turn LEDs OFF
unsigned char randomColor;
void setup()
{
ring.begin(); // start the NeoPixel display
ring.setBrightness(16); // brightness range: 0-255
randomSeed(analogRead(0)); // feed random number generator
}
void loop()
{
int i,k,j;
while(1){
randomColor = random(1,3); // generate random number between 1-3 (maximum 3 is exluded)
for (int k = 0; k < ringLEDs; k++) {
if (randomColor==1) ring.setPixelColor(k, green);
else ring.setPixelColor(k, red);
ring.show();
}
delay(250);
}
}
Done uploading.
** Resetting Target **
shutdown command invoked
22
BBC micro:bit on COM46

Figure 5.20 Generate random (green and red) colors to the LED ring. *Source:* Arduino Software.

the maximum brightness of each LEDs is achieved by the value 255). The *randomSeed(analogRead(0))* resets the PRNG to a random initial value, picked up from random noise generated by the unconnected analog PINA0.

Inside the Arduino *loop()* the *random(1,3)* function assigns to the *randomColor* variable either number 1 or 2, as decided by the PRNG sequence. Then the *for()* loop, which is repeated 24 successive times, sets the 24 LEDs of *Zip Halo* board to either green or red color. The latter decision is accomplished by the *if()* statement inside the loop and it is in agreement with the number generated by the *random()* function. It should be noted that the *ring.setPixelColor()* function sets the color of each LED of the ring. Yet, the color does not appear on the corresponding LED until the execution of the *ring.show()* function. Before the repetition of the Arduino *loop()* function, the code delays 250 ms. Accordingly, the LED ring of *Zip Halo* board alternates the colors red and green, randomly, every 0.25 s.

A possible entertaining game for the children (grounded on this particular test code) is to ask them try to hit the virtual button, every time the green color appears on the device. Hereafter, we continue to the bottom-up design method with a firmware, which controls the virtual button of the prototype interactive game.

Arduino Ex.5.4

The firmware test code that controls the LED ring but also implements the functionality of the virtual button (through BMP280 barometric sensor acquiring the air pressure derived from a cooling blower fan) is presented in Figure 5.21a. The first seven code lines are used for the control of the LED ring, where in this particular example we have declared variables just for the colors used by the LED ring, that is, the red, green, and black (where the latter turns *OFF* the LEDs of the ring).

Next, the code lines within the red frame are used for the control of BMP280 sensor. Because the latter sensor is controlled via I2C, but from the Micro:bit pins other than the dedicated I2C pins (and, in particular, from pins P1 and P2 for the control of SCL and SDA lines, respectively), the foremost code line within the frame invokes the *SlowSoftWire* library. This library, which emulates the functionality of Arduino *Wire* library, can be found at: https://github.com/felias-fogg/SlowSoftWire. The subsequent code line invokes the library (i.e. through the header file *DFRobot_BMP280.h*) for the control of the BMP280, which is delivered by DFRobot[4] and is available from https://github.com/DFRobot/DFRobot_BMP280.

Because the latter library is designed to work with Arduino *Wire* library, we need to make a couple of amendments to the *DFRobot_BMP280.h* and *DFRobot_BMP280.cpp* files, so as to pass the control of BMP280 sensor to the *SlowSoftWire* library. In detail, we comment the command line, which includes the Arduino *Wire* library within the *DFRobot_BMP280.h* file, as presented by Figure 5.21b. Then, we include the *SlowSoftWire* library within the *DFRobot_BMP280.cpp* file as well as create an instance of the *SlowSoftWire* class using the exact name *Wire*, while also defining SDA and SCL lines (herein, in P2 and P2, respectively), as presented by Figure 5.21c.

The last two code lines within the red frame first create an instance of the *DFRobot_BMP280* class using the name *bmp*, and then declare the constant *th* that determines the pressure threshold between the open/close state of the virtual button. We will explore next how the value 35 is determined by particular pressure measurements, which are obtained from the sensor.

The Arduino *setup()* function initializes the LED ring, the UART port, as well as the BMP280 sensor device. The latter task is performed through the call of *bmp.begin()*. Subsequent to the Arduino *loop()* function we have defined a function named *ringColor()*, which is used to turn the LED ring *ON*, using the particular color determined by the function argument. The function body is the same as the corresponding code used by the previous example.

4 It is here noted that a useful list of Arduino libraries is available at: https://www.arduinolibraries.info/.

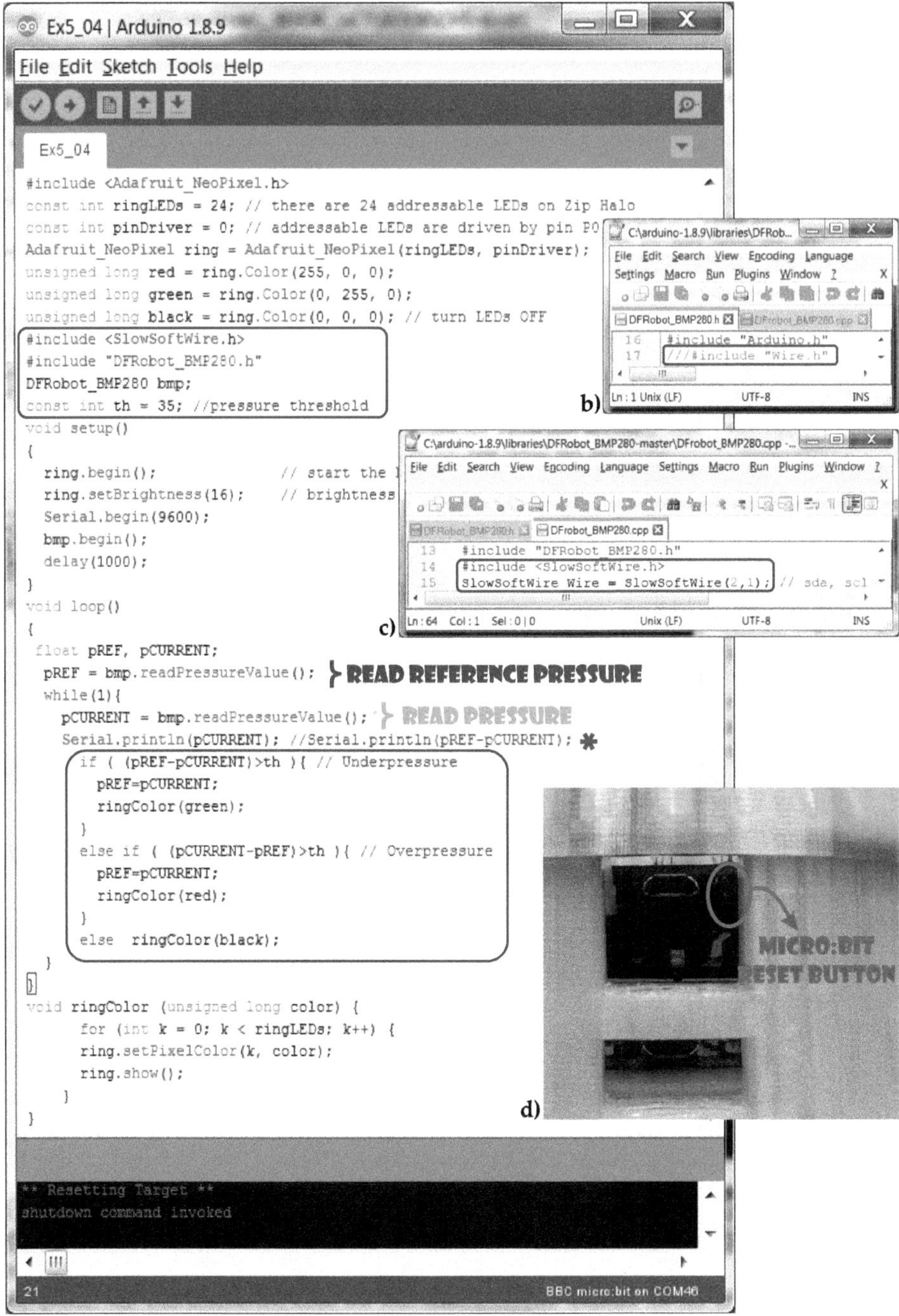

Figure 5.21 Emulate a push-button via a barometric pressure sensor (firmware). (a) application firmware, (b) requisite amendments to the *header (.h)* file of the library, (c) requisite amendments to the *source code (.cpp)* file of the library, (d) reset button location on the system. *Source:* Arduino Software.

The first code line within the Arduino *loop()* function declares to *float* variables that holds the pressure data (acquired by the sensor) in *Pascal (Pa)*, where 1 Pa = 10^{-5} bar. Accordingly, the subsequent code line acquires a pressure sample and assigns the result to a *reference* variable (of identical name *pREF*). Next, in the code, is found a *while()* loop where the foremost instruction of the loop acquires again a pressure sample. This time the measurement data are assigned to the variable named *pCURRENT*. The *Serial.println()* of the subsequent code lines sends to the serial port either the content *pCURRENT* variable or the difference *pREF – pCURRENT*.

The *if()*. . .*else* statement first evaluates if the latter difference is higher than the pressure threshold value an if the condition is *true*, the current pressure value is assigned to the reference pressure while also the green color appears on the LED ring. To clarify how this condition is associated with the user's interaction to the system apparatus, we explore the graph of by Figure 5.22a. The graph illustrates the pressure measurements as the user

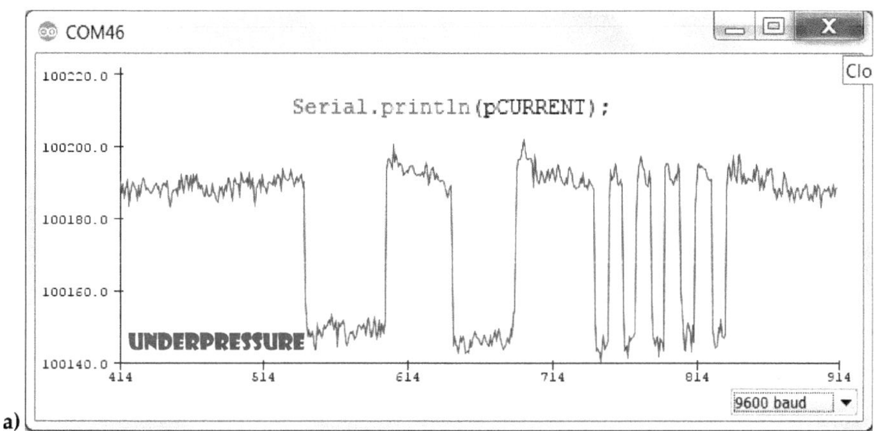

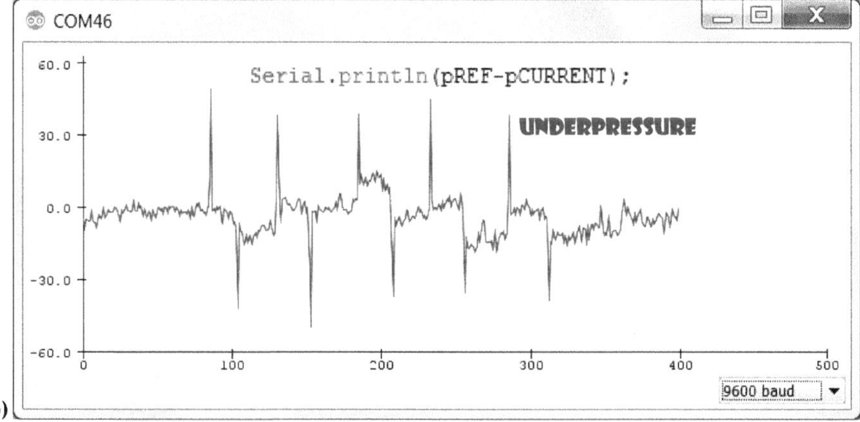

Figure 5.22 Emulate a push-button via a barometric pressure sensor (plots). (a) plot of pressure measurements: changes in the pressure level, (b) plot of pressure measurements: identify underpressure by the pressure difference. *Source:* Arduino Software.

presses/releases the virtual button, i.e. as he/she blocks/unblocks the air flow (see Figure 5.18b). When the user blocks the air flow with his/her palm, the identified pressure is reduced (referred to as *underpressure* in Figure 5.22a). The unblocking of the air flow results in an increment of the pressure (to the initial pressure level).

If we examine carefully the difference between the pressure values (in Figure 5.22a) as the user blocks/unblocks the air flow, we observe that the pressure is alternated between the values 100 140 Pa and 100 180 Pa. The difference is approximately 40 Pa, which is higher than the predefined pressure threshold (i.e. th = 35). This is also verified by the positive pulse spikes of Figure 5.22b, which plots the difference of the two pressure samples (i.e. *pREF* − *pCURRENT*).

What we expect to observe in regard to the response of the barometric pressure after releasing the virtual button (i.e. after unblocking of the air flow), is an increment of the acquired pressure. For that reason, the second condition of the *if()...else* statement evaluates if the difference *pCURRENT* − *pREF* is higher than the pressure threshold. Because the content of the reference pressure (i.e. *pREF*) has been redefined the moment of pressing the virtual button, and since the *pCURRENT* value is constantly refreshed on each repetition of the *while()* loop, the condition *pCURRENT* − *pREF* > *th* is verified as soon as the virtual button is released. In the case where the *else if ()* condition is met, the red color appears on the LED ring and the content of *pREF* is redefined again. The redefinition of *pREF* variable allows us to expect the subsequent press of the virtual button (i.e. the subsequent underpressure identification by the sensor device).

Finally, the latter *else* condition of the statement turns the LED ring *OFF*. In summary, when the user presses the virtual button the LED ring flashes for a moment (i.e. in less than 1 s) with green color, and as soon as the user releases the virtual button, the LED ring flashes (for a moment) with red color. The user can blow through the holes on the top of the prototype system to test an equivalent overpressure condition, as well. It is here noted that it might be needed to reset the Micro:bit board in order to run the application code. The reset can be performed with a small screwdriver through the reset button of the board, which can be accessed by the upper opening of the prototype apparatus (as illustrated in Figure 5.21d). Hereafter, we continue the bottom-up design method with the addition of the code part, which controls the speaker of prototype interactive game, so as to trigger the children's *hearing* sense.

Arduino Ex.5.5

To understand how to reproduce sounds from a speaker driven by a microcontroller device, we explore the frequencies associated to each note found on the piano keys. For the record, there are seven music notes Do(C), Re(D), Mi(E), Fa(F), Sol(G), La(A), and B(Si), where adjacent notes are one tone apart, despite the distance between B and C as well as E to F, which are one semitone apart. As depicted by the piano sketch in Figure 5.23 (where the notes are grouped in nine *octaves*), the notes found in the interval between a tone are referred *diesis*; that is, the notes of the piano's black keys. The *diesis,* which is notated by the *sharp (#)* symbol, indicates a higher (in pitch) semitone. For instance in octave 4, the black

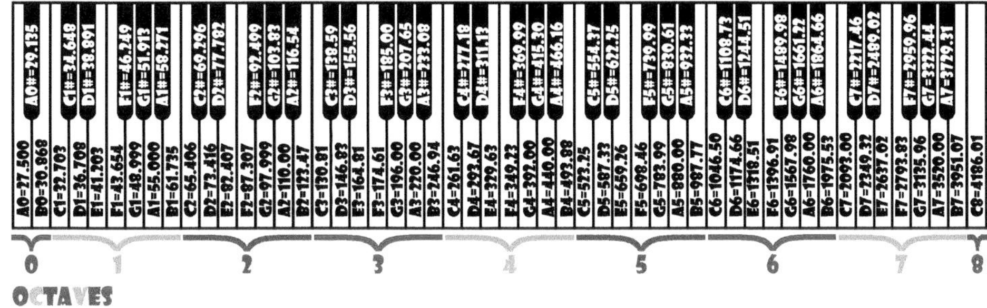

Figure 5.23 Notes and frequencies of the piano keys.

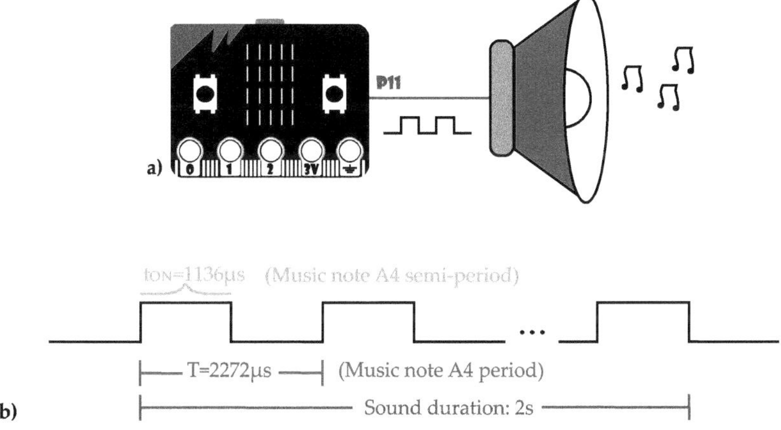

Figure 5.24 Generating by the microcontroller (and playing for 2 s) A4 music note. (a) connection diagram of the Micro:bit board to a speaker, (b) square waveform of frequency proportional to A4 note (440Hz).

key found immediately after the note A4 (and immediately before the B4 note) is referred to as A4#. To indicate a lower (in pitch) semitone we make used of the *flat* term, denoted by the *b* symbol. Subsequently, the music note between A4 and B4 can be denoted either A4# or B4b.

All semitones feature a frequency that is herein represented in Hz. For instance, the A4 music note is identical to the frequency 440 Hz. One possible (and simplified) way to reproduce a music note from a speaker controlled by the μC device (Figure 5.24a), is to generate a square pulse of a particular frequency (Figure 5.24b). For example, the period of A4 music note is equal to $T = 1/440\,\text{Hz} = 0.002\,272\,\text{s}$. Hence, if we need to play that music

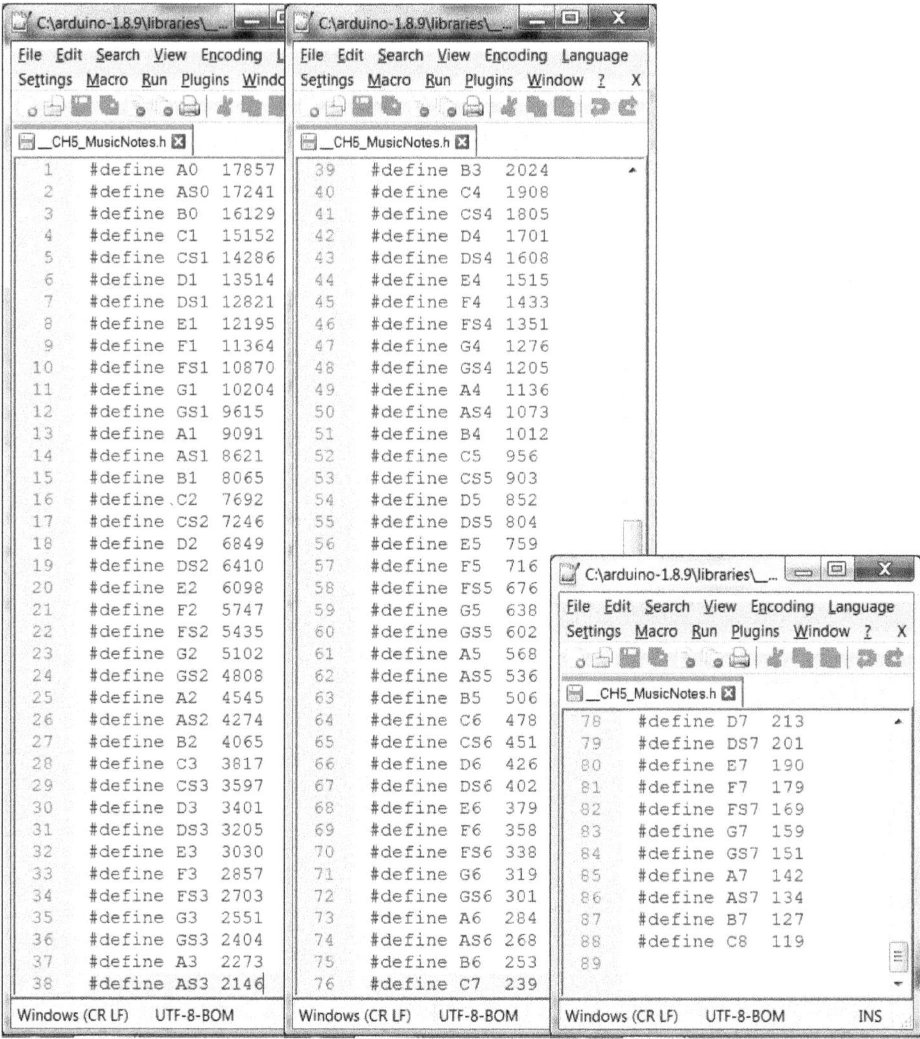

```
#define A0  17857
#define AS0 17241
#define B0  16129
#define C1  15152
#define CS1 14286
#define D1  13514
#define DS1 12821
#define E1  12195
#define F1  11364
#define FS1 10870
#define G1  10204
#define GS1 9615
#define A1  9091
#define AS1 8621
#define B1  8065
#define C2  7692
#define CS2 7246
#define D2  6849
#define DS2 6410
#define E2  6098
#define F2  5747
#define FS2 5435
#define G2  5102
#define GS2 4808
#define A2  4545
#define AS2 4274
#define B2  4065
#define C3  3817
#define CS3 3597
#define D3  3401
#define DS3 3205
#define E3  3030
#define F3  2857
#define FS3 2703
#define G3  2551
#define GS3 2404
#define A3  2273
#define AS3 2146
#define B3  2024
#define C4  1908
#define CS4 1805
#define D4  1701
#define DS4 1608
#define E4  1515
#define F4  1433
#define FS4 1351
#define G4  1276
#define GS4 1205
#define A4  1136
#define AS4 1073
#define B4  1012
#define C5  956
#define CS5 903
#define D5  852
#define DS5 804
#define E5  759
#define F5  716
#define FS5 676
#define G5  638
#define GS5 602
#define A5  568
#define AS5 536
#define B5  506
#define C6  478
#define CS6 451
#define D6  426
#define DS6 402
#define E6  379
#define F6  358
#define FS6 338
#define G6  319
#define GS6 301
#define A6  284
#define AS6 268
#define B6  253
#define C7  239
...
#define D7  213
#define DS7 201
#define E7  190
#define F7  179
#define FS7 169
#define G7  159
#define GS7 151
#define A7  142
#define AS7 134
#define B7  127
#define C8  119
```

Figure 5.25 Header file declaring the semi-period (in μs) of each music note. *Source:* Arduino Software.

note for 2 s (for instance) we should keep generating the A4 period for as long as $2/0.002272 = 880$ successive iterations. Then setting the voltage on the μC's pin to zero volts causes the termination of any sound being produced by the speaker.

In the header file depicted by Figure 5.25 (i.e. *__CH5_MusicNotes.h*), we have associated the music notes to the semi-period of each note, given in μs (of integer datatype). The S letter denotes the *sharp* symbol (e.g. the A# note of *octave 4* is denoted AS4).

The firmware example of Figure 5.26 constitutes an upgrade of the Ex5_03, which was used earlier so as to display randomly the red and green colors to the LED ring. The current example encompasses the header file *__CH5_MusicNotes.h* and is enriched with the *MakeSomeNoise()* function. The latter function (at the end of the code in Figure 5.26b)

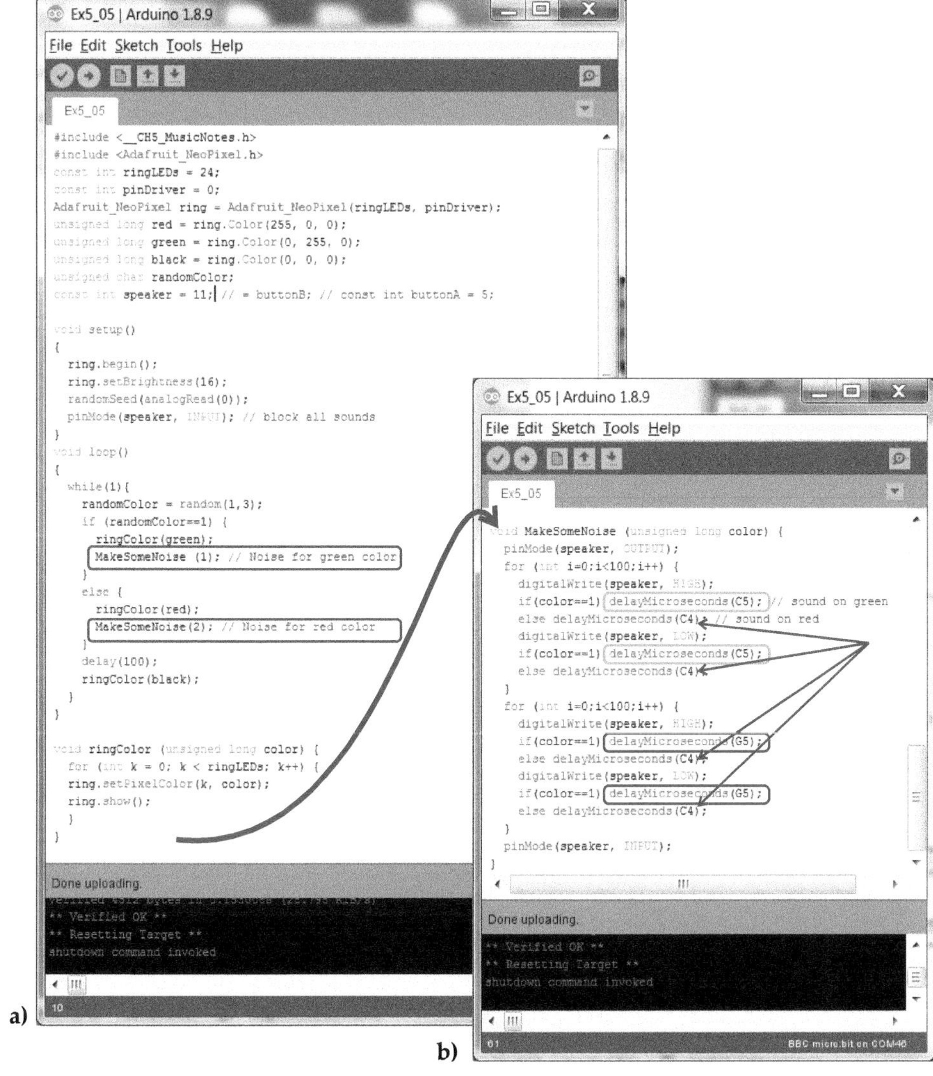

Figure 5.26 Generate random colors to the LED ring and associate sound to each color. (a) application firmware (1 of 2), (b) application firmware (2 of 2). *Source:* Arduino Software.

plays, additionally to the visual colors, an indicative sound for the green, as well as an indicative sound for the red color. The red color is acoustically accompanied by merely one music note, that is, the C4, while the green color is accompanied by two notes, that is, the C6 and G6.

The *MakeSomeNoise()* function generates a square pulse relative to the music note that is outputted by the speaker. The corresponding square pulse is repeated within a *for()* loop so as to increase the duration of each acoustic note. Because we need two music notes for the green color (instead of the one used for the red color) the *MakeSomeNoise()* function incorporates two different *for()* loops. The music notes for the sound accompanying the green color, are obtained from *octave 5* and, hence, they are of higher frequency (i.e. lower period) compared to the note accompanying the red color, which is obtained from *octave 4*.

To calculate the total time C4 note is heard, we need multiply the total period of C4 by 200 (which are the iterations of the loops). That is, $200 \times C4 = 200 \times (2 \times 1908\,\mu s) = 763\,200$ μs, provided that the rest of instructions within the two loops generate no additional delay. Equivalently, $100 \times C5 = 100 \times (2 \times 956\,\mu s) = 191\,200\,\mu s$ is the total period the C6 is heard, while $100 \times G6 = 100 \times (2 \times 638\,\mu s) = 127\,600\,\mu s$ is the total period the G6 is heard. Hence, the sound of green color is heard for approximately $191\,200 + 127\,600 = 318\,800\,\mu s$ (about half the time for the sound of the red color).

Arduino Ex.5.6

Having completed the control of the three individual subunits (incorporated by the prototype system), we are now ready to conclude the bottom-up design method with the implementation of the overall firmware of the proposed interactive game, which engages the children's *sight*, *hearing*, and *proprioception* senses. The three firmware codes for the control of the (i) LED ring, (ii) virtual button, and (iii) speaker, are linked together for the implementation of the system's firmware, which works as follows: The red and green colors are randomly displayed by the LED ring. The children are trying to hit the green color, while the displayed (red and green) colors are alternated in relative short time. On each hit of the virtual button, the color that is displayed by the LED ring remains stuck for a while on the system, while the indicative sound of the corresponding color is, at the same time, heard.

Because the prototype system employs the Micro:bit motherboard, the template firmware can be enriched with a Bluetooth interface, which could count the "correct" hits performed by the player, through a mobile device. The latter task is left as an exercise for the reader. Figure 5.27a,b presents the proposed (template) firmware of the prototype interactive game for engaging children's *sight*, *hearing*, and *proprioception* senses, while Figure 5.27c illustrates some printed results generated by the firmware execution.

The declarations at the very beginning of the code (i.e. in Figure 5.27a) incorporate all the information (from the three aforementioned firmware codes), which are needed to control the three individual subunits (i.e. LED ring, virtual button, and speaker). Accordingly, the Arduino *setup()* performs the necessary initializations to those subunits, as well as to the μC's serial port and the *random number generator*. Moreover, the main *loop()* function does not differ much from the previous example codes. In detail, the reference pressure (i.e. *pREF*) is acquired before the execution of the *while()* loop, while inside the loop the following tasks are performed:

a) The red and green colors are randomly displayed on the LED ring;
b) A delay (of 100 ms) determines the speed of which the colors are (randomly) displayed;
c) The current pressure is obtained again (and transmitted to the serial port);
d) The two *if()* statements within the black frame emulate the operation of the push button; the former *if()* is executed upon the detection of a pressure change to a lower (than threshold) level, while the latter is *if()* is executed upon the detection of a pressure change to a higher (than threshold) level;

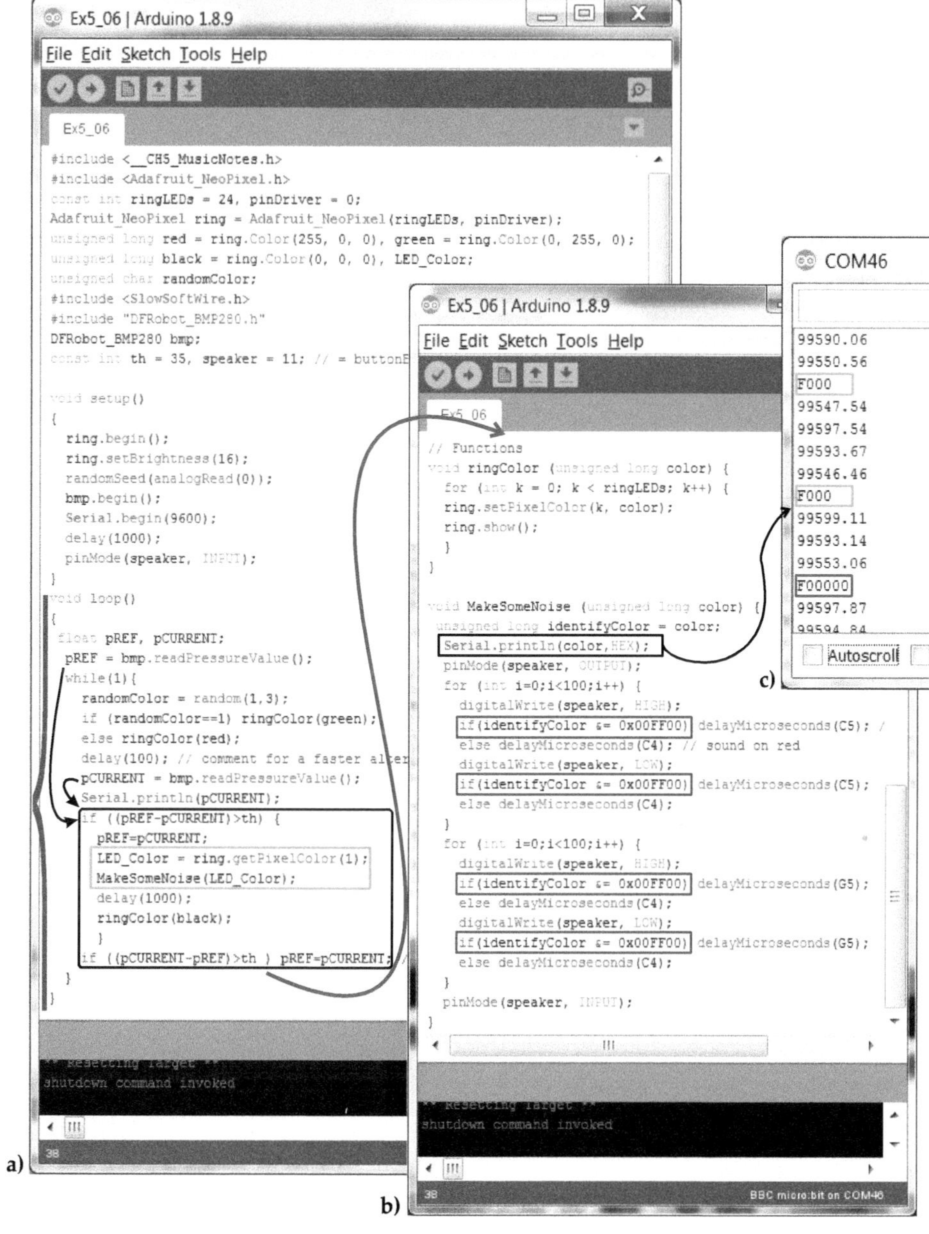

Figure 5.27 Template firmware of the prototype interactive game. (a) application firmware (1 of 2), (b) application firmware (2 of 2), (c) printed results. *Source:* Arduino Software.

The new part in this code is the *ring.getPixelColor(1)* instruction (within the green frame), which reads the color currently displayed by LED1 of the ring, in order to decide which note(s) to play via the speaker. While there is another way to decide which music note(s) the speaker reproduces, this method allows us to explore another useful function of *Adafruit_NeoPixel* library, used for acquiring the color currently displayed by a LED. Because in this particular example all LEDs of the ring display the same color, it is enough to obtain that information merely from a single LED.

The functions that print a color on the LED ring and play music notes to the speaker are given in Figure 5.27b. The *ringColor()* is the exact same function that was used by the previous firmware examples, while the *MakeSomeNoise()* function features some amendments that are traced within the colored frames. In detail, the *Serial.println()* sends to the serial port the argument of *MakeSomeNoise()* function, which is the color obtained from the LED1 of the ring. The *if()* statement with the blue frames perform a bitwise *AND (&)* to the content of a local variable, named *identifyColor*, with the hex value 0×00**FF**00. The *identifyColor* has been previously assigned to the color obtained from LED1, while the hex value clears the content of all information except the one that holds the information for the green color (i.e. 0×RR**GG**BB). Hence, the condition evaluated by the *if()* statement determines whether the LED features a green color and proceeds to the selection of the music notes that should be played by the speaker.

In Figure 5.27c, we observe the current pressure value acquired by the sensor as well as the information of the LED1 color when the user hits the virtual button. That is, the hex value 0×F000 when the user hits the virtual button upon the display of green color, or the hex value 0×F00000 when the user hits the virtual button upon the display of the red color.

3D Printing

Nowadays, makers have the opportunity to build things at home with a low-cost 3D printer, using freeware tools. We herein explore the 3D printing processes for building the enclosure(s) of μC products. The overall procedure can be separated into three distinctive tasks: (i) the modeling/design of 3D objects with the utilization of 3D *computer aided design* (*CAD*) software; (ii) the conversion of the 3D design file into *computer numerical control* (*CNC*) programming language for the automated control of the 3D printer; and (iii) the proper calibration of the 3D printer toward a successful build of the 3D printed part.

The examples presented in this subchapter apply to the 3D printed enclosure explored earlier by the prototype interactive game. In consideration of the 3D modeling as well as the conversion of the 3D design file into CNC programming language, we address the freeware tools FreeCAD and Ultimaker Cura, respectively, while for the production of the 3D printed parts we employ the PrimaCreator P120 3D printer.

Modeling 3D Objects with FreeCAD Software

Up until this point we have made clear that the process of building microcontroller prototypes entails particular tasks from different fields of study. We herein are challenged to get involved with another working area in order to model/design the 3D object(s) of the enclosure of a μC product. To ensure *simplicity in design*, the 3D modeling tasks, presented hereafter, apply to merely the utilization of *primitive shapes* (such as cube, cylinder, etc.), using consecutive Boolean operations to the primitive shapes (e.g. cutting shapes, making an intersection of shapes, and so forth). This particular modeling technique is used by many CAD systems for solid modeling and it is known as *constructive solid geometry* (*CSG*).

a) 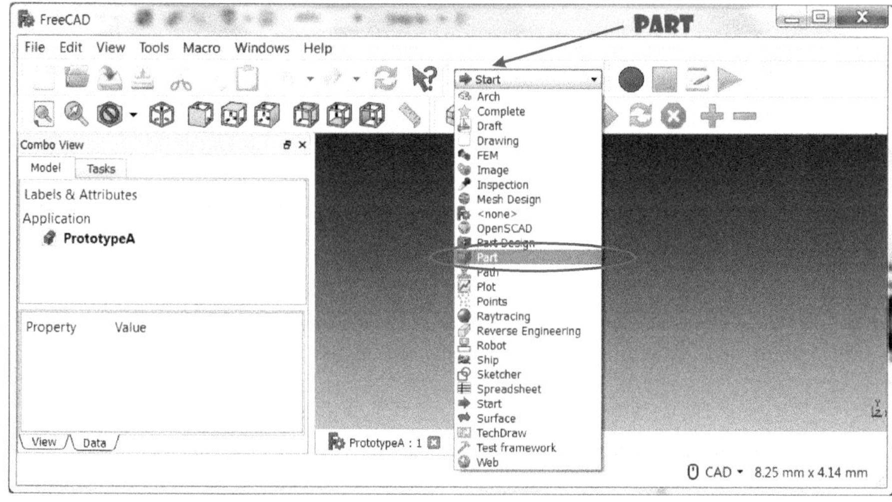

b) 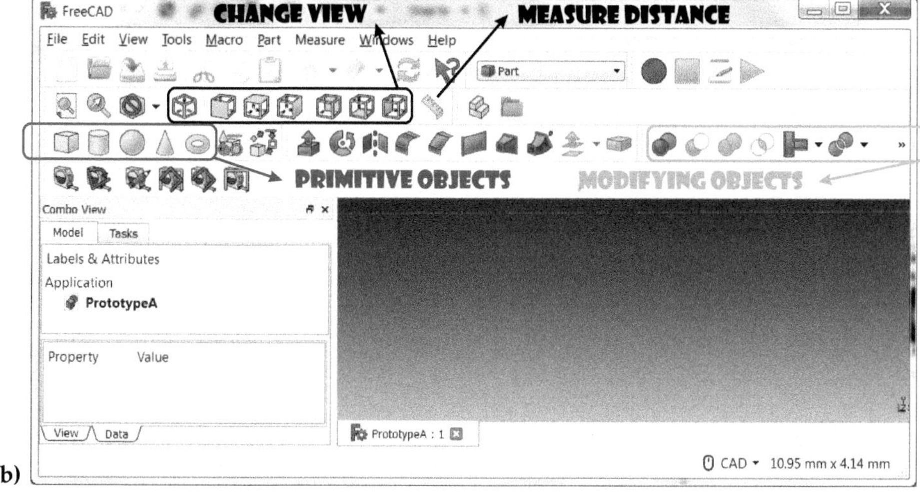

Figure 5.28 FreeCAD: getting started with primitive shapes and Boolean operations. (a) select the *Part* Workbench on the FreeCAD, (b) toolbars of *primitive objects* for *modifying objects* appear on the FreeCAD. *Source:* Arduino Software.

The FreeCAD software tool can be found at: www.freecadweb.org. As soon as the user downloads and installs the software, he/she may start a new application from the menu *'File → New'*. Through the menu *'File → Save As..'* the user saves the current project of a given name, that is, *<file-name>.FCStd* (we, hereafter, use the name *PrototypeA.FCStd*). To apply the CSG modeling technique, the user should first select the *Part* Workbench from the available list as presented in Figure 5.28a. As soon as selecting the *Part* Workbench, the toolbar of the available *primitive objects* and the toolbar for *modifying objects* appear on the FreeCAD environment, as highlighted by Figure 5.28b (with red and green frames, respectively).

Some useful buttons of FreeCAD are highlighted with black color. In detail, the black frame encloses buttons, which allow us to change the view of our models (e.g. top view,

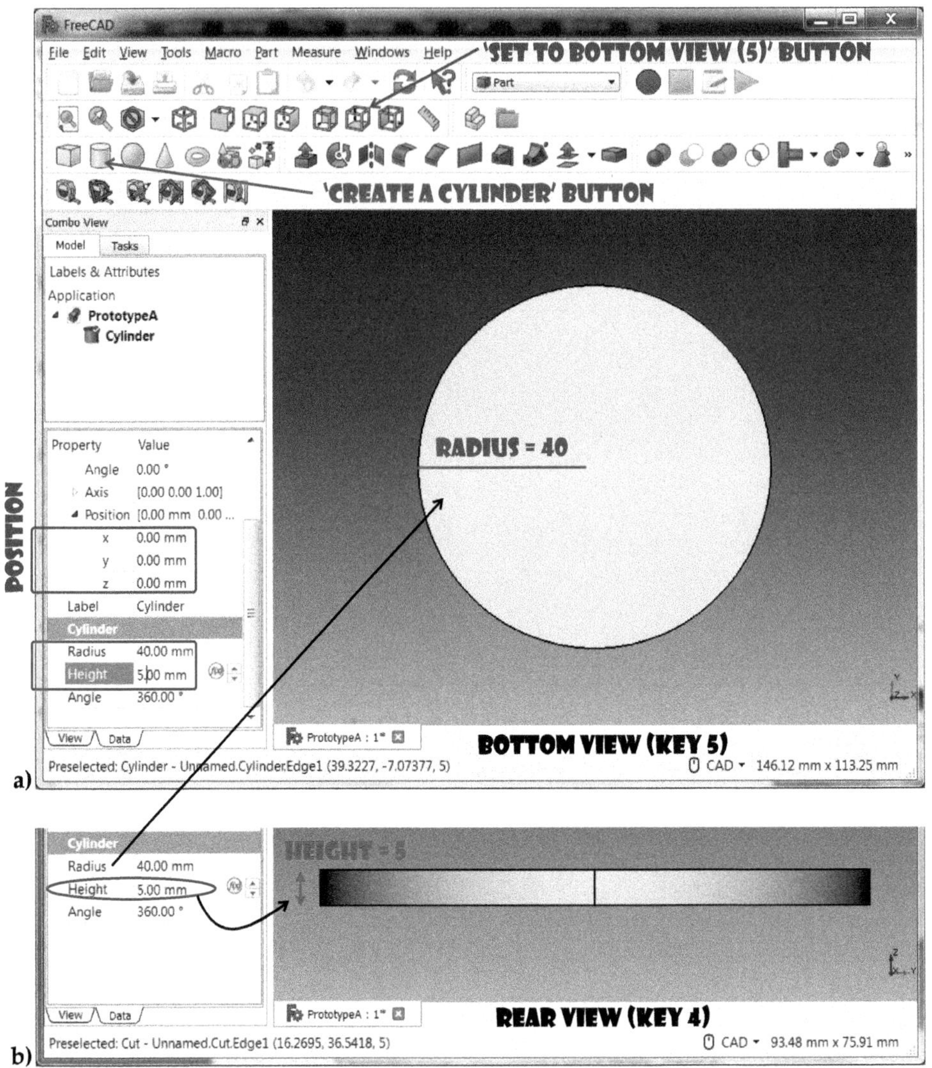

Figure 5.29 FreeCAD: modeling the *floor part* of the prototype interactive game (step 1). (a) *bottom view* activated (key 5), (b) *rear view* activated (key 4). *Source:* Arduino Software.

rear view, etc., which can also be invoked by the keyboard keys: 1 to 6), while the ruler tool (pointed by the black arrow) measures the distance between two user-defined points of a selected part. Next, we present the CSG modeling process of the *floor* 3D printed part used by the prototype interactive game (presented earlier in this chapter).

To get started the CSG modeling process of 3D printed *floor* of the enclosure (of the prototype interactive game), we initiate the design with a cylinder by pressing the *'Create a Cylinder'* button, as pointed by the red arrow of Figure 5.29a. Though the *Model* window in the left area of the figure, we are able to change the *radius (R)* and *height (H)* of the cylinder, which are herein defined 40 mm and 5 mm, respectively (as highlighted by the red frame). From the same window we may change the position (i.e. *X*, *Y*, and *Z*) of the 3D object on the working sheet. In Figure 5.29a, we have selected to observe the *bottom view* of the 3D object through the corresponding button pointed by the blue arrow (or by pressing number 5 on the keyboard). In order to explore the height of the cylinder, we may change to the *rear view* through the corresponding button (or by pressing number 4 on the keyboard), as illustrated in Figure 5.29b.

Having determined the radius and height of the cylinder we proceed to design of the holes of the *floor* part, through the six successive steps of Figure 5.30. In detail, in Figure 5.30a we design another cylinder of $R = 3.6$ mm and $H = 3$ mm. In Figure 5.30b, we change the *X* position of the cylinder (i.e. $X = 30$ mm). Because the holes of the *floor* part (explored herein) intend to "hide" the head of the screw, the present cylinder is slightly bigger than the head of the screw.

In regard to the body of the M3×80 mm screw that pass through the *floor* part, we design another smaller cylinder in the same position of $R = 1.6$ mm and $H = 2$ mm, as presented in Figure 5.30c. Then we change the *Z* position of the latter cylinder to $Z = 3$ mm, and hence, the outline of the smaller holes is not any more visible in the *bottom view* of the working sheet, as illustrated by Figure 5.30d. Yet, it becomes visible in the *top view* of the working sheet, as illustrated by Figure 5.30e. The design of the latter cylinder is repeated in Figure 5.30f, where this time the cylinder is moved to the $X = -30$ mm location.

The final cylinder (of $R = 3.6$ mm and $H = 3$ mm), which is meant for the head of the second M3×80 mm screw, is designed in Figure 5.31a and placed in the $X = -30$ mm (where, in this particular figure, we have changed again the *view* area of the working sheet to *bottom*). Using the CTRL key on the PC keyboard we select the four cylinders that are meant for the holes of the *floor* part, as presented by Figure 5.31b. Then the four cylinders are fused together (i.e. are all attached to a union) by clicking the *'Make a union of several shapes'* button, pointed by the red arrow of Figure 5.31c. Then, in Figure 5.31d we select the *main* cylinder of the *floor* part, and by holding the CTRL key on the PC keyboard, we subsequently select the *Fusion* of the four cylinders (meant for the *holes* of the part), as presented in Figure 5.31e. Having selected the *main* cylinder and the *Fusion* of the four holes, with that particular order, we click on the *'Make a cut of two shapes'* button, highlighted by the red arrow of Figure 5.31f.

We have now completed the final model of the *floor* part, which is presented in Figure 5.32. In detail, the *top* view of the model is depicted by Figure 5.32a, the *bottom* view is depicted by Figure 5.32b, while in Figure 5.32c we have rotated the object in order to reveal the cavity that is meant for the head of the screws. The rotation is achieved by clicking and holding pressed the middle button of the mouse, and then, clicking and holding pressed the left button.

It is worth mentioning that we could continue the modeling of the rest of 3D parts of the prototype interactive game, in the same application of the FreeCAD software (i.e. the *PrototypeA.FCStd*). In that case, it would be useful to rename the final *cut* of the current part to a more descriptive name, herein renamed to *floor*, as presented in Figure 5.33a. To do so, the user should press right click on the final *cut* and select the option *Rename*, and then give the desired name to that part, as presented in Figure 5.33a,b. Another useful option when modeling a 3D part is to change its appearance to *Transparency*, by selecting the part and then clicking on the menu *'View → Appearance'*. Through that particular option we may observe, in our design, the implementation of the holes of the *floor* part, as

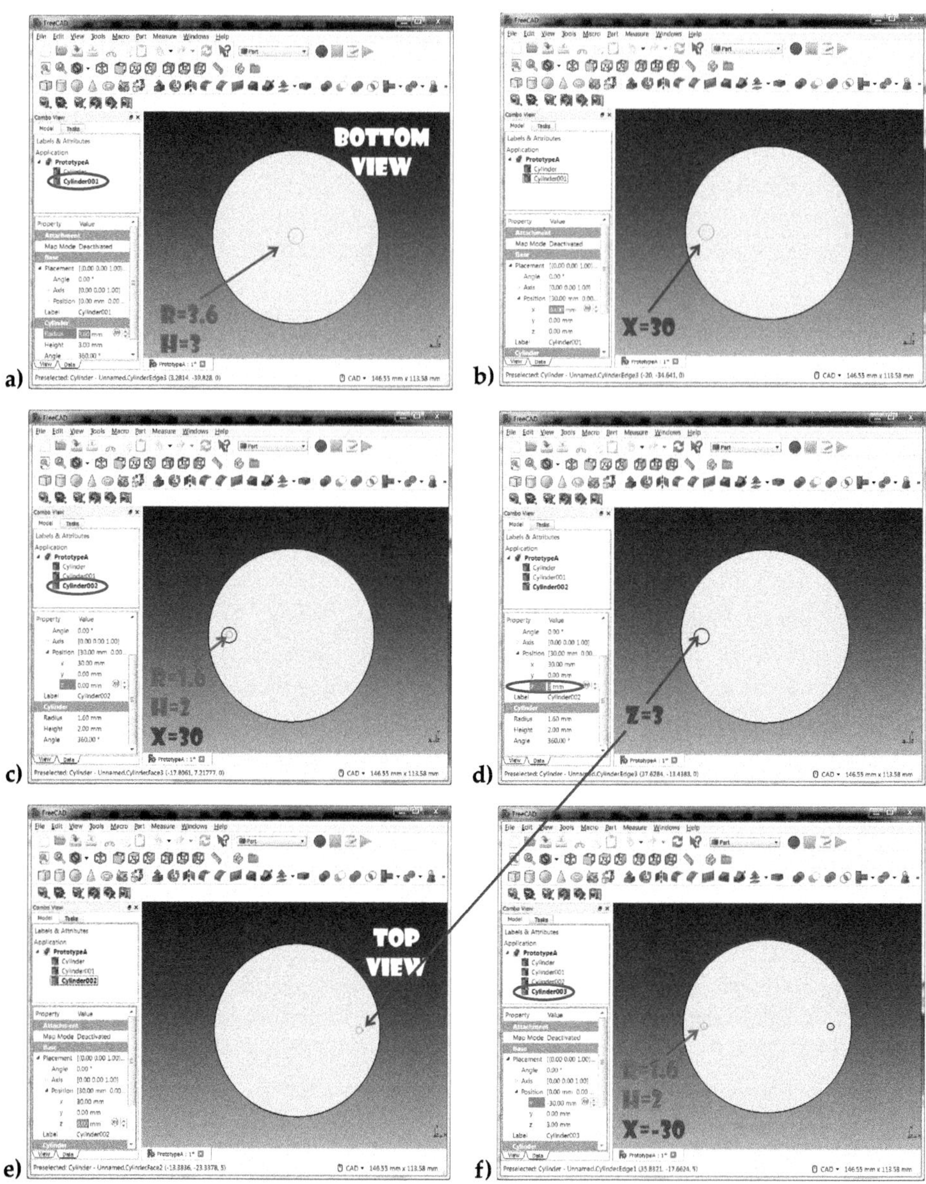

Figure 5.30 FreeCAD: modeling the *floor part* of the prototype interactive game (steps 2–7). (a) Design cylinder (R=3.6mm, H=3mm), (b) Change the X position of cylinder (X=30mm), (c) Design cylinder (R=1.6mm, H=2mm), (d) Change the Z position of the latter cylinder (Z=3mm), (e) *Top* view activated, (f) Repeat the latter design and change X position of the new cylinder (R=1.6mm, H=2mm, X=-30mm). *Source:* Arduino Software.

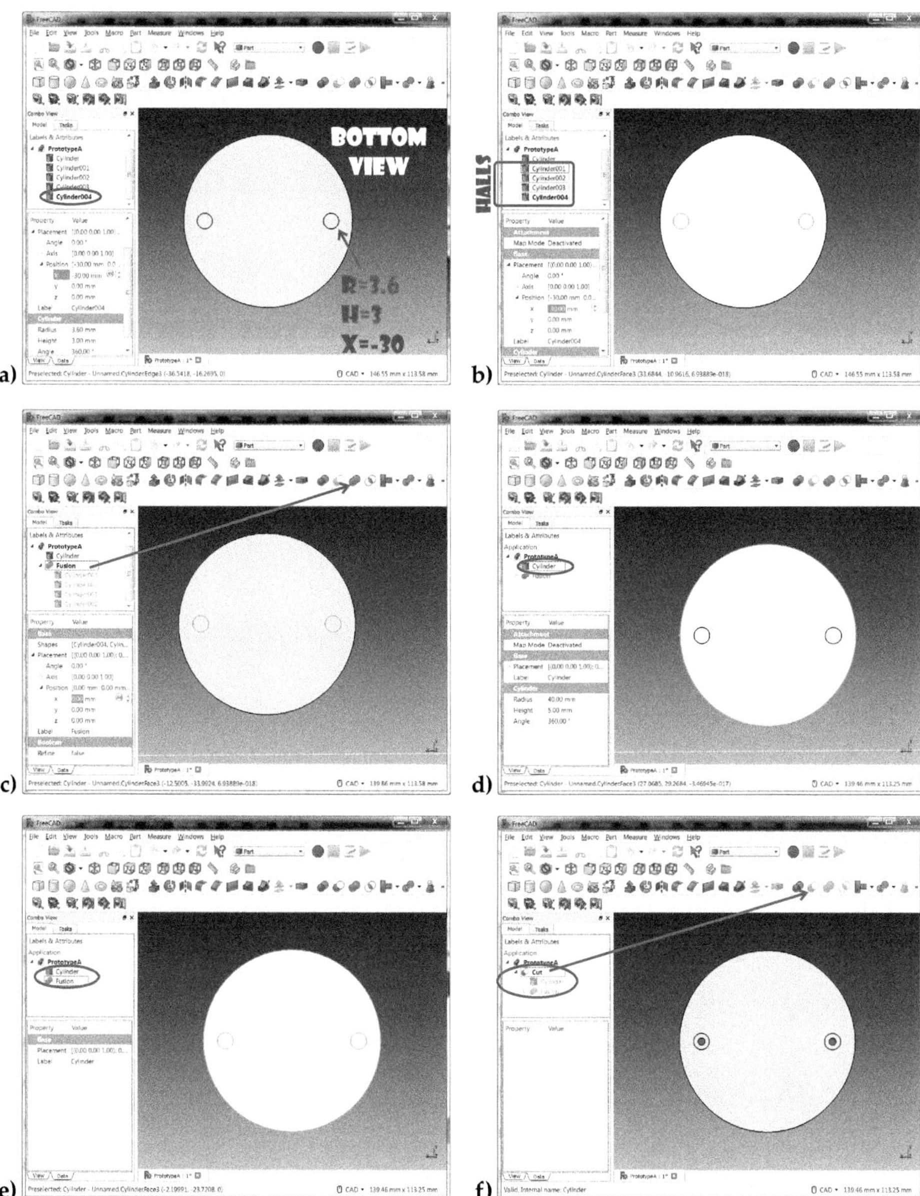

Figure 5.31 FreeCAD: modeling the *floor part* of the prototype interactive game (steps 8–13). (a) design cylinder RxH=3.6x3 mm and modify its position to X=-30mm, (b) select the four cylinders meant for the holes of the *floor* part, (c) fuse the selected parts to a union, (d) Select the main cylinder of the floor part, (e) select the fusion of the holes, (f) make a cut of two shapes to create the holes to the main cylinder. *Source:* Arduino Software.

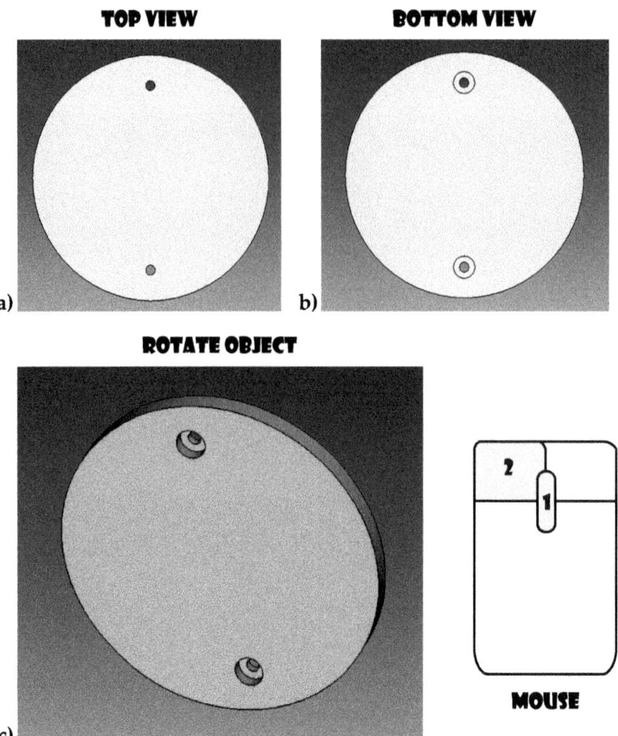

Figure 5.32 Top/bottom view and rotation of the *floor* part of the prototype interactive game. (a) top view, (b) bottom view, (c) rotation of the *floor* part by pressing together the highlighted buttons of the mouse.

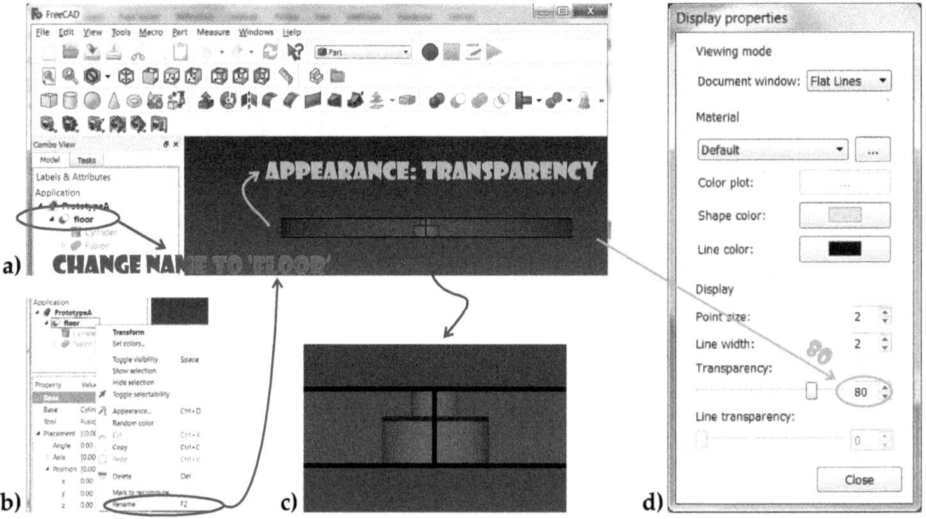

Figure 5.33 FreeCAD: rename a 3D part and change its appearance to *transparency*. (a) select the previous cut of the selected parts, (b) press right click on the mouse and choose rename, (c) modify appearance of the *floor* part to transparency, (d) set transparency to 80%. *Source:* Arduino Software.

presented in Figure 5.33c. To achieve this particular output, we have changed the corresponding number of *Transparency:* from 0 to 80, as presented in Figure 5.33d.

The 3D modeling process of the rest parts of the enclosure used by the prototype interactive game is based on the same design philosophy. All models are available on the WEB and can be found at the link: www.mikroct.com. We next explore how to reform the current 3D model (i.e. the *floor* part) to the appropriate format for the automated control of the 3D printer.

Preparing the 3D Prints with Ultimaker Cura Software

Before preparing the *floor* 3D model to the appropriate format for the automated control of the 3D printer, we need to export the current model to an output *.stl* file (usually denoting the abbreviation of *stereolithography*). This file format is commonly used by the 3D printing software packages so as to describe the geometry of a three-dimensional object. The generation of the *floor.stl* file via FreeCAD software is achieved from the menu *'File → Export'*. As soon as finishing the exporting process, the user may quickly explore the generated *.stl* using an online STL viewer, such as the www.viewstl.com presented in Figure 5.34.

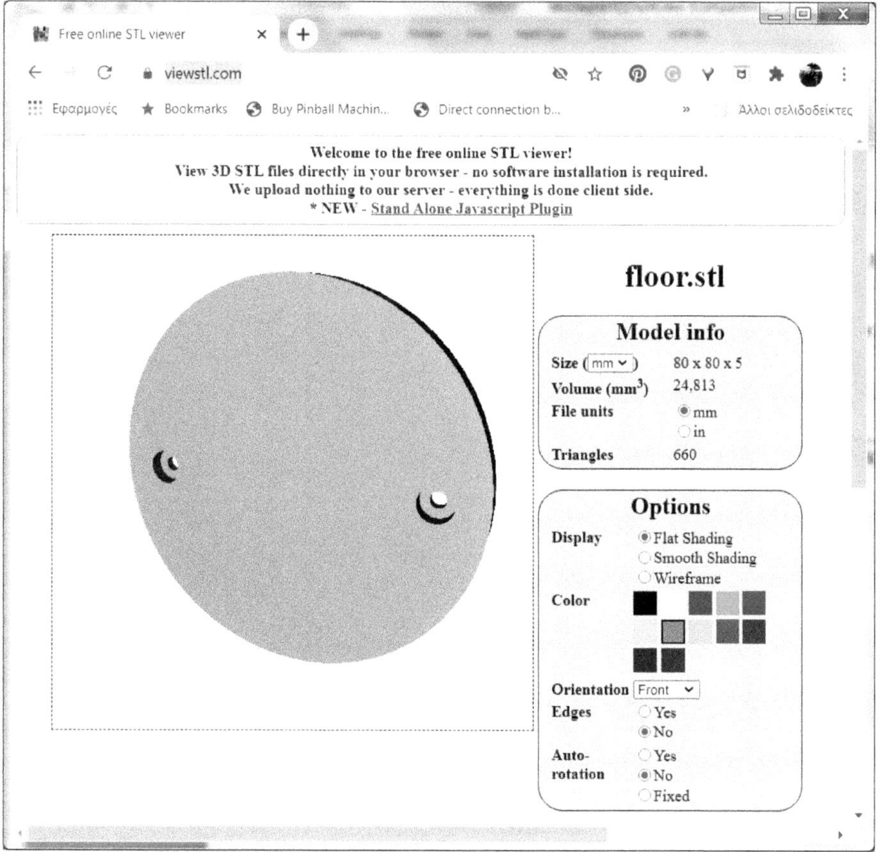

Figure 5.34 Free online STL viewer. *Source:* Arduino Software.

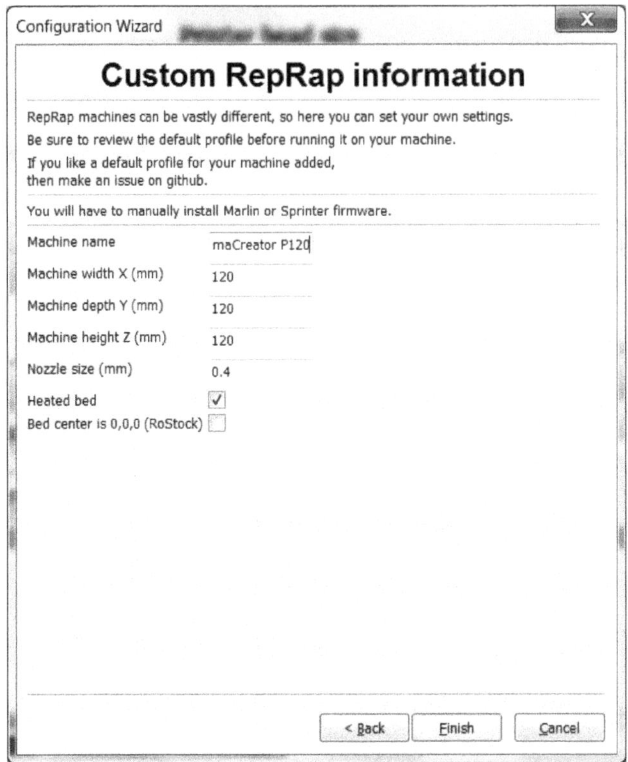

Figure 5.35 Cura: define settings for Prima Creator P120 3D printer. *Source:* Arduino Software.

Next the *floor.stl* file will be imported to the Cura software, and through particular configurations made by Cura, it will be converted into *.gcode* file format (which constitutes the CNC programming language for the automated control of the 3D printer). The Cura software can be downloaded for free from the website https://ultimaker.com/software/ultimaker-cura.

When opening Cura software for the first time, the user needs to add information about the model of the 3D printer that will be used for building 3D objects. In the version Cura 15.04.6 introduced herein, this information is accessed through the menu *'File → Machine settings'* and then clicking on the icon *Add new machine*. The user should then configure the following: (i) click *Next* to the *Add new machine wizard* pop-up window, (ii) select the option *Other* to the *Select your machine* window and click again *Next*, (iii) without selecting any of the available options, the user should click *Next* to the *Other machine information* window, and (iv) in the final pop-up window the user should specify the printer's settings and click the *Finish* button. The settings for the 3D printer used by the present example, that is, the Prima Creator P120, are illustrated in Figure 5.35.

The Prima Creator P120 has a print volume of $X{\cdot}Y{\cdot}Z = 120{\cdot}120{\cdot}120\,\text{mm}$ and, hence, the machine width, depth, and height are determined by those particular values. Prima Creator uses the default nozzle diameter for most of today's 3D printers, that is, the 0.4 mm. Finally, the P120 employs a heated bed (thereby it can use a wider range of filaments) and that particular information is declared below the nozzle diameter.

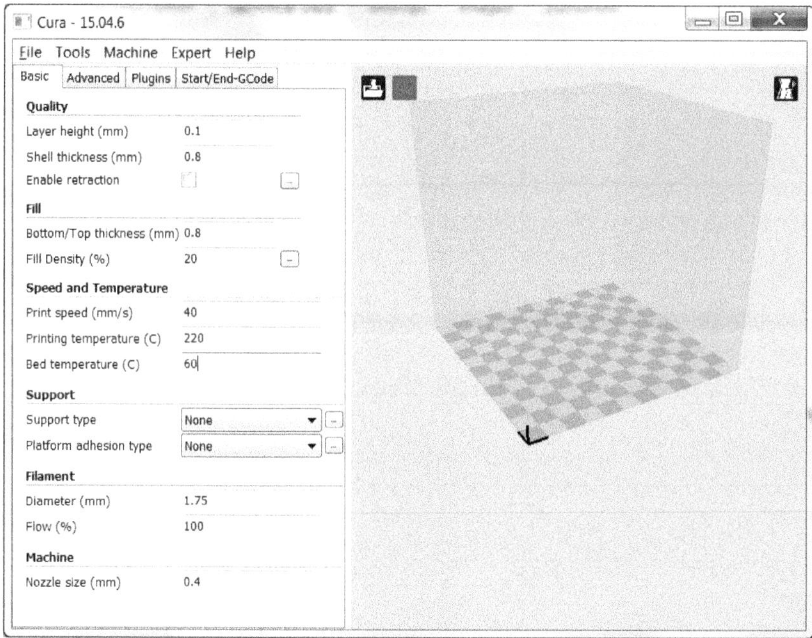

Figure 5.36 Cura: define settings for setting of the desired 3D printing procedure. *Source:* Arduino Software.

Having added the model of our 3D printer we return to the main screen of Cura software, as presented in Figure 5.36, where we define some additional information about the 3D printing procedure. In detail, the *Layer height (mm),* which determines the quality of prints, is set to 0.1 (as it constitutes the layer resolution of the Prima Creator P120). Lower values of *Layer height (mm)* determine better quality of prints but they cannot be achieved by the current printer device. Next, the *Shell thickness (mm),* which is used in combination with the nozzle size so as to define the thickness of the outside shell (i.e. the number of perimeter lines), is set to 0.8. As a rule of thumb, the shells are a multiple of the nozzle diameter and regularly are set to 2 nozzle diameter (herein, $2 \cdot 0.4 = 0.8\,\text{mm}$).

The *Bottom/Top thickness (mm)*, which is herein set to 0.8, is usually expressed as a multiple of the *layer height*. A higher value of the *Bottom/Top thickness* ensures that all gaps on the top and bottom layers are closed, yet, it can increase the print time and amount of utilized filament. The *Fill Density (%)*, which is set to 20 (usually enough value), controls how densely are filled the insides of the printed part (i.e. 100% for a solid part and 0% for an empty part), while it does not affect the outside of the print.

The *Speed and Temperature* values are determined in reference to the technical data of Creator P120 printer, where the maximum *Print speed (mm/s)* and *Printing temperature (C)* as defined by the technical data of the printer, are $55\,\text{mm}\,\text{s}^{-1}$ and 250 °C, respectively (we herein use $40\,\text{mm}\,\text{s}^{-1}$ and 220 °C, instead). Yet, we have determined the maximum *Bed temperature (C)* of the printer (that is, 60 °C).

None 3D printing support structure is used in the current settings. In general, the support structures are not part of the printed part, and if selected, the user should remove the support structures as soon as the printing process ends. Such an option would be useful when printing more complicated models than simple enclosures. The filament diameter for the current printer is 1.75 mm, while the amount of material extruded is set to 100%. Finally, the nozzle size of the current printer is (as mentioned earlier) 0.4 mm.

After the essential configurations of the printing process, the *floor.stl* file can be loaded to the Cure software via the menu *'File → Load model file'*, as presented in Figure 5.37a. At the bottom of the working sheet we may find three useful buttons (i.e. the rotate, scale, and mirror), in order to further configure the 3D printing process of the part. In Figure 5.37b, we have rotated the part 180° so as the bigger holes (which are meant for the head of the screws) to be printed after the smaller holes. In general, it is advised to place the 3D part in the most convenient position that would assist the printing process. For instance, if we rotate the current part 90° left or right, then the printing process would be trickier (and would normally need printing support structures).

The final step after the aforementioned configurations is to generate the CNC programming language for the automated control of the 3D printer; that is, the *floor.gcode* file. This task is performed through the menu *'File → Save GCode'*. The generated *floor.gcode* file may now be loaded to the 3D printer in order to initiate the printing process of the *floor* part of the enclosure.

3D Printing with Prima Creator P120

We herein introduce the Prima Creator P120 3D printer, which has been used for printing the enclosure of the prototype interactive game. The 3D printer is presented in Figure 5.38. As mentioned earlier, the P120 incorporates a heated build plate (i.e. heated bed) able to print materials of maximum dimensions 120·120·120 mm. The heated-bed feature provides flexibility on the possible filaments used for producing the 3D parts. Hence, despite the commonly used polylactic acid (PLA) filament, which can be printed cold, the P120 printer accepts other filaments, such as ABS, HIPS, polyethylene terephthalate glycol (PETG), and similar materials.

In the *fused deposition modeling* (*FDM*) 3D printers (as the P120 used by this chapter), a filament material is fed into an extruder. The extruder heats and melts the material and, then, it pushes it out through a nozzle. As the extruder moves in *X,Y,Z* axes, via directions obtained from a *.gcode* file, it produces the 3D printed part.

The most regular filament for 3D printers is the *PLA,* which requires no heated plate. Yet, in this example make use of a *PETG* filament, delivered by Prima Creator. The main purpose for selecting PETG relies on the fact that it is easiest to achieve transparency. In the current project, transparency is required because the children's *sight* is triggered by particular colors arisen from the electronics found inside the enclosure of the interactive game. However, the PETG material has other improved features over PLA, such as the higher physical strength. The selected PETG filament requires print temperature 210–235 °C and bed temperature 60 °C. It is worth mentioning that 3D printers of no heated plate cannot work with PETG and similar thermoplastic materials, as the filament cannot be adhered to the printer's bed.

The adhesion of the filament on the printer's bed is very important toward creating successful prints. For that reason, we hereafter explore a couple of tips that should be followed

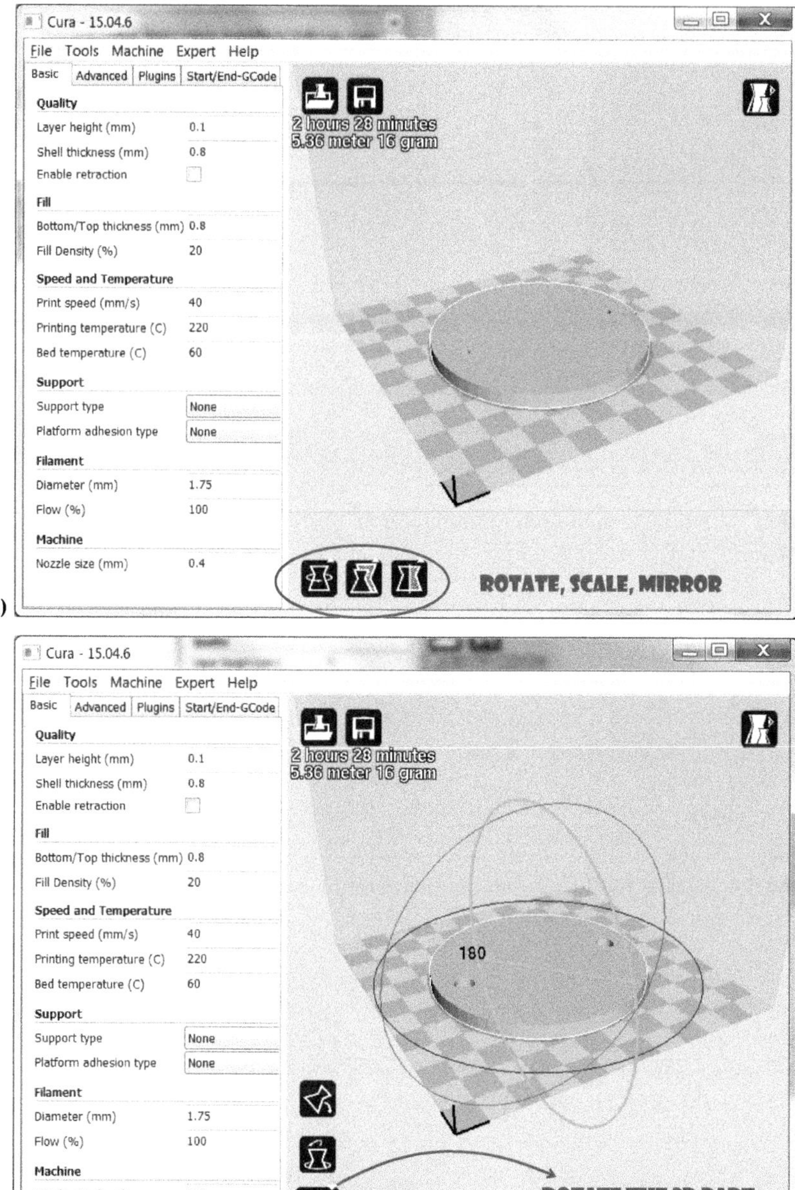

Figure 5.37 Cura: load the *.stl* file. (a) Cura software buttons for *rotating*, *scaling*, and *mirroring* the 3D model, (b) rotate 180° the 3D model. *Source:* Arduino Software.

before the printing process. The first is to place an adhesive tape above the printer's bed. The Prima Creator P120 is delivered with a *polyetherimide* (*PEI*) sheet on the build plate for better adhesion. Another solution is to remove the PEI sheet (which could be damaged after a few prints as we try to remove the printed part from the printer's bed) and replace it with a

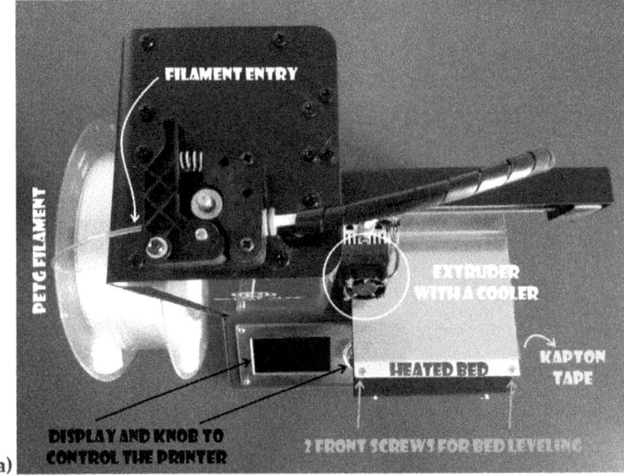

Figure 5.38 The *Prima Creator P120* 3D printer. (a) front view, (b) rear view.

Kapton tape, as the one presented in Figure 5.38b. The Kapton tape constitutes a polyester film tape with a high-temperature–resistant adhesive, which also provides an easy release of the printed part from the build plate. Hence, it protects the printer's bed as well.

Another important tip for creating successful prints is to occasionally level the printer's bed. In P120 printer the bed leveling is performed through the two screws presented in Figure 5.38a, as well as the two screws presented in Figure 5.38b (using the Allen tool that is included in the package of P120 printer). Figure 5.39a, b illustrate the menu for moving manually the extruder in the X,Y,Z axes. The process for leveling the printer's bed is as follows:

a) Move the extruder close to the one of the four screws, relative to the printing boundaries of the printer;
b) Place the nozzle one click above the building plate and insert a regular sheet of paper (e.g. a single A4 page of 80 g);
c) Move the nozzle one click down on the Z axis so as to touch the bed of the printer;
d) Turn counterclockwise/clockwise the corresponding screw for moving the bed up/down in the Z axis, until the paper sheet is pulled out and off the bed with some kind of strength, but without being damaged (as illustrated by Figure 5.39c);

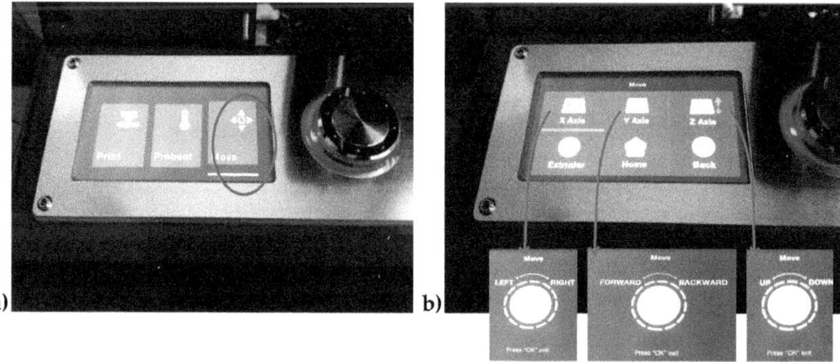

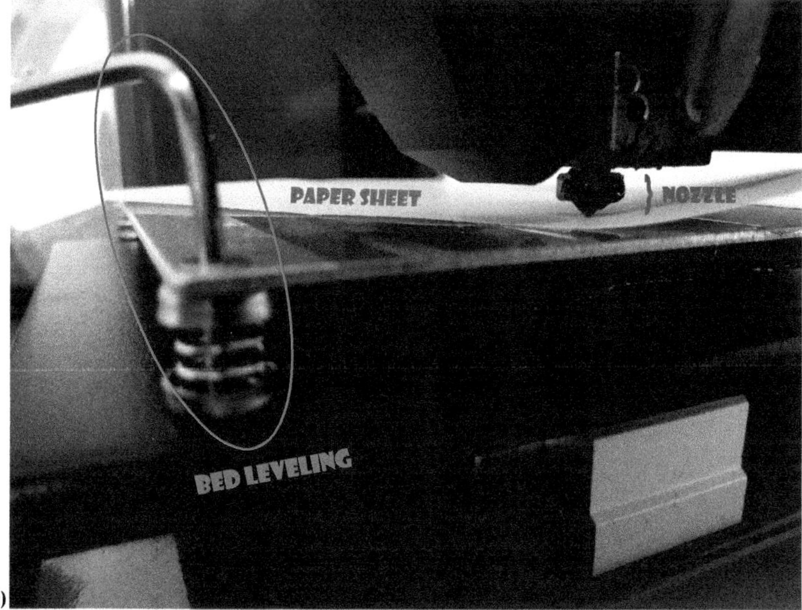

Figure 5.39 Bed leveling of the *Prima Creator P120* 3D printer. (a) menu for moving manually the bed and extruder of the 3D printer, (b) menu for moving the bed left/write, forward/backward, and the extruder up/down, (c) bed leveling using a paper sheet.

e) As soon as finishing the bed leveling in that particular corner, place the nozzle one click above the building plate and direct it to another corner;
f) Repeat the process with the screw found in the new corner and repeat the overall process, i.e. steps from (a) to (d), for all four corners of the printer's bed (this way the nozzle will have equal distance from the build plate at all points and prints will not fail).

After placing a Kapton tape (or similar material) on the printer's build plate and performing bed leveling, the P120 device is ready to begin the printing process.

Next we present the 3D printing process of the *floor.gcode* file with Prima Creator P120 3D printer. To initiate the procedure, we first need to load into a microSD card the *.gcode* file, and then, insert the card into the corresponding socket of the 3D printer (depicted by Figure 5.38b). Next we prepare the P120 printer through the round knob and via the menu appearing on the printer's display.

In Figure 5.40a, we select (by pressing the printer's round knob) the *Preheat* option so as to prepare the desired nozzle as well as bed temperature. By clicking on the *Nozzle* and *Bed* options of Figure 5.40b we set the corresponding temperature. Then we select the *Start Preheat* option to force the printer to preheat the nozzle and bed before the printing process. In Figure 5.40c, the P120 printers increases the nozzle's temperature up to the 220 °C and the bed's temperature up to 60 °C. As soon as the temperature reaches the desired value in both elements, we select the *Back* option. In Figure 5.40d, we select *Print* and in Figure 5.40e, we select the *floor.gcode* file traced inside the microSD card. In Figure 5.40f, the 3D printer initiates the process by printing a bullet in the lower left corner of the building plate, so as to remove the filaments that remain from the nozzle. Then the P120 start printing the *floor* part in Figure 5.40g, while Figure 5.40h illustrates the progress of the printed part in approximately 25% of the overall procedure.

It is worth noting how densely filled the insides of the printed part are, in about 50% of the overall printing process, as depicted by Figure 5.41. This particular result arises from the fact that we have previously set 20% *Fill Density* in Cura software.

Nowadays, 3D printers can be purchased at a surprising small cost and render a particularly useful tool for the hobbyists who intend to build their own μC-based prototypes. Despite building a custom-designed enclosure, the reader can find a plethora of free 3D models, over the internet, for miscellaneous applications. It is worth making a reference to the website www.thingiverse.com where the reader can download digital designs and build them with his/her 3D printer.

Presentation of the Rest 3D Models of the Prototype Interactive Game

Hereafter, we present the rest of the 3D parts of the prototype game, the design of which is based on the CSG modeling technique (as in the previous example of the *floor* model). To print those parts, the user should follow the procedure explored before, that is, (i) to export the 3D models into *.stl* file format via FreeCAD, (ii) to insert each *.stl* file into Cura software and after the necessary configuration, to generate the *.gcode* file format for the printer, and (iii) to prepare the printer via its menu and possibly place Kapton tape on the printer's building plate, as well as perform bed leveling, if needed.

PrototypeB (Modeling the battery.stl Part)

Figures 5.42–5.44 illustrate the partial steps toward the design of the 3D part, which is addressed for holding the battery of the system. Having set the FreeCAD to bottom view we place, in Figure 5.42a, a cylinder of $R = 32$ mm and $H = 2$ mm. In Figure 5.42b, we insert another cylinder of same height but with $R = 27$ mm. We then apply a *cut* to the homocentric cylinders to create the ring depicted in Figure 5.42c (which is given the indicative name *Cut_RING*). In Figure 5.42d, e, we successively place two homocentric cylinders at the left side (i.e. at $X = 30$ mm location) of the ring. The same process is repeated for the generation of two homocentric cylinders at the right side of the ring (i.e. at $X = -30$ mm), as presented by Figure 5.42f. Each pair of the homocentric cylinders is addressed for a hole of the screw,

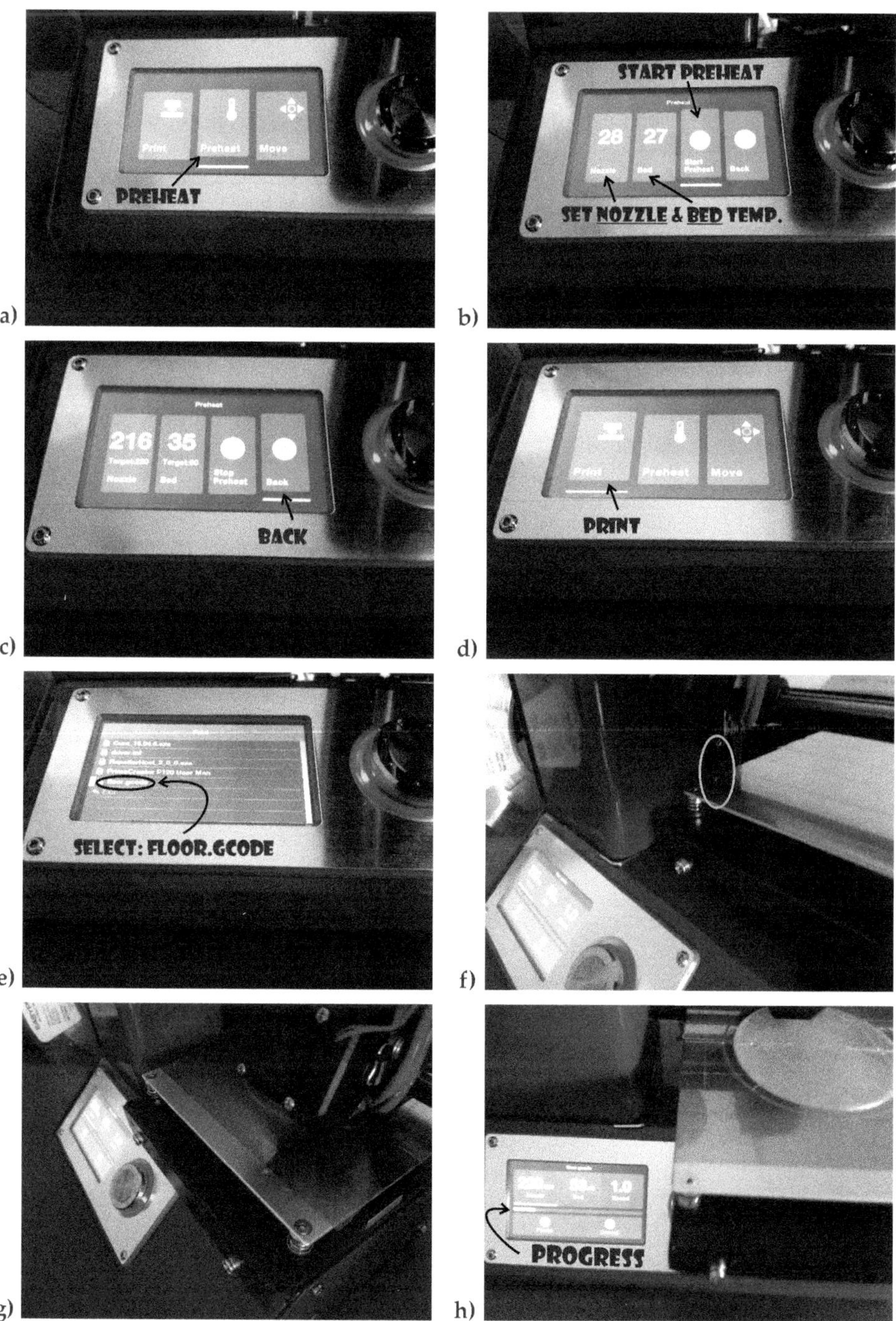

Figure 5.40 Prepare the printing process in *Prima Creator P120*. (a) preheat menu, (b) set nozzle and bed temperature, (c) return to the previous view, (d) print menu, (e) select the .gcode file to be 3D printed, (f) the 3D printer leaves a bullet of filament in the lower left corner so as to remove the remains of filament extruded during the preheat process, (g) the printer starts the printing process of the 3D model, (h) menu illustrating the overall 3D printing progress.

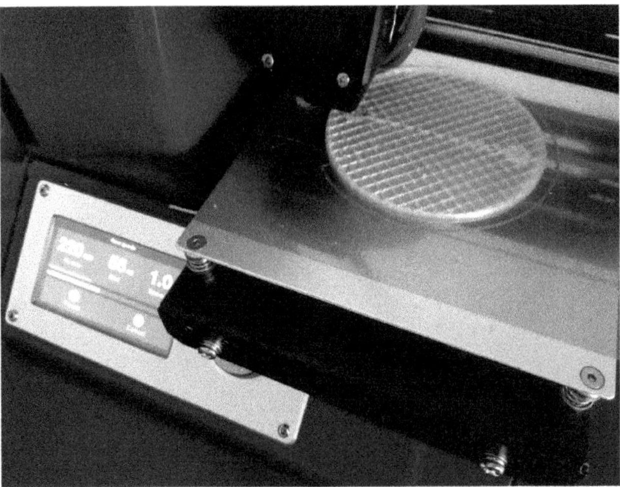

Figure 5.41 Fill density of the printed part (i.e. 20%).

which holds the overall apparatus, but also for a spacer to leave the necessary distance between the adjacent parts. That is why the two pairs feature $H = 8.75\,\text{mm}$. In Figure 5.42g, we generate a cut on the pair of cylinders on the right side of the ring, and hence, we generate the hole/spacer illustrated in the right side of the figure (i.e. highlighted by the red frame). The same process is repeated for the pair of cylinders on the left side of the ring, in Figure 5.42h (where the holes/spacers have been named *Cut_SPACER1* and *Cut_SPACER2*).

The aforementioned holes/spacers feature a hole up to the surface of the ring (i.e. only within the spacer). To create the hole within the ring, as well, we place two more cylinders at the position depicted in Figure 5.43a. Having applied a *union* of those two cylinders, we cut them from the ring so as to conclude the hole for the screw, as illustrated in Figure 5.43b. The overall design is grouped together on the *Fusion* named *Ring,Holes,Spacers,* which is depicted (in axonometric view) in Figure 5.43c. Then in Figure 5.43d, we utilize the *'Create a cube solid'* button to create the cuboid of *length (L)* equal to 40 mm, *width (W)* equal to 55 mm and $H = 2\,\text{mm}$. To cut the four angles of the cuboid that extend beyond the ring, we generate four small cubes, which are placed in the four corners of the cuboid, as illustrated in Figure 5.43e, f, and cut them from the cuboid, as illustrated in Figure 5.43g.

In Figure 5.43h, we create one of the four cuboids (named *Cube_Wall1*) indented to prevent the battery from sliding *left and right* or *back and forth*. The design of the three walls are presented in Figure 5.44a–c, while an axonometric view of the four walls grouped together, is illustrated in Figure 5.44d. The three separated models, i.e. the (i) *Fusion_Walls* of Figure 5.44d, (ii) *Cut_Cuboid* of Figure 5.44e, and (iii) *Fusion_Ring,Holes,Spacers* of Figure 5.44f, are grouped together to generate the final model of Figure 5.44g, named *Battery*. The latter fusion should be exported into *.stl* file, such as the *battery.stl* depicted by the online *.stl* viewer of Figure 5.44h.

PrototypeC (Modeling the booster.stl Part)

The *battery.stl* model features only few differences from the *booster.stl* part, presented herein. The latter part is meant for holding the PowerBoost 1000 charger. For that reason, we save the previous FreeCAD project as *PrototypeC.FCStd* and perform the following

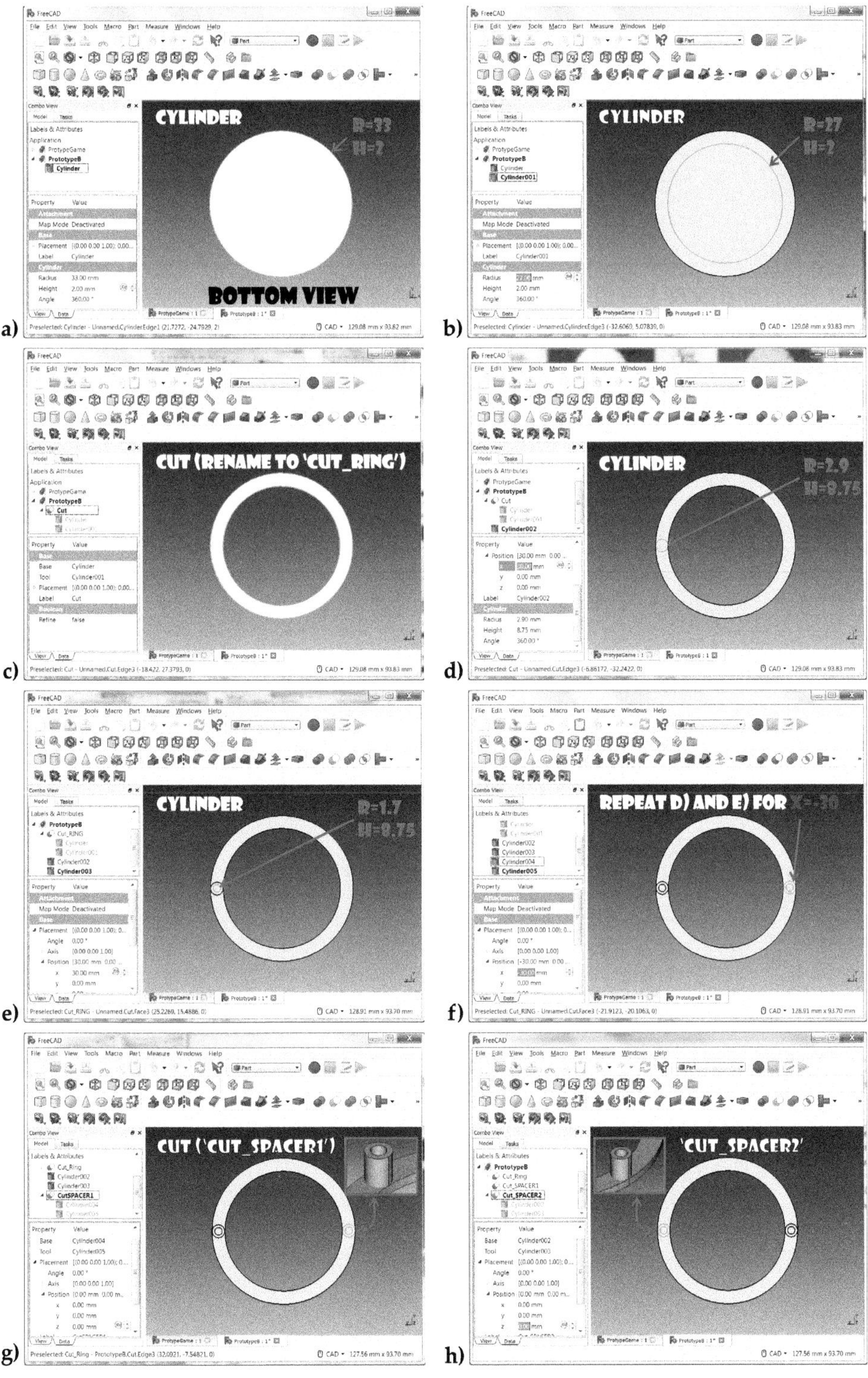

Figure 5.42 FreeCAD: modeling the *Battery.stl* part of the prototype game (1 of 3). (a) place cylinder RxH=32x2 mm, (b) place cylinder RxH=27x2 mm, (c) apply a cut to the homocentric cylinders (Cut_RING), (d) place cylinder RxH=2.9x8.75 mm @X=30mm, (e) place cylinder RxH=1.7x8.75 mm @X=30mm, (f) repeat the latter two steps to create the same homocentric cylinders @X=-30mm, (g) generate a cut on the pair of cylinders on the right side of the ring to generate hole and spacer (Cut_SPACER1), (h) repeat the cut on the left side of the ring to generate the second hole and spacer (Cut_SPACER2). *Source:* Arduino Software.

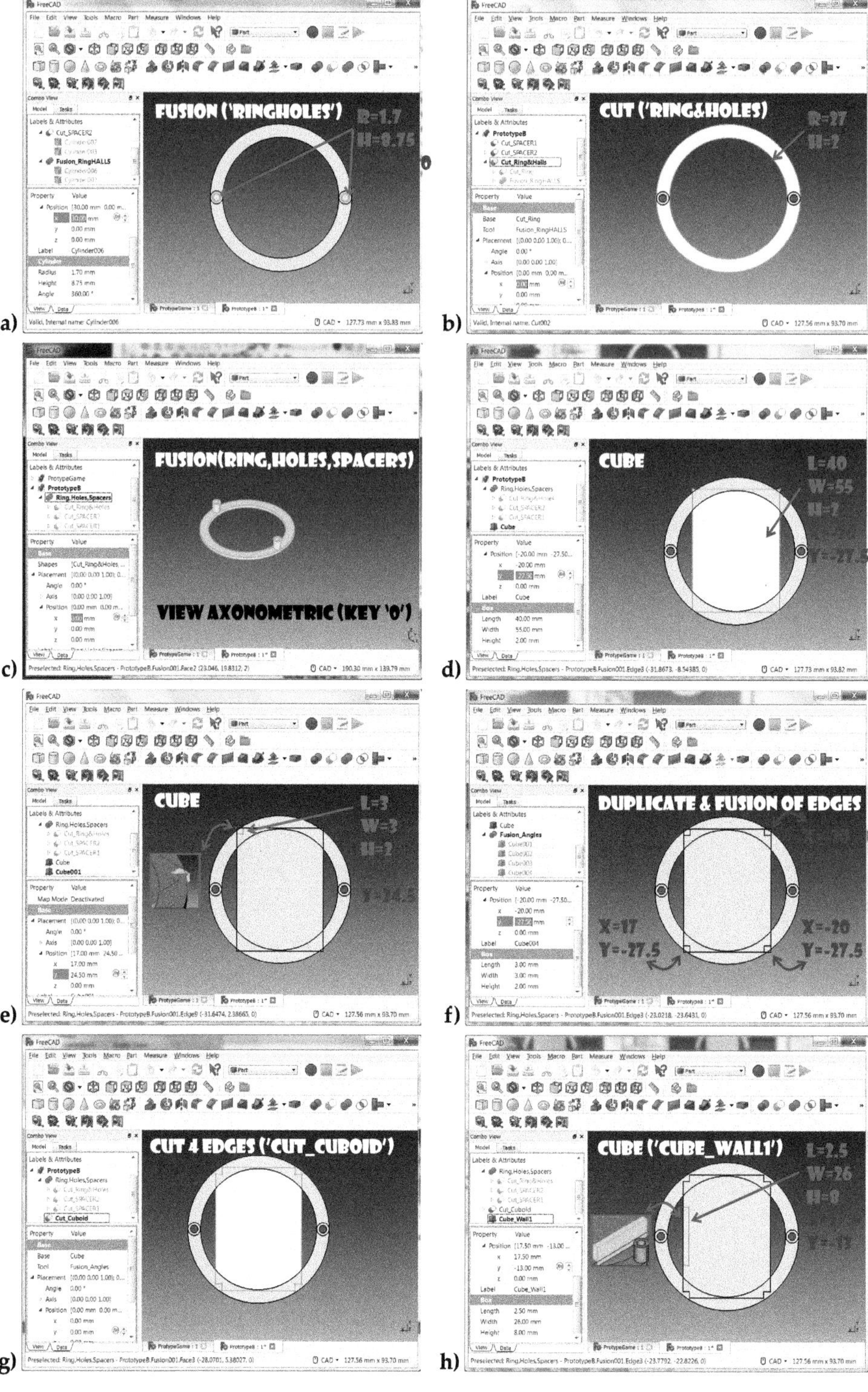

Figure 5.43 FreeCAD: modeling the *Battery.stl* part of the prototype game (2 of 3). (a) place two more cylinders RxH=1.7x8.75 mm @X=30mm and X=-30mm, (b) apply a cut from the ring to finish the holes for the two screws, (c) the overall design is grouped together to a fusion named Ring,Holes,Spacers, (d) create a cuboid of LxWxH=40x55x2 mm @X=-20/Y=-27.5 mm, (e) create a cuboid of LxWxH=3x3x2 mm @X=17/Y=24.5 mm, (f) repeat the latter creation three successive times and place the cubes @X2/Y2=-20/24.5 mm, @X3/Y3=-20/-27.5 mm and @X4/Y4=17/-27.5 mm, (g) cut the four small cubes from the large cuboid (i.e. cut edges of the cuboid), (h) create a cuboid of LxWxH=2.5x26x8 mm @X=17.5/Y=-13 mm. *Source:* Arduino Software.

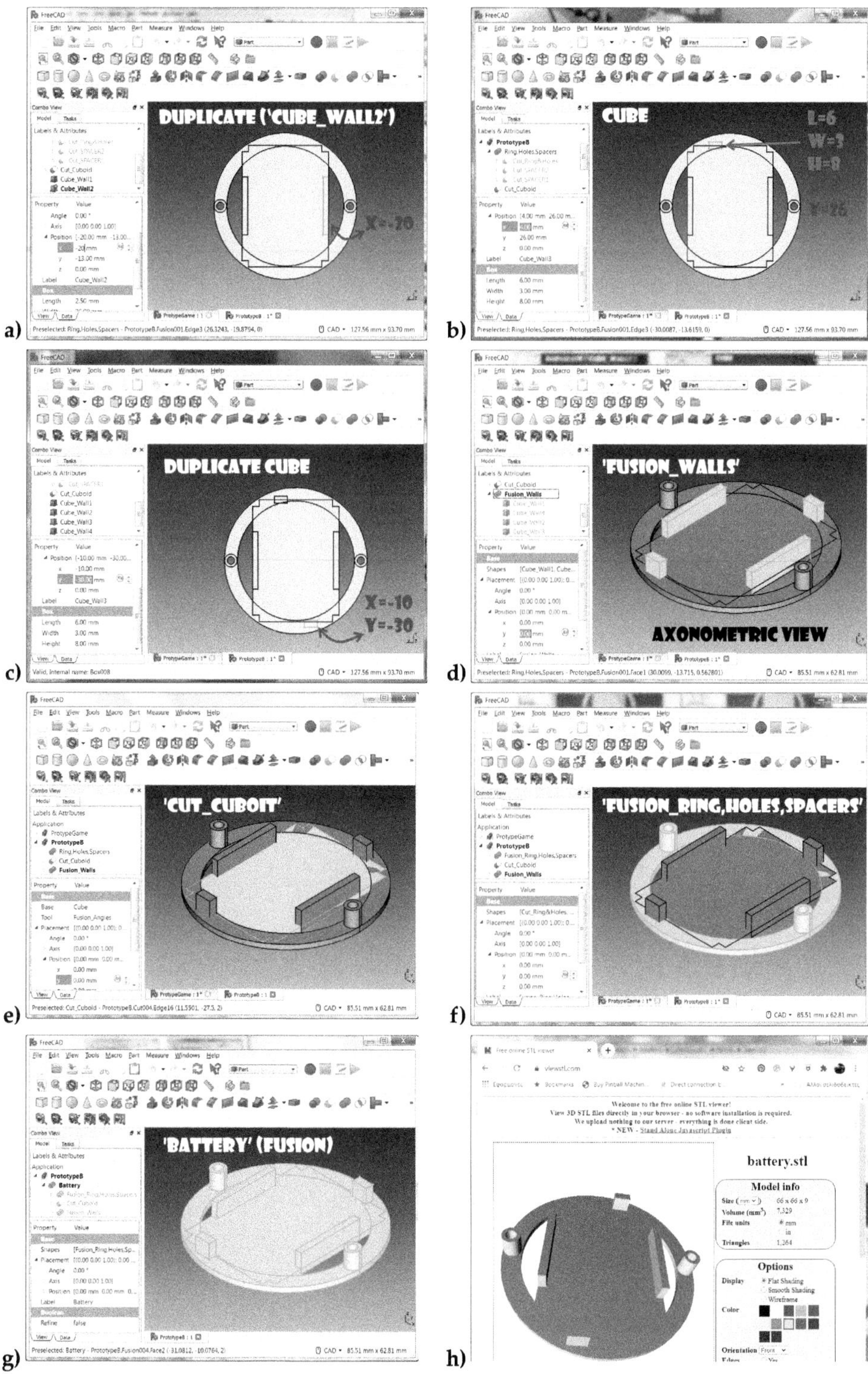

Figure 5.44 FreeCAD: modeling the *Battery.stl* part of the prototype game (3 of 3). (a) duplicate the latter cuboid @X=-20mm location, (b) create a cuboid of LxWxH=6x3x8 mm @X=4/Y=-26 mm, (c) duplicate the latter cuboid @X=-10/Y=-30 mm, (d) create a fusion of the last four cuboids (axonometric view activated), (e) the overall parts of the design are fused together and export the 3D model, (f) the exported file (*battery.stl*) is depicted by an online *.stl viewer* application. *Source:* Arduino Software.

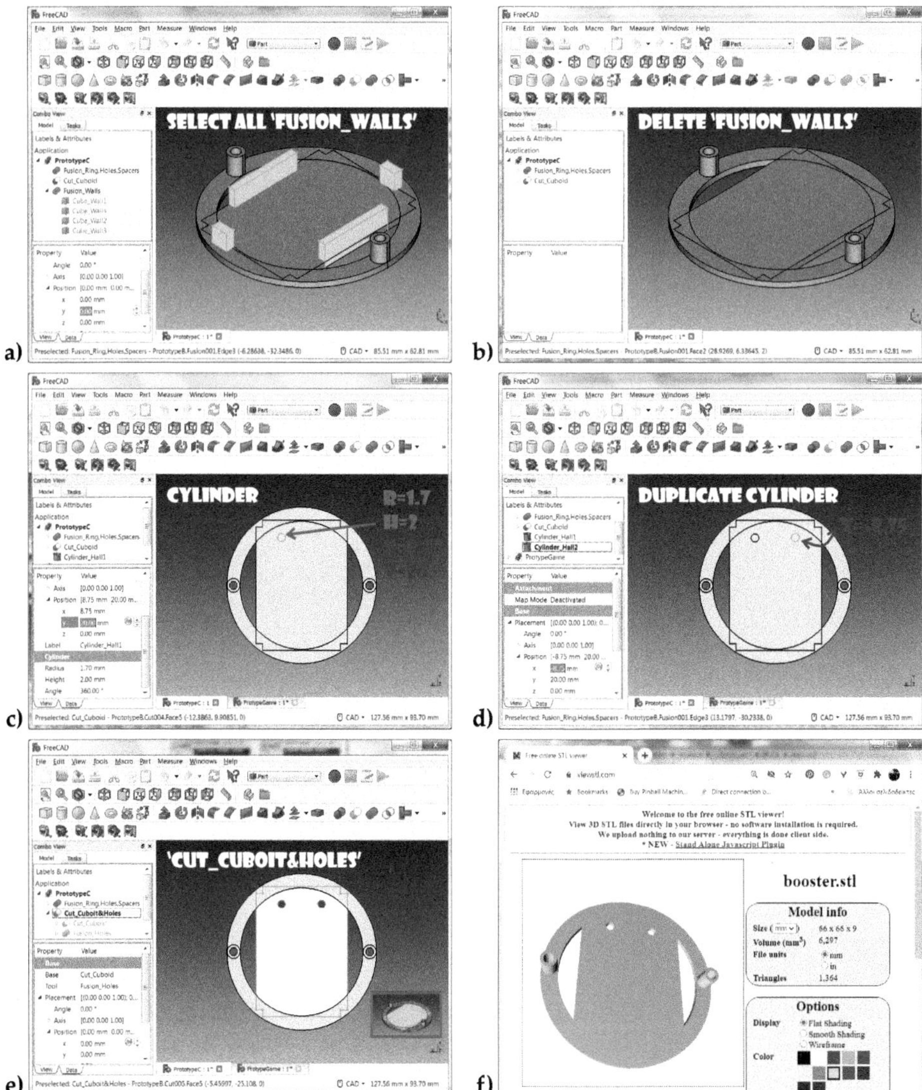

Figure 5.45 FreeCAD: modeling the *Booster.stl* part of the prototype game (1 of 1). (a) the previous design is duplicated and its four walls are selected, (b) the four walls are deleted from the current design, (c) place cylinder RxH=1.7x2 mm @X=8.75/Y=20 mm, (d) duplicate the latter cylinder @X=−8.75, (e) cut the latter two cylinders from the cuboid (to generate two holes) and export the design, (f) the exported file (*booster.stl*) is depicted by an online *.stl viewer* application. *Source:* Arduino Software.

adjustments. In Figure 5.45a, we select all the components incorporated by the *Fusion_Walls* design and in Figure 5.45b, we delete them. We, then, successively place two cylinders in the position illustrated by Figure 5.45c, d, respectively, while the cylinders are grouped together in Figure 5.45e. Then we cut from the cuboid the aforementioned *Fusion* of cylinders so as to generate holes where the PowerBoost 1000 module is screwed. The overall design is exported as *booster.stl* and the latter illustrated in Figure 5.45f.

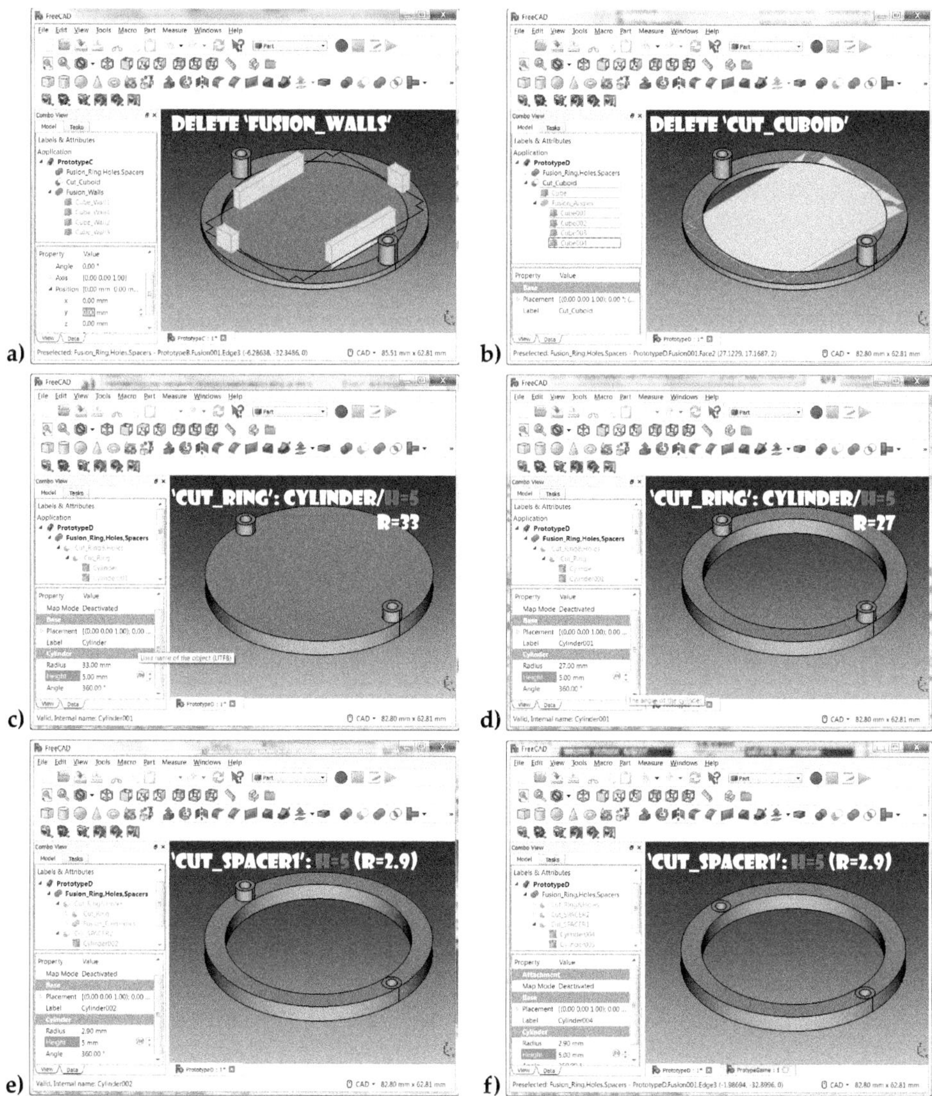

Figure 5.46 FreeCAD: modeling the *Speaker.stl* part of the prototype game (1 of 2). (a) the previous design is duplicated and its four walls are selected, (b) the four walls are deleted from the current design and the cuboid is selected, (c) delete the cuboid and increase the outer cylinder of the ring (H=5mm), (d) increase also the height of inner cylinder of the ring (H=5mm), (e) decrease the height of the spacer on the right bottom corner (H=5mm), (f) decrease the height of the spacer on the left top corner (H=5mm). *Source:* Arduino Software.

PrototypeD (Modeling the speaker.stl Part)

To create the *speaker.stl* we repeat the modification of the *battery.stl* model, as some designs are repeated. In detail, we delete the battery's walls as well as the cuboid at which the battery is touched, as presented by Figure 5.46a, b, respectively. In Figure 5.46c, we increase the height (i.e. $H = 5$ mm) of the outer cylinder of the ring, which causes the disappearance of the inner hole. However, in Figure 5.46d we also increase the inner cylinder to $H = 5$ mm and, hence, the hole of the ring is revealed. In Figure 5.46e, f, we decrease the height of the two *spacers* to a value equal to the height of the ring (i.e. 5 mm).

In Figure 5.47a, we delete the union (i.e. *Fusion*) of the sub-designs in order to straightforwardly modify the latter models. Accordingly, in Figure 5.47b we increase the radius of the two *spacers* to $R = 3.5$ mm. In Figure 5.47c, we first move each *spacer* 7.5 mm closer to the ring's center, and then, we create a cuboid of length, width, and height equal to $L = 9$ mm, $W = 7$ mm, and $H = 5$ mm. The cuboid is transferred to the position $X = 22.5$ mm and $Y = -3.5$ mm, so that the right edge of the cuboid reaches the middle of the left *spacer*. In Figure 5.47d, we duplicate the inner cylinder of the left *spacer* and move the duplicated part to the position $X = 22.5$ mm (i.e. the same position where the cuboid was previously transferred). We then generate a *cut* of the hole from the cuboid and rename it "Cut_Cuboid2." Then in Figure 5.47e we duplicate "Cut_Cuboid2," rename the duplicated part as "Cut_Cuboid1," and change its angle to 180°.

In Figure 5.47f, we first make a union of each cuboid with each spacer (e.g. "Cut_Cuboid1" with "Cut_SPACER1") so as to prepare the so-called *bearing* (e.g. "Fusion_Bearing1"). Then the two bearings are grouped together. The latter *Fusion* is used to hold (via two screws) the module of the speaker, while its angle[5] has been modified to 45°). The bearings as well as the ring are also grouped together to shape the *speaker* 3D model. In Figure 5–47g, we make the final corrections to the position (X,Y) of the two bearings, and finally, we generate the *speaker.stl* model, which is presented in Figure 5.47h.

PrototypeE (Modeling the cover.stl Part)

For the *cover.stl* part, we initiate the design from two cylinders of height 38 mm and radius 40 mm for the outer, as well as 34 mm for the inner cylinder, as presented by Figure 5.48a. Then we perform a *cut* of the two homocentric cylinders to generate the ring of Figure 5.48b. Afterward, we copy[6] from *PrototpeD.FCStd* (FreeCAD) project the "Fusion_Bearings" sub-part and we paste it in the current project, as illustrated in Figure 5.48c. To adjust the *bearings*, we change the FreeCAD's view to *bottom*, as in Figure 5.48d. In Figure 5.48e, we increase the in-between gap of *bearings*, by changing X position of each one of them.

In Figure 5.48f, we create a cuboid of length, width, and height equal to 16 mm, 13 mm, and 16.5 mm, respectively, and move the cuboid to $X = -8$ mm and $Y = -42$ mm. The cuboid will be used to make a cut to the cover, at the position where the USB connectors of the system are traced. For the same reason, we duplicate the cuboid in order to make a new one of $L \times W \times H = 16 \times 13 \times 7.5$ mm. The latter cuboid is relocated to the position $X,Y,Z = -8,-42,21.5$ as illustrated in Figure 5.48g. The two cuboids are grouped together in Figure 5.48h to generate the "Fusion_Cubes," which are subsequently cut from the cover, in Figure 5.49a. The completed *cover.stl* is presented in Figure 5.49b.

5 To change the angle of another axis, choose an object and go to the menu '*Edit → Placement...*' and select the desired axis (i.e. X, Y, or Z) in the *Rotation* tab.

6 When copying select '*Yes*' to the pop-up window inquiring to select the objects that have dependency to the current copy.

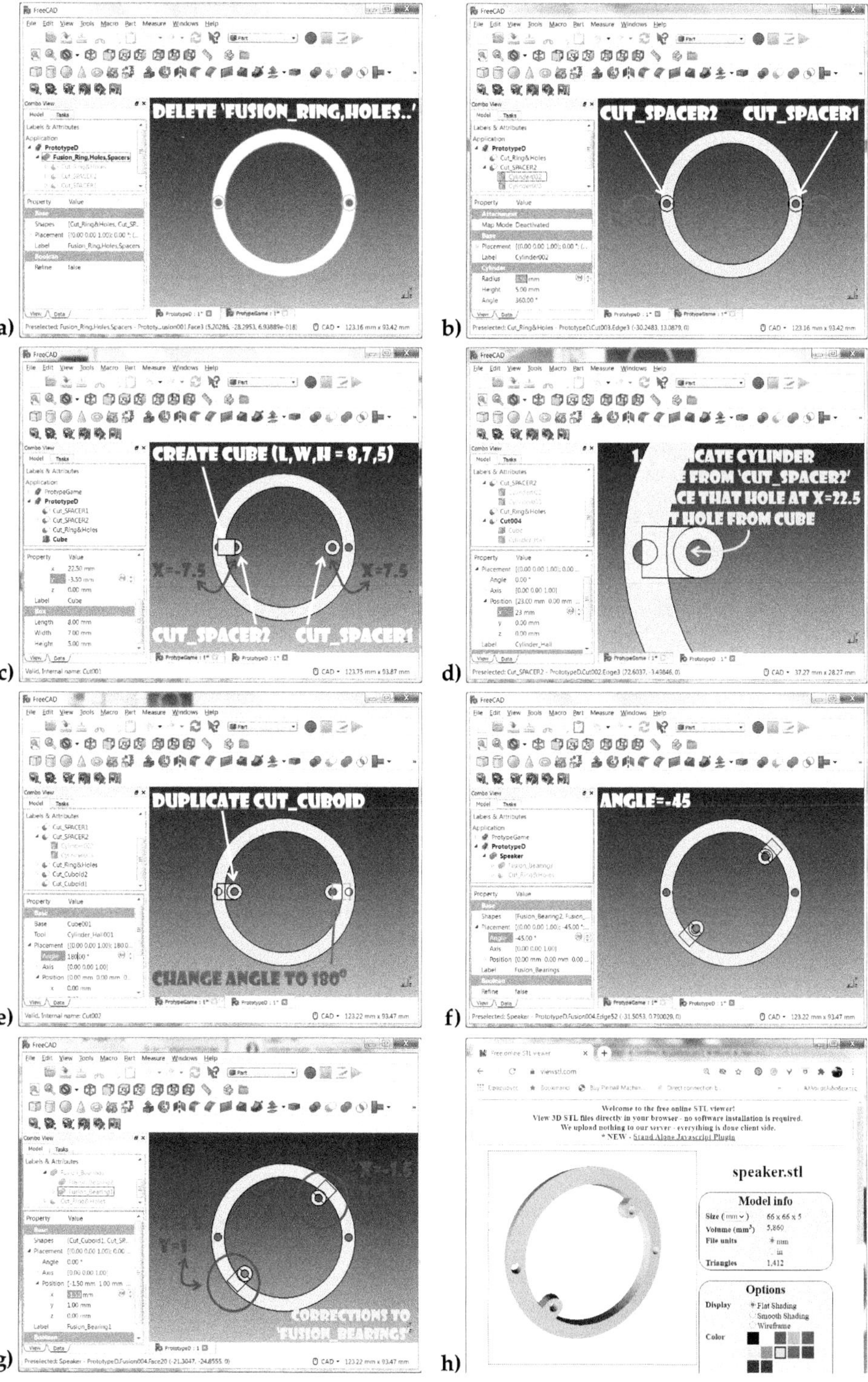

Figure 5.47 FreeCAD: modeling the *Speaker.stl* part of the prototype game (2 of 2). (a) delete the union of the sub-designs in order to modify the part models, (b) increase the radius of the two spacers to R=3.5mm, (c) move each spacer 7.5mm closer to ring center (X1=7.5/X2=-7.5 mm) and create a cuboid of LxWxH=8x7x8 mm, (d) create hole to the latter cuboid (consider the process described by the text), (e) duplicate the Cut_Cuboid and change its angle to 180°, (f) fuse together cuboids, holes and spacers and change the angle of the union to 45°, (g) apply correction to the X,Y position of the two bearings and export the 3D model, (h) the exported file (*speaker.stl*) is depicted by an online *.stl viewer* application. *Source:* Arduino Software.

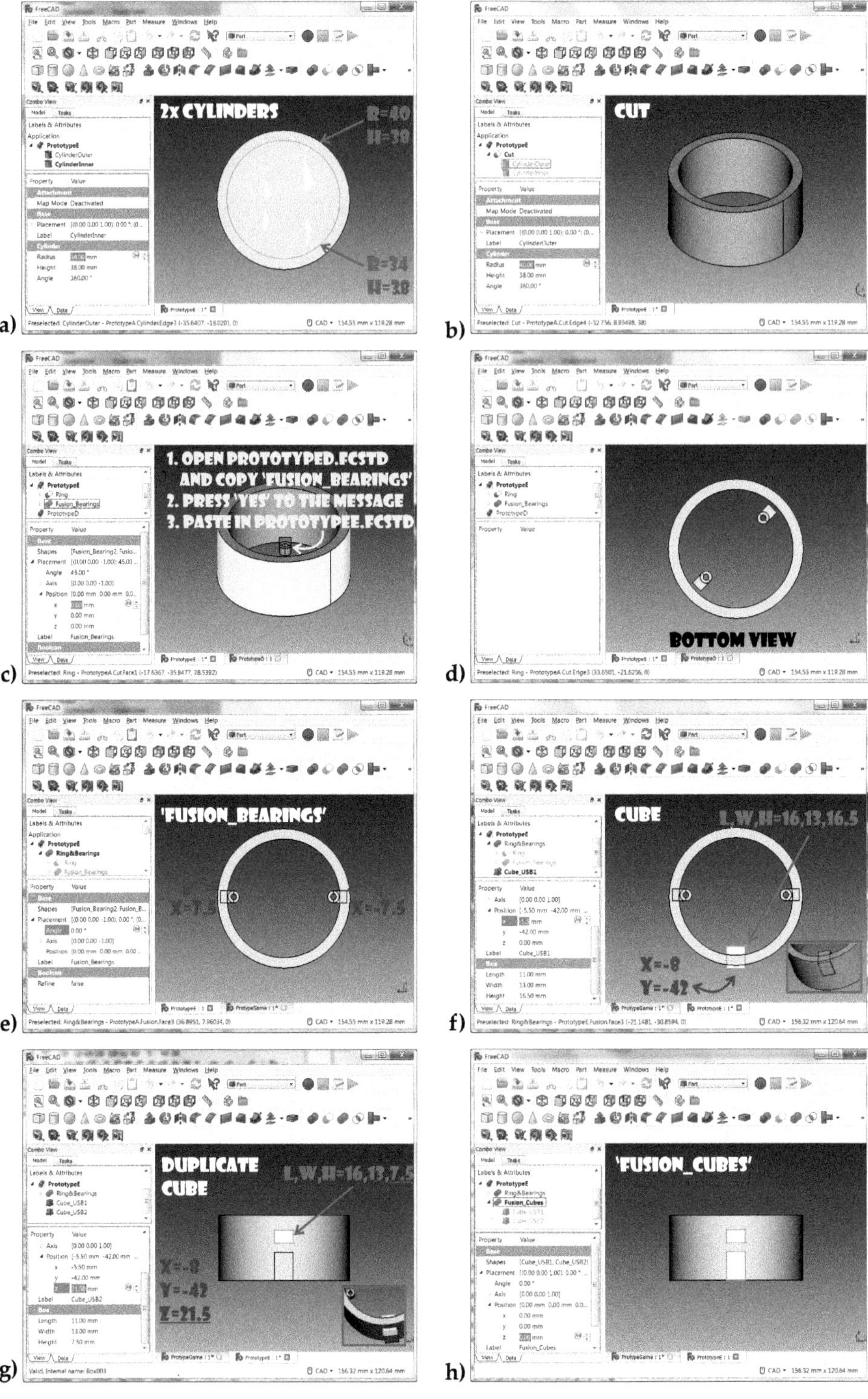

Figure 5.48 FreeCAD: modeling the *cover.stl* part of the prototype game (1 of 2). (a) create two cylinders of R1xH1=40x38 mm and R2xH2=34x38 mm, (b) cut the two cylinders to generate a ring, (c) copy Fusion_Bearings sub-part from the previous design and paste it to the current, (d) activate bottom view and increase the in between gap of bearings (X1=7.5/X2=-7.5 mm), (e) change angle of bearings to zero, (f) create a cuboid of LxWxH=16x13x16.5 mm @X=-8/Y=-42 mm, (g) duplicate cuboid and modify as follows: LxWxH=16x13x7.5 mm @X=-8/Y=-42/Z=21.5 mm, (h) fuse cubes together. *Source:* Arduino Software.

a) 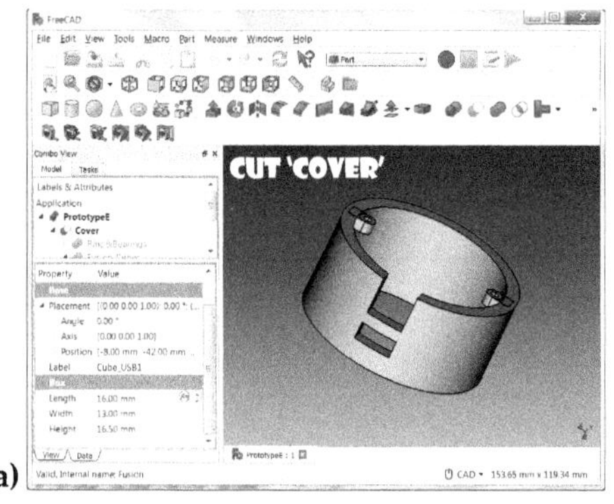

b) 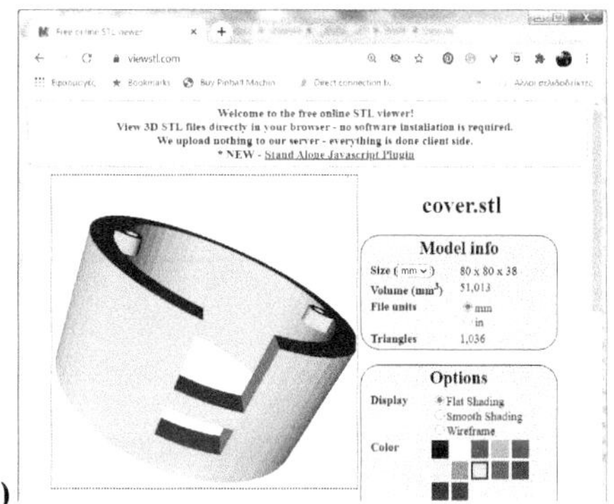

Figure 5.49 FreeCAD: modeling the *cover.stl* part of the prototype game (2 of 2). (a) cut the previous fusion of cubes from the ring and export the 3D model, (b) the exported file (*cover.stl*) is depicted by an online *.stl viewer* application. *Source:* Arduino Software.

PrototypeF (Modeling the button.stl Part)

For the *button.stl* part, we are going to need the bearings of the previous design, and hence, we copy the "Fusion_Bearings" sub-part from *PrototpeE.FCStd* and paste it into the current project, as illustrated in Figure 5.50a. In Figure 5.50b, we create a *sphere* solid, and in Figure 5.50c, d we modify Angle1 and Angle2 of that sphere. In Figure 5.50e we create a cylinder of $R = 21.5$ mm and $H = 15$ mm, while in Figure 5.50f we create a second smaller cylinder $R = 16.5$ mm and of same height. In Figure 5.50g, we perform a *cut* of the two homocentric cylinders so as to create a ring, and then we modify further the two angles of the sphere so as the latter fits perfectly on the top of the ring. The latter modifications are illustrated in Figure 5.50h, where the sphere properties are defined as follows: Radius = 24.54 mm, Angle1 = 29°, Angle2 = 70°, and *no transparency*.

In Figure 5.51a, we prepare the rotation of the *bearings* copied from the previous project. First, we select the desired part (herein, the "Fusion_Bearing2") and then, we change the workbench from *Part* to *Draft* from (from the menu denoted be the red arrow). The latter configuration applies a grid to our working project. In Figure 5.51b, we press left click on the mouse, on the exact point around of which we would like to perform the rotation of the selected object. As we move the mouse around the desired selection, the FreeCAD reveals a white mark on the potential position to be selected (where the red frame of Figure 5.51b illustrates the point we have actually selected).

A second mouse click causes a rotation the selected object (until the final mouse click), and as the object rotates with the motion of our mouse, a black shadow of rotated object appears on the screen (highlighted by the yellow frame Figure 5.51b). The exact degrees of rotation can be defined from the field pointed by the yellow arrow of the figure. After the rotation of the first bearing we perform the same operation for the second bearing around the point highlighted by Figure 5.51c. This way, the holes of the two *bearings* remain on the exact position where the two screws (holding the overall apparatus) pass through. Because the 3D part designed herein is smaller than the previous one, we increase the length of the cuboid to 11 mm, for each *bearings*, as illustrated by Figure 5.51d, e. It is here reminded that the current 3D part is placed above the fan and, in order to allow the air to pass through this

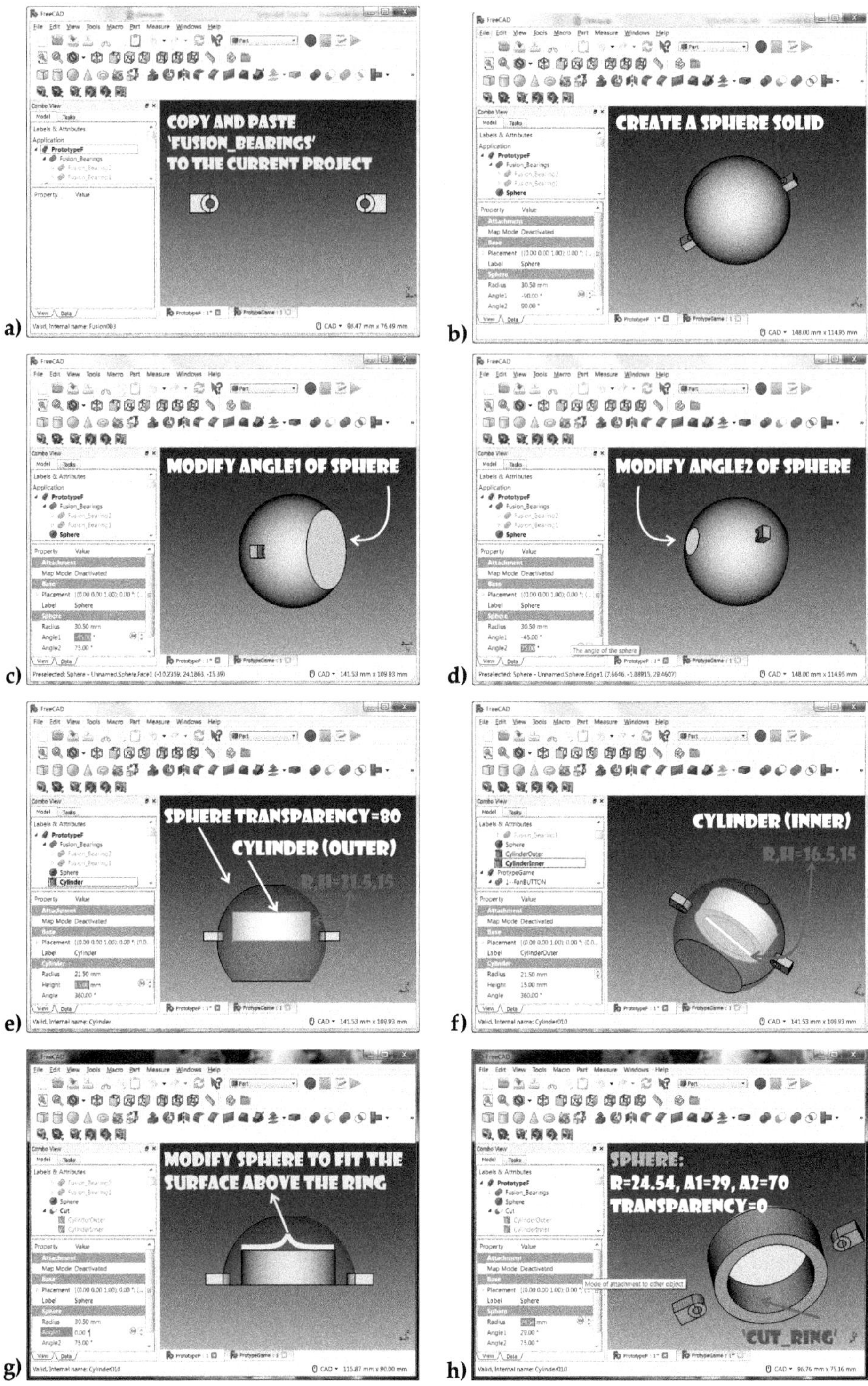

Figure 5.50 FreeCAD: modeling the *button.stl* part of the prototype game (1 of 2). (a) copy again the Fusion_Bearings sub-part to the current design, (b) create a sphere solid of R=30.5 mm, (c) modify Angle1 of the sphere to -45°, (d) modify Angle2 of the sphere to 75°, (e) create a cylinder of RxH=21.5x15 mm, (f) create a cylinder of RxH=16.5x15 mm, (g) cut homocentric cylinders, (h) modify further the angles and radius of the sphere so as to fit perfectly on the top of the ring (R=24.54, Angle1=29°, Angle1=70°). *Source:* Arduino Software.

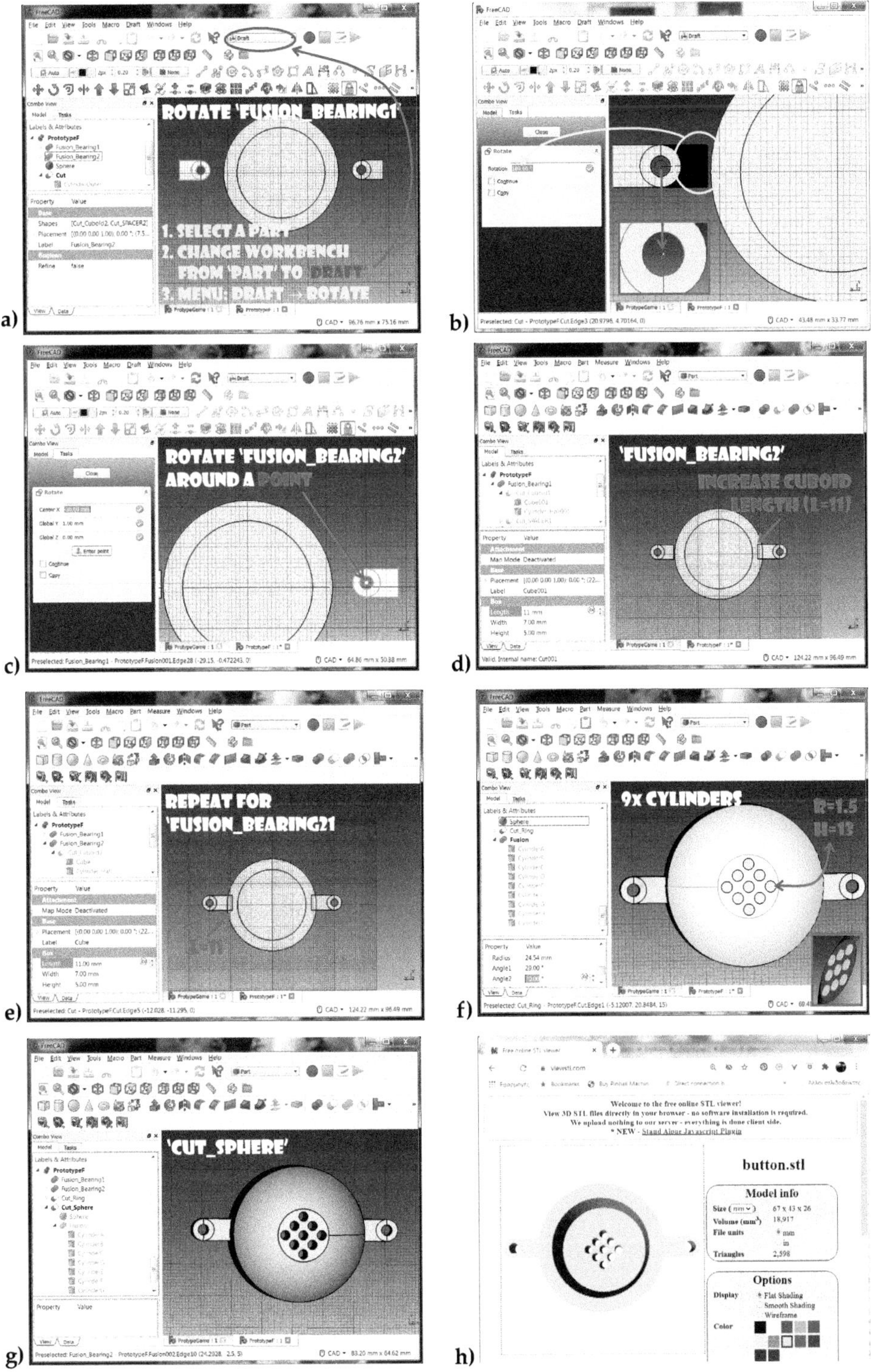

Figure 5.51 FreeCAD: modeling the *button.stl* part of the prototype game (2 of 2). (a) prepare the rotation of bearings copied from the previous project (consider the process described by the text), (b) rotate 180° the left bearing (consider the process described by the text), (c) repeat the process for the bearing on the right side, (d) increase the length of cuboid of the right bearing so as it touches the ring (L=11 mm), (e) repeat the process for the left bearing so that it also touches the ring, (f) insert 9 cylinders on the top of the sphere of RxH=1.5x13 mm each and fuse them together, (g) cut cylinders from the sphere (to generate holes) and export the 3D model, (h) the exported file (*cover.stl*) is depicted by an online *.stl viewer* application. *Source:* Arduino Software.

particular object, we insert nine cylinders (of 1.5 mm radius and 13 mm height) on the top of the sphere, as illustrated by Figure 5.51f. The nine cylinders are grouped together and the corresponding *Fusion* is cut from the sphere to shape the finalized design, depicted by Figure 5.51g. The *button.stl* part is presented in Figure 5.51h.

PrototypeG (Modeling the sensor.stl Part)

The *sensor.stl* part, which is used for holding the sensor the DFRobot module, constitutes a smaller design from the previous parts, and hence, the location of each sup-part (e.g. holes) is quite critical. We initiate the design from the main surface, which constitutes a cuboid of $L \times W \times H = 32 \times 27 \times 1.5$ mm, as presented in Figure 5.52a. We continue the design with the cuboid which fits within the exit gate of the blowing fan. Accordingly, in Figure 5.52b we design the outer cuboid of $L \times W \times H = 16.9 \times 12.35 \times 7.5$ mm and in Figure 5.52c, we design the inner cuboid of $L \times W \times H = 13.9 \times 9.35 \times 7.5$ mm. The latter is moved to the location $X = 1.5$ mm and $Y = 1.5$ mm in Figure 5.52d, and in Figure 5.52e we generate a cut of the two cuboids. The latter *cut* is moved at the middle of the main surface (i.e. $Y = 7.33$ mm) in Figure 5.52f, while in Figure 5.52g we duplicate the inner cuboid of this particular *cut* in order to create, later in the design, a *cut* in the main surface (so as the air from the blowing fan reaches the barometric pressure sensor of the module). Hence, in Figure 5.52h we move the duplicated cuboid in the same position, as highlighted within the red frame of the figure.

In Figure 5.53a, we prepare another *cut* in the position where the header of the module is traced, and hence, we create a cuboid of $L \times W \times H = 7.5 \times 13.5 \times 6$ mm at the position $X = 24.5$ mm and $Y = 6.83$ mm. In Figure 5.53b, we prepare another cylinder-style of *cut* of $R \times H = 1.5 \times 1.5$ mm at $X = 3.7$ mm and $Y = 3$ mm, which represents one of the two holes that attach the module on the 3D part (via screws). Therefore, in Figure 5.53c we duplicate the cylinder and move it at the position $X = 3.7$ mm and $Y = 23$ mm and in Figure 5.53d, we group the holes and provide a correction to the position of the "Fusion_holes" part, (i.e. $Y = 0.5$ mm). In Figure 5.53e, we group the holes and cuboids that should be removed from the main surface of the 3D part and, in Figure 5.53f, we apply that particular *cut* that shapes the final design. In Figure 5.53g, we group all parts, rename fusion as *sensor*, and export the latter to the *.stl* format of Figure 5.53h.

PrototypeH (Modeling the front.stl Part)

The *front.stl* part holds the ON/OFF switch device as well as the distance (i.e. time-of-flight [ToF]) sensors and constitutes a "thin" design, so as to let light (generated by the LED ring) spread. With the usual method depicted by Figure 5.54a, we design two homocentric cylinders of radius $R1 = 41$ mm and $R2 = 40$ mm and of height equal to 36 mm. Then we cut the two cylinders and generate the ring illustrated by Figure 5.54b. We repeat the latter two steps to generate the front side of the 3D part, as represented by Figure 5.54c, using two homocentric cycles of $R1 = 41$ mm and $R2 = 22$ mm, respectively, and of $H = 2$ mm. The two generated rings are then grouped together (i.e. "Fusion_Cylinders").

In Figure 5.54d, we create two cylinders of $R = 1.2$ mm and $H = 2$ mm so as to generate the two holes that attach the 3D object to the electronic system. The cylinder on the left is moved to the location $X = 30$, while the cylinder on the right is moved to position $X = -30$. We group the two cylinders together and we set the angle of the latter fusion (i.e. the "Fusion_FrontHoles") to 45°. In the lower part of the figure, we illustrate how the *button.stl* is going to fit with the current design.

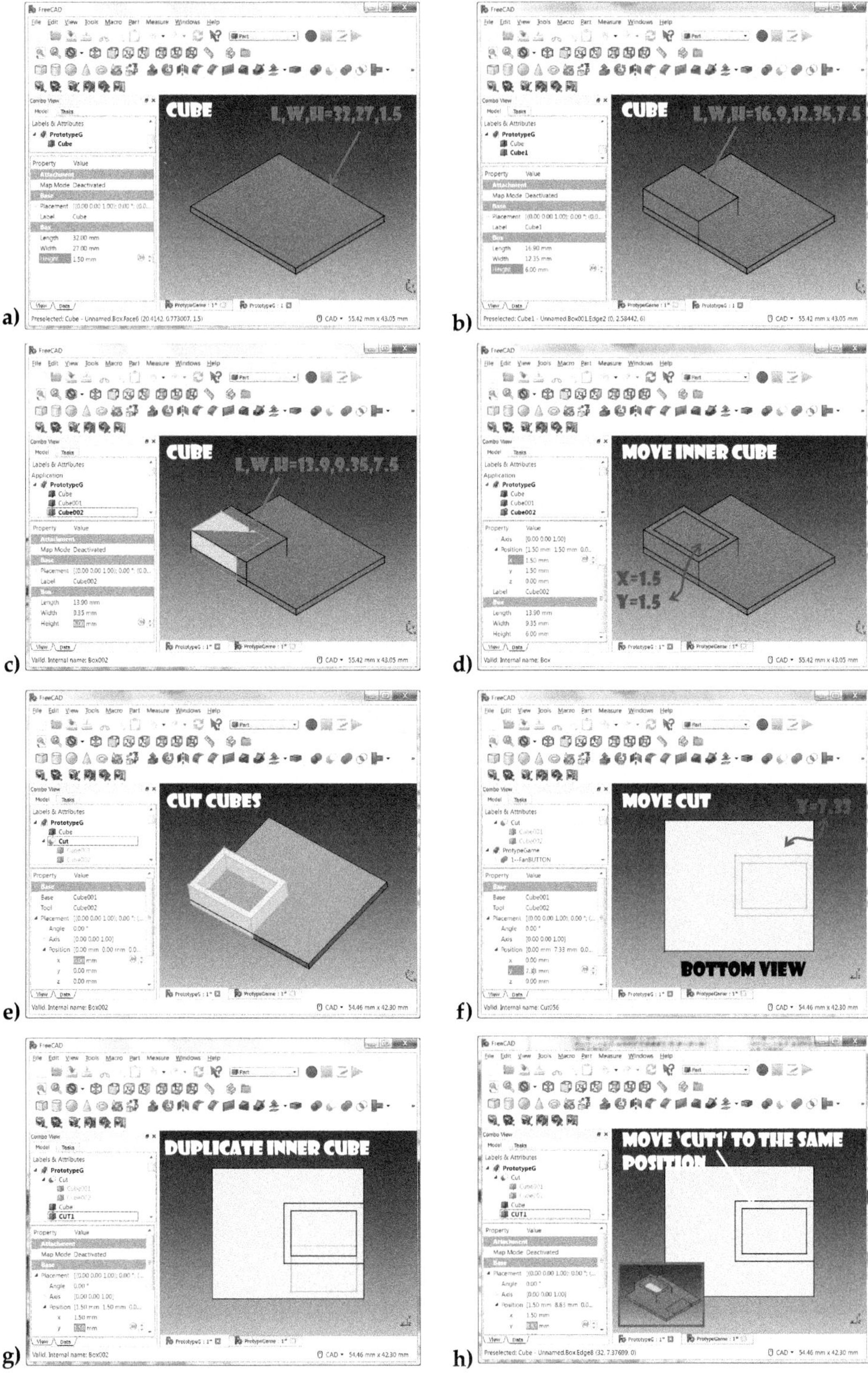

Figure 5.52 FreeCAD: modeling the *sensor.stl* part of the prototype game (1 of 2). (a) create a cuboid of LxWxH=32x27x1.5 mm, (b) create a cuboid of LxWxH=16.9x12.35x7.5 mm, (c) create a cuboid of LxWxH=13.9x9.35x7.5 mm, (d) move inner cuboid @X=Y=1.5 mm, (e) cut the latter two cuboids, (f) move the cut @Y=7.33 mm, (g) duplicate inner cube, (h) move the duplicated part at Y=8.83 mm (as depicted by the figure). *Source:* Arduino Software.

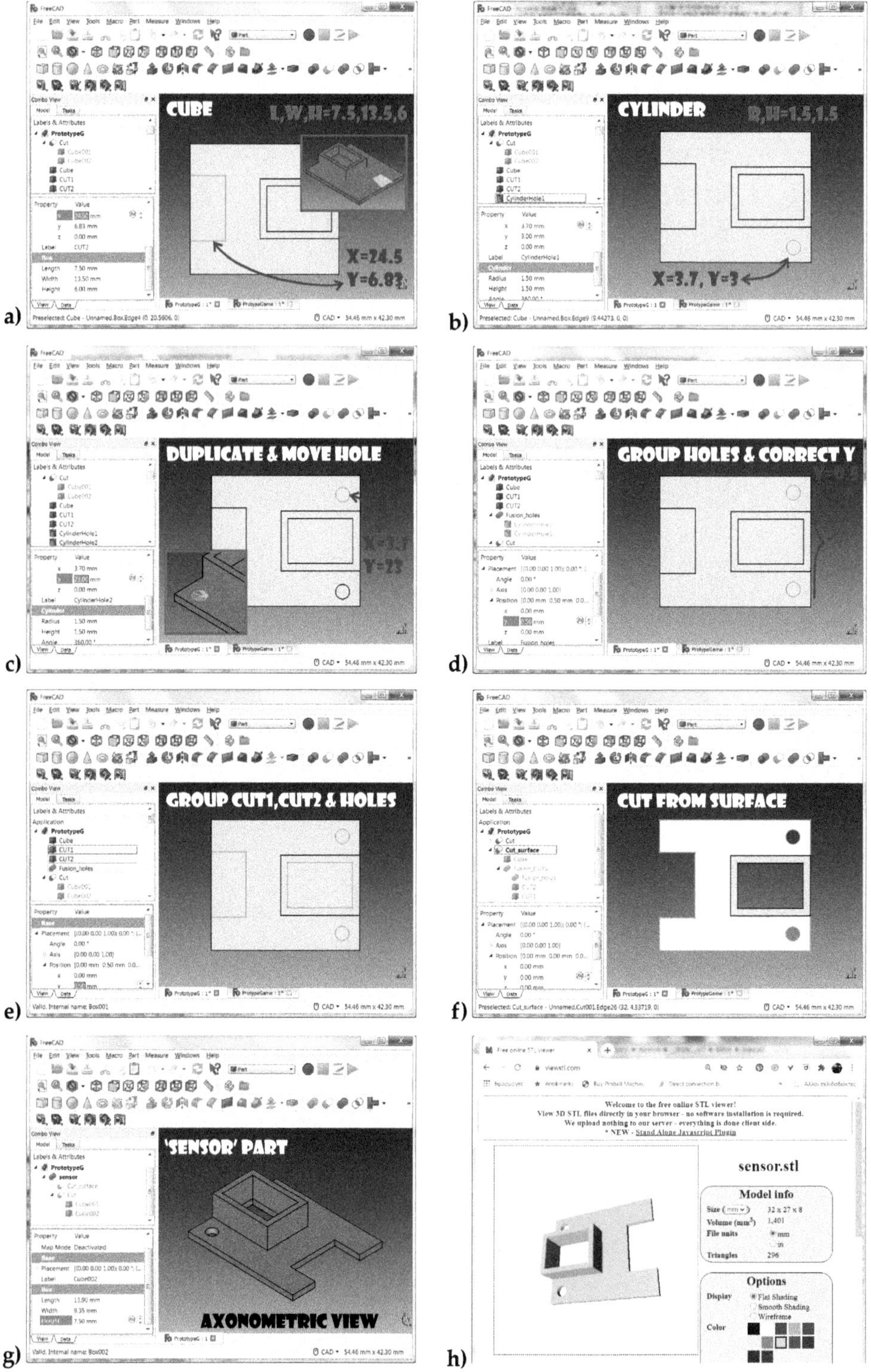

Figure 5.53 FreeCAD: modeling the *sensor.stl* part of the prototype game (2 of 2). (a) create a cuboid of LxWxH=7.5x13.5x6 mm, (b) create a cylinder of RxH=1.5x1.5 mm and place it @ X=3.7,Y=3 mm, (c) duplicate and move cylinder @X=3.7,Y=23 mm, (d) group the two cylinders and correct their position (Y=0.5 mm), (e) group together the selected parts, (f) cut the selected parts from the main surface cuboid, (g) activate axonometric view and export the 3d model, (h) the exported file (*sensor.stl*) is depicted by an online *.stl viewer* application. *Source:* Arduino Software.

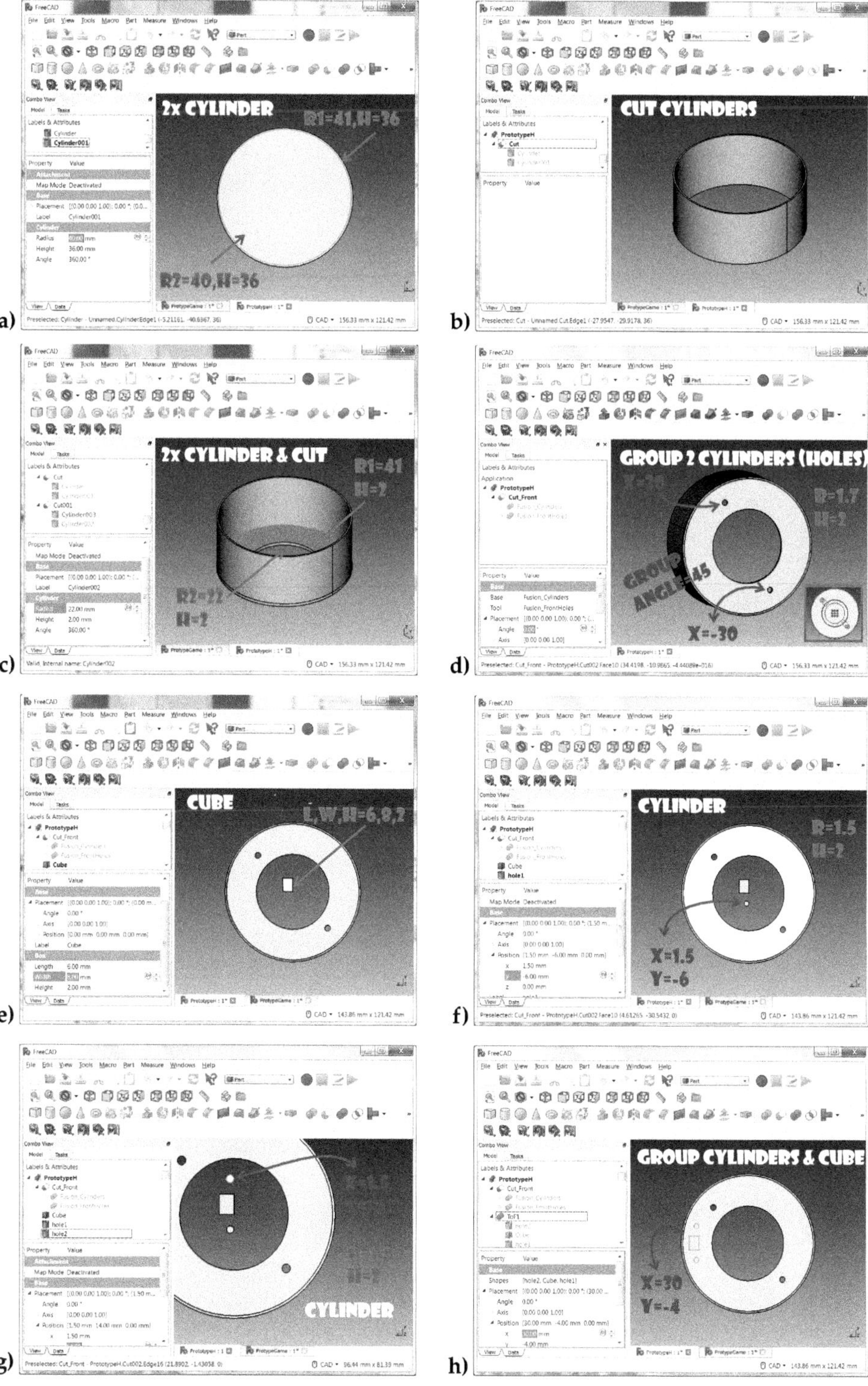

Figure 5.54 FreeCAD: modeling the *front.stl* part of the prototype game (1 of 2). (a) create two cylinders of R1xH1=41x36 mm and R2xH2=40x36 mm, (b) cut the two cylinders to generate a ring, (c) create two cylinders of R1xH1=41x2 mm and R2xH2=22x2 mm and cut them to generate a ring, (d) create two cylinders of RxH=1.7x2 mm and place them at X=30 mm and X=-30 mm, respectively, then change the group angle to 45° and cut them from the latter ring, (e) create a cuboid of LxWxH=6x8x2 mm, (f) create a cylinder of R1xH1=1.5x2 mm @X=1.5 mm, Y=-6 mm, (g) duplicate the latter cylinder @Y=14 mm, (h) group together the cylinders and cuboid and place them @ X=30 mm, Y=-4 mm. *Source:* Arduino Software.

In Figure 5.54e, we initiate the design of a *cut* on the front side of the model, where the ToF (i.e. distance) sensor will be attached. Accordingly, we place a cuboid of $L \times W \times H = 6 \times 8 \times 2$ mm and in Figure 5.54f, we place a cylinder, which represents one of the two holes for holding the ToF on the current 3D part. The cylinder is of radial equal to 1.5 mm and height 2 mm, and it is moved to the location $X = 1.5$ mm and $Y = -6$ mm. The latter cylinder is duplicated in Figure 5.54g and it is moved to the location $X = 1.5$, $Y = 14$. In Figure 5.54h, we group together the two cylinder as well as the cuboid and move the fusion (named *ToF1*) to the location $X = 30$, $Y = -4$.

In Figure 5.55a, we duplicate "ToF1" fusion to generate the so called "ToF2", which is intended for the second distance sensor. The later fusion is moved on the right side of the model, and in particular, to the position $X = -30$ and $Y = -4$. This way, the user may test the gesture recognition techniques explored in the previous chapter. In Figure 5.55b, we start the design of a *cut,* which is intended for the ON/OFF switch of the prototype system. Therefore, we duplicate the "ToF2" sub-model and rotate it 90°, while also changing the dimensions of the cuboid to $L \times W \times H = 4 \times 10.6 \times 2$ mm. Then we change the radius of the two holes to 1.2 mm and reposition the left hole to $X = 2$, $Y = -2.3$, and the right hole to $X = 2$, $Y = 12.9$. The overall "ON/OFF" fusion (i.e. the grouped cylinders and cube) is moved to the position $X = 5.3$, $Y = 33$, as illustrated by Figure 5.55c. In Figure 5.55d, we group the fusions "ToF1", "ToF2", "ON/OFF" and in Figure 5.55e, we cut the latter fusion from the rest of the part. The latter task generates the finalized model, which is exported to *.stl* file format and represented by the online STL viewer of Figure 5.55f.

Conclusion

This chapter has dealt with real-world projects and provided a thorough examination on the tinkering tasks (i.e. repairing and/or improving electronic products) with Arduino, as well as on the prototype design issues with Arduino and 3D printing technology. *Creativity* and *simplicity in design* are two of the main features that were addressed by the carefully thought examples of this chapter. Hereafter, are some unsolved problems for the reader.

Problem 5.1 (Tinkering with a Low-cost RC Car)

Revise **Ex.5.2** as follows: Decode the commands of the Control Pad function (found in the Bluefruit app), which corresponds to the keys '1', '2', '3', '4' (see Figure 5.9) and implement *backward left-turn* as well as *backward right-turn* commands for the RC car. In addition, utilize the remaining two function keys of Bluefruit app to force the car in 360° rotation (either clockwise or counterclockwise). The latter task is performed by forcing the two motors in diverse motion (i.e. the first in clockwise motion and the second in counterclockwise motion, and vice versa).

Problem 5.2 (A Prototype Interactive Game for Sensory Play)

Revise **Ex.5.6** as follows: Change the colors Green and Red which appear on the LED ring to some other colors of your choice and also modify the sounds accompanying the two colors, by applying different acoustic notes, of your choice as well.

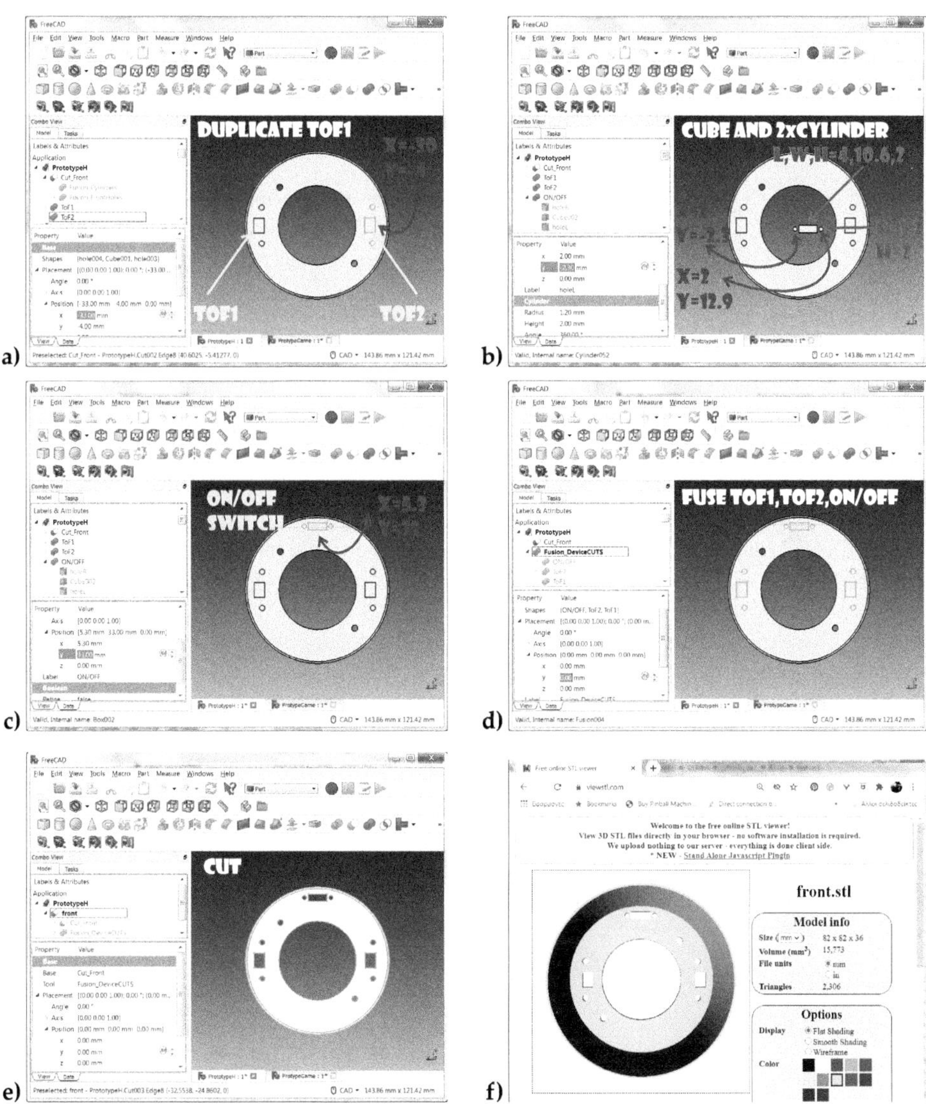

Figure 5.55 FreeCAD: modeling the *front.stl* part of the prototype game (2 of 2). (a) duplicate the latter group (meant to the hold the ToF sensor) and place it @X=-30 mm, (b) create a similar group of a cuboid and two cylinders for the ON/OFF switch (consider the process described by the text), (c) place the group meant for the ON/OFF switch to the position X=5.3,Y=33 mm, (d) fuse the two ToF sensors and the ON/OFF switch, (e) cut the latter group from the main part and export the 3D model, (f) the exported file (*front.stl*) is depicted by an online *.stl viewer* application. *Source:* Arduino Software.

Problem 5.3 (A Prototype Interactive Game for Sensory Play)

Revise the Arduino **Ex5-6** as follows: Enrich the template firmware with a Bluetooth interface, which would count the "correct" hits performed by the player, via a mobile device.

Problem 5.4 (A Prototype Interactive Game for Sensory Play)

Revise the Arduino **Ex5-6** as follows: Replace the virtual button function with 1D gesture recognition, using the two distance sensors at the top of the prototype game.

Tips: Because no I2C multiplexer is used by the current setup, exploit the XSHUT pin of the distance sensor module driven by Micro:bit P5 pin, in order to deactivate that particular sensor and modify the I2C address of the other sensor.

Problem 5.5 (3D Printing)

Redesign the *floor* 3D part of the prototype interactive game in order to add a wireless charging feature to the system.

Tips: (i) Buy a *wireless charging pad* along with a *wireless charger* (as the ones used in mobile phones). (ii) Solder a PH2.0 Male two-pin connector to the *wireless charger*, which will be connected to the corresponding female connector, driven by the USB pin of *PowerBoost 100 charger* module, as highlighted by green color in Figure 5.56. (iii) Given that the *wireless charger* fits within the *floor* part of the prototype game, remodel the latter 3D part along with a cavity determined by the outline of the *wireless charger*. (iv) Design the *floor* part as thin as possible so that when the prototype game is placed above the charging pad (at the left of the figure), the power is transmitted to the *wireless charger* (at the right of the figure) embedded within the new *floor* part.

Figure 5.56 Add wireless charging feature on the prototype interactive game.

References

1 Lye, S.Y. and Koh, J.H.L. (2014). Review on teaching and learning of computational thinking through programming: what is next for K-12? *Computers in Human Behavior* 41: 51–61.

2 Pears, A., Seidman, S., Malmi, L. et al. (2007). A survey of literature on the teaching of introductory programming. Working group reports on ITiCSE on innovation and technology in computer science education, 204–223.

3 Robins, A., Rountree, J., and Rountree, N. (2003). Learning and teaching programming: a review and discussion. *Computer Science Education* 13 (2): 137–172.

4 Ackovska, N. and Ristov, S. (2013). Hands-on improvements for efficient teaching computer science students about hardware. *2013 IEEE Global Engineering Education Conference (EDUCON)*, 295–302. IEEE.

5 Hamblen, J.O., Smith, Z.C., and Woo, W.W. (2013). Introducing embedded systems in the first C/C++ programming class. *2013 IEEE International Conference on Microelectronic Systems Education (MSE)*, 1–4. IEEE.

6 Lo, D.C.T., Qian, K., and Hong, L. (2012). The use of low cost portable microcontrollers in teaching undergraduate Computer Architecture and Organization. *IEEE 2nd Integrated STEM Education Conference*, 1–4. IEEE.

7 Gil-Sánchez, L., Masot, R., and Alcañiz, M. (2015). Teaching electronics to aeronautical engineering students by developing projects. *IEEE Revista Iberoamericana de Tecnologias del Aprendizaje* 10 (4): 282–289.

8 Su, S., Kerwin, J., Crowe, S. et al. (2013). Teaching embedded programming to electrical engineers, BioEngineers, and mechanical engineers via the escape platform. *2013 3rd Interdisciplinary Engineering Design Education Conference*, 87–92. IEEE.

9 Geth, F., Verveckken, J., Leemput, N. et al. (2013). Development of an open-source smart energy house for K-12 education. *2013 IEEE Power & Energy Society General Meeting*, 1–5. IEEE.

10 Gasser, M., Lu, Y.H., and Koh, C.K. (2010). Outreach project introducing computer engineering to high school students. *2010 IEEE Frontiers in Education Conference (FIE)*, F2E-1. IEEE.

11 Musa, L. (2008). FPGAS in high energy physics experiments at CERN. *2008 International Conference on Field Programmable Logic and Applications*, 2-2. IEEE.

Microcontroller Prototypes with Arduino and a 3D Printer: Learn, Program, Manufacture, First Edition. Dimosthenis Bolanakis.

12 Bolanakis, D.E., Rachioti, A.K., and Glavas, E. (2017). Nowadays trends in microcontroller education: do we educate engineers or electronic hobbyists? Recommendation on a multi-platform method and system for lab training activities. *2017 IEEE Global Engineering Education Conference (EDUCON)*, 73–77. IEEE.

13 Bolanakis, D.E., Kotsis, K.T., and Laopoulos, T. (2016). Ethernet and PC-based experiments on barometric altimetry using MEMS in a wireless sensor network. *Computer Applications in Engineering Education* 24 (3): 443–455.

14 Blikstein, P., Kabayadondo, Z., Martin, A., and Fields, D. (2017). An assessment instrument of technological literacies in makerspaces and FabLabs. *Journal of Engineering Education* 106 (1): 149–175.

15 Blikstein, P. and Krannich, D. (2013). The makers' movement and FabLabs in education: experiences, technologies, and research. *Proceedings of the 12th International Conference on Interaction Design and Children*, 613–616.

16 Bolanakis, D.E., Glavas, E., Evangelakis, G.A. et al. (2012). Microcomputer Architecture: Low-level Programming Methods and Applications of the M68HC908GP32. Self-publishing (Printed by Createspace).

17 Bolanakis, D.E., Kotsis, K.T., and Laopoulos, T. (2010). Switching from computer to microcomputer architecture education. *European Journal of Engineering Education* 35 (1): 91–98.

18 Bolanakis, D.E., Evangelakis, G.A., Glavas, E., and Kotsis, K.T. (2011). A teaching approach for bridging the gap between low-level and high-level programming using assembly language learning for small microcontrollers. *Computer Applications in Engineering Education* 19 (3): 525–537.

19 Bolanakis, D.E., Glavas, E., and Evangelakis, G.A. (2007). An integrated microcontroller-based tutoring system for a computer architecture laboratory course. *International Journal of Engineering Education* 23 (4): 785.

20 Shulman, L. (1987). Knowledge and teaching: foundations of the new reform. *Harvard Educational Review* 57 (1): 1–23.

21 Shulman, L.S. (1986). Those who understand: knowledge growth in teaching. *Educational Researcher* 15 (2): 4–14.

22 Pierson, M.E. (2001). Technology integration practice as a function of pedagogical expertise. *Journal of Research on Computing in Education* 33 (4): 413–430.

23 Mishra, P. and Koehler, M.J. (2006). Technological pedagogical content knowledge: a framework for teacher knowledge. *Teachers College Record* 108 (6): 1017–1054.

24 Koehler, M. and Mishra, P. (2009). What is technological pedagogical content knowledge (TPACK)? *Contemporary Issues in Technology and Teacher Education* 9 (1): 60–70.

25 Zappa, F. (2017). Microcontrollers: Hardware and Firmware for 8-bit and 32-bit Devices. Editrice Esculapio.

26 Pereira, F. (2009). HCS08 Unleashed: Designer's Guide to the HCS08 Microcontrollers. Booksurge.

27 Valdes-Perez, F.E. and Pallas-Areny, R. (2017). Microcontrollers: Fundamentals and Applications with PIC. CRC Press.

28 Bolanakis, D.E. (2017). *Microcontroller education: do it yourself, reinvent the wheel, code to learn. Synthesis Lectures on Mechanical Engineering* 1 (4): 1–193.

29 Rachioti, A.K., Bolanakis, D.E., and Glavas, E. (2016). Teaching strategies for the development of adaptable (compiler, vendor/processor independent) embedded C code. *2016 15th International Conference on Information Technology Based Higher Education and Training (ITHET)*, 1–7. IEEE.

30 Kordaki, M. (2010). A drawing and multi-representational computer environment for beginners' learning of programming using C: design and pilot formative evaluation. *Computers & Education* 54 (1): 69–87.

31 Palumbo, D.B. (1990). Programming language/problem-solving research: a review of relevant issues. *Review of Educational Research* 60 (1): 65–89.

32 Lee, C.S., Su, J.H., Lin, K.E. et al. (2009). A project-based laboratory for learning embedded system design with industry support. *IEEE Transactions on Education* 53 (2): 173–181.

33 Bruce, J.W., Harden, J.C., and Reese, R.B. (2004). Cooperative and progressive design experience for embedded systems. *IEEE Transactions on Education* 47 (1): 83–92.

34 Levin, J.R. (1981). On functions of pictures in prose. In: Neuropsychological and Cognitive Processes in Reading (eds. F.J. Pirozzolo and M.C. Wittrock), 203. New York;: Academic Press.

35 Wing, J.M. (2006). *Computational thinking. Communications of the ACM* 49 (3): 33–35.

36 Wing, J. (2011). Research notebook: computational thinking—what and why. *The Link Magazine* 6.

37 Guzdial, M. (2011). A definition of computational thinking from Jeannette Wing. *Computing Education Blog*.

38 Selby, C. and Woollard, J. (2013). Computational thinking: the developing definition.

39 K-12 Computer Science Framework Steering Committee (2016). K-12 Computer Science Framework. ACM.

40 Aho, A.V. (2012). Computation and computational thinking. *The Computer Journal* 55 (7): 832–835.

41 García Peñalvo, F.J., Reimann, D., Tuul, M. et al. (2016). An overview of the most relevant literature on coding and computational thinking with emphasis on the relevant issues for teachers.

42 Wing, J.M. (2008). Computational thinking and thinking about computing. *Philosophical Transactions of the Royal Society A: Mathematical, Physical and Engineering Sciences* 366 (1881): 3717–3725.

43 Brennan, K. and Resnick, M. (2012). Using artifact-based interviews to study the development of computational thinking in interactive media design. *Annual American Educational Research Association Meeting*, Vancouver, BC, Canada, 1–25.

44 Leens, F. (2009). An introduction to I2C and SPI protocols. *IEEE Instrumentation Measurement Magazine* 12 (1): 8–13.

45 Fox, A. and Patterson, D. (2012). Crossing the software education chasm. *Communications of the ACM* 55 (5): 44–49.

46 Koehler, M.J., Mishra, P., and Cain, W. (2013). What is technological pedagogical content knowledge (TPACK)? *Journal of Education* 193 (3): 13–19.

47 Papert, S. (1980). Mindstorms: Children, Computers, and Powerful Ideas. New York;: Basic Books. Inc.

48 Bolanakis, D.E. (2019). A survey of research in microcontroller education. *IEEE Revista Iberoamericana de Tecnologias del Aprendizaje* 14 (2): 50–57.

49 Wicker, S.B. (2018). Smartphones, contents of the mind, and the fifth amendment. *Communications of the ACM* 61 (4): 28–31.

50 Butler, M. (2010). Android: changing the mobile landscape. *IEEE Pervasive Computing* 10 (1): 4–7.

51 Lee, J.C. (2008). Hacking the nintendo wii remote. *IEEE Pervasive Computing* 7 (3): 39–45.

52 Pell, R. (2018). Computer the size of grain of salt embeds in everyday objects. eeNews Europe report.

53 Bosch Sensortech (2015). BME280 combined humidity and pressure sensor.

54 Bosch Sensortech (2015). BMP280 digital pressure sensor.

55 Bolanakis, D.E. (2017). MEMS barometers toward vertical position detection: background theory, system prototyping, and measurement analysis. *Synthesis Lectures on Mechanical Engineering* 1 (1): 1–145.

56 Bosch Sensortech (2016). BNO055 Intelligent 9-axis absolute orientation sensor.

57 STMicroelectronics (2018). VL53L0X World's smallest Time-of-Flight ranging and gesture recognition sensor.

58 Gainsley, S. (2011). Look, listen, touch, feel, taste: the importance of sensory play. *Extensions. Curriculum Newsletter from HighScope* 25 (5): 1–4.

Index

Microcontroller Prototypes with Arduino and a 3D Printer: Learn, Program, Manufacture, First Edition.
Dimosthenis E. Bolanakis.